Sounds of the Pandemic

Sounds of the Pandemic offers one of the first critical analyses of the changes in sonic environments, artistic practice, and listening behaviour caused by the Coronavirus outbreak.

This multifaceted collection provides a detailed picture of a wide array of phenomena related to sound and music, including soundscapes, music production, music performance, and mediatisation processes in the context of COVID-19. It represents a first step to understanding how the pandemic and its by-products affected sound domains in terms of experiences and practices, representations, collective imaginaries, and socio-political manipulations.

This book is essential reading for students, researchers, and practitioners working in the realms of music production and performance, musicology and ethnomusicology, sound studies, and media and cultural studies.

Maurizio Agamennone is Full Professor of Ethnomusicology at the University of Florence. His interests span theoretical issues in ethnomusicology; poetic improvisation and other forms of sung poetry; living polyphonies; the activities and productions of migrant musicians; the compositional and performance practices in the European musical avant-garde; and intercultural exchanges in contemporary music.

Daniele Palma is a post-doctoral research fellow at the University of Bologna, working on evidence of Giuseppe Verdi's operas' performance practice in 19th-century music periodicals. His research concerns operatic vocality in the 19th and 20th centuries, early sound media, cultural imaginaries of opera, and amateur music practices, from children's records to mental patients.

Giulia Sarno is a post-doctoral research fellow in Ethnomusicology at the University of Florence. Her current interests span a variety of topics in contemporary musical and sonic practices, from avant-garde electronics, to oral traditions, and popular music. She is investigating the relationship between sound and football, and the history of the centre for music research *Tempo Reale* (Florence).

Sounds of the Pandemic
Accounts, Experiences, Perspectives in Times of COVID-19

Edited by Maurizio Agamennone,
Daniele Palma, and Giulia Sarno

LONDON AND NEW YORK

Cover image: Claudio Palma

First published 2023
by Routledge
4 Park Square, Milton Park, Abingdon, Oxon OX14 4RN

and by Routledge
605 Third Avenue, New York, NY 10158

Routledge is an imprint of the Taylor & Francis Group, an informa business

© 2023 selection and editorial matter, Maurizio Agamennone, Daniele Palma, and Giulia Sarno; individual chapters, the contributors

The right of Maurizio Agamennone, Daniele Palma, and Giulia Sarno to be identified as the authors of the editorial material, and of the authors for their individual chapters, has been asserted in accordance with sections 77 and 78 of the Copyright, Designs and Patents Act 1988.

All rights reserved. No part of this book may be reprinted or reproduced or utilised in any form or by any electronic, mechanical, or other means, now known or hereafter invented, including photocopying and recording, or in any information storage or retrieval system, without permission in writing from the publishers.

Trademark notice: Product or corporate names may be trademarks or registered trademarks, and are used only for identification and explanation without intent to infringe.

British Library Cataloguing-in-Publication Data
A catalogue record for this book is available from the British Library

Library of Congress Cataloging-in-Publication Data
Names: Agamennone, Maurizio, editor. | Palma, Daniele, editor. | Sarno, Giulia, editor.
Title: Sounds of the pandemic : accounts, experiences, perspectives in times of COVID-19 / edited by Maurizio Agamennone, Daniele Palma, and Giulia Sarno.
Description: Abingdon, Oxon ; New York : Routledge, 2022. |
Includes bibliographical references and index. |
Identifiers: LCCN 2022027734 (print) | LCCN 2022027735 (ebook) |
ISBN 9781032060231 (paperback) | ISBN 9781032060248 (hardback) |
ISBN 9781003200369 (ebook)
Subjects: LCSH: Music–Social aspects–History–21st century. |
COVID-19 Pandemic, 2020–Social aspects.
Classification: LCC ML3916 .S695 2022 (print) |
LCC ML3916 (ebook) | DDC 306.4/842–dc23/eng/20220725
LC record available at https://lccn.loc.gov/2022027734
LC ebook record available at https://lccn.loc.gov/2022027735

ISBN: 978-1-032-06024-8 (hbk)
ISBN: 978-1-032-06023-1 (pbk)
ISBN: 978-1-003-20036-9 (ebk)

DOI: 10.4324/9781003200369

Typeset in Times New Roman
by Newgen Publishing UK

Contents

List of Illustrations	viii
List of Contributors	x
Acknowledgments	xiv

Introduction: Understanding a Pandemic through Sound   1
MAURIZIO AGAMENNONE, DANIELE PALMA, AND GIULIA SARNO

PART I
Accounts: Sounds from a World under Lockdown   13

1 Listening to the First Lockdown: The Auditory Experience of
Wrocław's Inhabitants   15
ROBERT LOSIAK, RENATA TAŃCZUK, AND SŁAWOMIR WIECZOREK

2 Together in Discipline and Turmoil: Remembering Public
Sounds during the COVID-19 Pandemic in the Czech Republic
and Slovakia   30
DOMINIKA MORAVČÍKOVÁ

3 Listening to the Hustle and the Hush: Sound, City, and the Pandemic   40
NAKSHATRA CHATTERJEE AND SRIJITA BISWAS

4 Applauses and Banners, Horns and Fireworks: Tracing the Sonic
Expression of French Social Movements during Lockdown   50
ALESSANDRO GREPPI AND DIANE SCHUH

5 Pandemic Soundscaping: Rediscovering a New Aura in the
Mediatised Sonic Reality   65
LUDOVICO PERONI

6 Not People but a Sound: Virtual Audio and the Appropriation of
Fandom Practices in Pandemic Football   76
GIULIA SARNO

vi *Contents*

7 A Digital Archive of Participatory Location Rhythm Performances:
   Listening as a Way of Attending to the Pandemic   87
   MARCEL ZAES SAGESSER

PART II
Experiences: Musicking in the Face of the Pandemic   97

8 *Huapanguitos pa seguir aguantando en cuarentena*: Mexican
   SonTube Channels as Emergent Digital Spaces of Music and
   Community during COVID-19   99
   DANIEL MARGOLIES AND J.A. STRUB

9 'WHY DO THEY DANCE IN THE MIDDLE OF THE
   PANDEMIC?' Post-Pandemic Cumbia, Mediated Live Music, and
   Digital Heritage from Mexico City   110
   MICHAËL SPANU

10 Sardinian Traditional Music during the COVID-19 Pandemic   125
    MARCO LUTZU AND IGNAZIO MACCHIARELLA

11 Becoming Visible: Proud Roma and Sinti Musicians in Italy during
    the Pandemic   136
    ANTONELLA DICUONZO

12 Rethinking Intermedia Practices during the Pandemic: Staging
    and Conception of Alexander Schubert's Virtual Reality Video
    Game *Genesis*   148
    LUCA BEFERA

13 Musicians in the Brazilian Pandemic: Facing COVID-19 during the
    Bolsonaro Regime and the Aldir Blanc Emergency Bill   163
    SUZEL ANA REILY AND REGINA MACHADO

14 Musical Performance during and after the COVID-19 Pandemic:
    Days of Future Passed?   176
    ALESSANDRO BRATUS, ALESSANDRO CALIANDRO, FULVIA CARUSO,
    FLAVIO ANTONIO CERAVOLO AND MICHELA GARDA

PART III
Perspectives: Rethinking Sound and Music against the Backdrop of a
Global Crisis   191

15 Coronamusic(king): Types, Repertoires, Consolatory Function   193
    MELANIE WALD-FUHRMANN

Contents vii

16 The Pandemic as a Catalyst for Remotivity in Music   213
MARK THORLEY

17 Music in Lockdown: On Sonic Spaces during the COVID-19
Pandemic, March–June 2020   226
ESTEBAN BUCH

18 What a Blackbird Has Told Me: Latent Acoustic Learning in the
Times of COVID-19   237
THEODOROS LOTIS

19 The Sounds and Silence of COVID-19 Quarantine: Media
Representation, Debility, and Neoliberal Biopolitics   249
JAMES DEAVILLE

20 Four Sounds against Capitalocene: Lockdown, Music, and the Artist
as Producer   263
MAKIS SOLOMOS

Afterword: Coping with Crisis through Coronamusic   272
NIELS CHR. HANSEN

Index   285

Illustrations

Figures

4.1	The slogan reads 'Money for the public hospital, solidarity with the cashiers'	52
4.2	The banner reads 'The corona will disappear; The commune will bloom again'	52
4.3	The banner reads 'We're here even if there's corona; We are here; Support for caregivers; For the hospital; + money'	54
7.1	Screenshot of #*otherbeats* website with several colour dots on a dark background, with mouse cursor	91
7.2	Still from video recording showing a tree leaf in nature (courtesy of Annie Rüfenacht)	93
7.3	Graphic representation of the rhythm produced by the tree leaf irregularly shaking in the wind	94
9.1	Richard TV's YouTube video 'FROM THE ROOFTOP HE ORGANIZES HIS DANCES AND THIS IS HOW HIS NEIGHBORS RESPONDED'	114
9.2	Richard TV's official YouTube page	114
9.3	Sonidero Latino TV's YouTube video 'THIS IS HOW WE DANCE CUMBIA IN MEXICO CITY'S NEIGHBORHOODS – ((SAMPUESANA SIBONERA)) SIBONEY – LA MERCED.' Siboney is the name of the *sonidero*, 'Sampuesana Sibonera' is the name of the song, and La Merced the name of the neighbourhood	116
9.4	Richard TV's YouTube video '(((THE LAST SAMPUESANA))) FROM SONIDO CONDOR IN LA MERCE[D] 62th ANNIVERSARY EQUIPO RETRO'	117
9.5	Richard TV's YouTube video 'THE SONG THAT TOOK OVER THE NIGHT PERUVIAN CUMBIA ((SIN TU AMOR)) SONIDO CARIBE 66 SAN JUAN DE ARAGON.' San Juan de Aragon is the name of a neighbourhood in the north-eastern part of Mexico City	118
9.6	Richard TV's list of videos that emphasise women's assertive bodies	120
12.1	*Genesis* game interface (courtesy of Alexander Schubert)	149
12.2	Prominent features of Schubert's works over the last four years	151
12.3 a–e	Homepages of selected pieces by Schubert, in order of appearance: *Genesis, Control, Wiki-Piano. Net, Av3ry,* and *Crawlers* (courtesy of Alexander Schubert)	154
12.4	Items pictures digital editing, from the original photos to their appearance in the game inventory (courtesy of Alexander Schubert)	158
12.5	Different ways of joining *Genesis* virtual reality regarding gamers, avatars, and items	158
15.1	User comments expressing very positive aesthetic evaluations of a *Sound of Music* parody by the singer Dovelybell on YouTube	204

Illustrations ix

15.2 Top user comments to Kathy Mak's parody of Natalie Imbruglia's
'Torn' on YouTube   205
18.1 Spectrogram of birds chirping and the salami sound   241
20.1 8 March 2020, Paris. Feminist demonstration as part of the social
movement against pension reform (photo taken by the author,
Makis Solomos)   264
A.1   The putative characteristics of coronamusic – open for future investigation.
Overall, coronamusic seems to differ from pre-pandemic music in terms
of its active, multimodal, creative, and attentive modes of engagement;
its temporal focus on the here and now; the agency it assigns to the
collective with amateur and co-creator identities; its positive and humorous
sentiment; its informal, authentic, and intimate interaction patterns;
its location in home and outdoors settings; and its topical focus on the
simplicities and trivialities of life in lockdown   278

Tables

10.1 'Su Baballoti' lyrics (courtesy of Antonio Pani)   131
14.1 Top 10 most viewed videos   186

Contributors

Maurizio Agamennone is Full Professor of Ethnomusicology at the University of Florence. His interests span theoretical issues in ethnomusicology; poetic improvisation and other forms of sung poetry; living polyphonies; the activities and productions of migrant musicians; the compositional and performance practices in the European musical avant-garde; and intercultural exchanges in contemporary music.

Luca Befera graduated in Musicology from the University of Pavia. After studying classical piano and composition, he developed interests in sound-based music and informatics mediation. He is currently a doctoral candidate at the University of Turin, where he studies the aesthetics of algorithmic and intermedial performances.

Srijita Biswas is a PhD Research Scholar and Teaching Assistant at the Indian Institute of Science Education and Research, Bhopal. Her doctoral research area is literary studies, with specific emphasis on gastronomy and city space in Bengali literature and culture. Her interests lie in cultural studies and literature written in Indian English and Bengali.

Alessandro Bratus is Associate Professor of Popular Music at the University of Pavia (Cremona). His preferred research area is the analysis of popular music recorded audio and audiovisual artifacts as mediatisation of performative acts related to the construction of the relationship between performers and their listeners/viewers.

Esteban Buch, Professor of Musicology at the Ecole des Hautes Etudes en Sciences Sociales (EHESS) in Paris, is the author of *Playlist. Musique et Sexualité* (Editions MF, 2022), *Trauermarsch. L'Orchestre de Paris dans l'Argentine de la dictature* (Éditions du Seuil, 2016), and *Beethoven's Ninth: A Political History* (University of Chicago Press, 2003), among other books. He has co-edited *Composing for the State: Music in Twentieth Century Dictatorships* (Routledge, 2016), *Finding Democracy in Music* (Routledge, 2021), and other volumes.

Alessandro Caliandro is Associate Professor in Sociology of Culture and Communication at the Department of Political and Social Sciences at the University of Pavia. His current research focuses on digital methods, digital consumer culture, platformisation of culture, surveillance capitalism, and smartphone use.

Fulvia Caruso is Associate Professor in Ethnomusicology at the University of Pavia (Cremona). Since 2014 she has been researching music and migration in the central Po Valley. Her fieldwork also addresses the processes of heritagisation of Italian intangible cultural heritage, with a focus on religious rites, vocal styles, and craftsmanship.

Flavio Antonio Ceravolo teaches Sociology at the Department of Political and Social Sciences at the University of Pavia. He is Director and co-Coordinator of the Master's Degrees in MUST and in Digital Communication at the University of Pavia, and also Rector of

Contributors xi

Collegio Benvenuti Griziotti. His research focuses on universities' third mission, science communication, and digital communication.

Nakshatra Chatterjee is currently writing his PhD on contemporary Indian literature at the Indian Institute of Science Education and Research, Bhopal. His area of interest includes the ways of understanding the intersections between aurality and literary studies. His research approaches visual epistemology from critical and interdisciplinary perspectives.

James Deaville teaches music at Carleton University, where he carries out research on news music and sound. He recently published, with Chantal Lemire, 'Latent Cultural Bias in Soundtracks of Western News Coverage from Early COVID-19 Epicentres,' in *Frontiers in Psychology* (2021). His other publications address news music, the Vietnam War, and the Iraq War.

Antonella Dicuonzo is a PhD candidate in History of the Performing Arts at the University of Florence. She obtained a MA in Musicology at the University 'La Sapienza' – Rome. Her research interests range from the relationships between music, trance, and new therapeutic devices, to those between music and religion, and music and minorities.

Michela Garda is Professor for Musical Aesthetics and Sociology of Music at the Department of Musicology and Cultural Heritage, University of Pavia. Her recent publications include *The Female Voice in the Twentieth Century* (ed. with Serena Facci, Routledge, 2021). Her research addresses the history of music aesthetics, the performativity of the voice, and the theory of vocal gesture.

Alessandro Greppi holds degrees in law, political science, and musicology. Trained as a classical pianist, his research explores the relationship between sound, music, and armed violence in the Sahel. He is currently completing an assignment as a Protection of Civilians Advisor with the United Nations in Mopti, Mali.

Niels Chr. Hansen (PhD, MSc, MMus) is Assistant Professor and Marie Skłodowska-Curie Fellow at Aarhus Institute of Advanced Studies & Center for Music in the Brain, Aarhus University, Denmark. He is a member of the Danish Young Academy and co-founded the global MUSICOVID research network in 2020.

Robert Losiak is a musicologist and sound ecologist. He holds a PhD and is affiliated with the University of Wrocław, where he founded there the Soundscape Research Studio. He has designed a research project on the soundscape of Wrocław, and co-edited Audiosfera miasta (Wydawnictwo Uniwersytetu Wrocławskiego, 2012), Audiosfera Wrocławia (Wydawnictwo Uniwersytetu Wrocławskiego, 2014), and other publications.

Theodoros Lotis is a composer working in the field of electroacoustic and mixed music. He has taught composition at the University of London (Goldsmiths), the Aristotle University of Thessaloniki, and the Technical Educational Institute of Crete. He is currently Associate Professor at the Ionian University. His music has been released by Empreintes Digitales.

Marco Lutzu is Research Associate of Ethnomusicology at the University of Cagliari. He has conducted field research in Sardinia, Cuba, and Equatorial Guinea, focusing on the relationship between music and religion, improvised poetry, multipart music, ethnoorganology, audiovisual ethnomusicology, performance analysis, and hip-hop culture in Sardinia.

xii *Contributors*

Ignazio Macchiarella is Full Professor of Ethnomusicology at the University of Cagliari. He has carried out fieldwork in Sardinia and Corsica focusing on multipart music, the relationship between music and religion, formal analysis of orally transmitted music, music as performance, and music making in the landscape of contemporary social life.

Regina Machado is a singer, composer, guitarist, and Professor of Popular Music at the State University of Campinas (UNICAMP). She holds a PhD in Semiotics and Linguistics from the University of São Paulo. She has published widely on Brazilian popular music, alongside an extensive performance career, including as part of the trio Líricas Modernas, which toured more than 40 cities in Brazil.

Daniel Margolies (PhD) is Director of Strategic Initiatives for the Guadalupe Cultural Arts Center in San Antonio, Texas. He has written widely on historical topics, Mexican migrant music, and cultural sustainability in Texas-Mexican music. He is founder and Artistic Director of the Festival of Texas Fiddling, which features *son huasteco* music each year.

Dominika Moravčíková is a postgraduate researcher in the Institute of Musicology at Charles University in Prague. She conducts ethnographic research on music education of Roma children in Slovakia. Her research interests include nationalism, folklore revival movement, urban soundscapes, voice culture, and racial constructions in listening.

Daniele Palma is a post-doctoral research fellow at the University of Bologna, working on evidence of Giuseppe Verdi's operas performance practice in 19th-century music periodicals. His research concerns operatic vocality in the 19th and 20th centuries, early sound media, cultural imaginaries of opera, and amateur music practices, from children's records to mental patients.

Ludovico Peroni is a musicologist, improviser, and composer. He holds a PhD in History of Performing Arts – Musicology at the University of Florence. His interests include improvisation, aesthetic studies and musical analysis, mainly applied in the repertoire of contemporary and audiotactile music.

Suzel Ana Reily is Titular Professor of Ethnomusicology in the Music Department of the State University of Campinas (UNICAMP). She is currently the coordinator of the FAPESP-funded framework project 'Local Musicking – New Pathways for Ethnomusicology.' Her publications include *The Routledge Companion to the Study of Local Musicking* (with K. Brucher, 2018), and *The Oxford Handbook of Music and World Christianity* (with J. Dueck, 2017).

Marcel Zaes Sagesser is Assistant Professor at SUSTech (Shenzhen, China), holding a PhD in Computer Music and Multimedia from Brown University. His work is located at the intersection of sonic materiality, the technologies of sound (re)production, digital rhythm machines, and popular culture. He has been awarded a number of grants and art prizes, and repeated artist residencies.

Giulia Sarno is a post-doctoral research fellow in Ethnomusicology at the University of Florence. Her current interests span a variety of topics in contemporary musical and sonic practices, from avant-garde electronics, to oral traditions, and popular music. She is investigating the relationship between sound and football, and the history of the centre for music research *Tempo Reale* (Florence).

Diane Schuh is an experimental landscape architect and musician. She is studying ways of making gardens as a composer. Her research explores the pedagogical and operative potential of the symbiosis model for compositional and listening devices that invite attention to the living. She is preparing a PhD under an EDESTA contract, within the CICM team of Musidanse laboratory at University of Paris VIII.

Makis Solomos is Professor of Musicology at the University Paris 8 and director of the research team Musidanse. He has published many books and articles about recent music. His latest book is *From Music to Sound* (Routledge, 2019).

Michaël Spanu holds a PhD in Sociology (University of Lorraine). He is a fellow of the Post-doctoral Fellowship Program of the Universidad Nacional Autónoma de México at the Centro de Investigaciones sobre América del Norte, under Dr Alejandro Mercado's supervision. His work focuses on the live music industry in North America and its links to digital platforms, in addition to nightlife in Mexico City.

J.A. Strub is completing a PhD in Ethnomusicology at the University of Texas-Austin and is the founding Director of the *Gran Foro de Cultura Huapanguera en Texas* music festival, in addition to doing public ethnomusicology work presenting traditional music at festivals and symposia.

Renata Tańczuk is Professor of Cultural Studies at the University of Wrocław. Her main research areas are material culture and collecting, and the urban soundscape. She is the author of Ars colligendi. Kolekcjonowanie jako forma aktywności kulturalnej (Wydawnictwo Uniwersytetu Wrocławskiego, 2011), Kolekcja – pamięć – tożsamość. Studia o kolekcjonowaniu (Wydawnictwo Uniwersytetu Wrocławskiego, 2018), and co-editor of several books, including Audiosfera miasta (Wydawnictwo Uniwersytetu Wrocławskiego, 2012, Audiosfera Wrocławia (Wydawnictwo Uniwersytetu Wrocławskiego, 2014), Sounds of War and Peace (Peter Lang Verlag, 2018).

Mark Thorley's research focuses on the impact of emerging technology on the creative industries, and draws upon his background as a classically-trained musician, technologist, and entrepreneur. He has developed and managed several academic programmes in the UK and is Senior Fellow of the UK Higher Education Academy, and a past Director of the Music Producers' Guild.

Melanie Wald-Fuhrmann is a musicologist and has been the Director of the Music Department at the Max Planck Institute for Empirical Aesthetics in Frankfurt since 2013. Before this, she held professorships for Musicology at the Music University in Lübeck and Humboldt University Berlin.

Sławomir Wieczorek holds a PhD in Musicology. He works at the Institute of Musicology at the University of Wrocław. He is the author of the book *On the Music Front* (Peter Lang Verlag, 2020), and co-editor of the book *Sounds of War and Peace* (Peter Lang Verlag, 2018). His research interests focus on the history of 20th-century music and soundscapes.

Acknowledgments

This volume originates from an academic conference that the editors conceived and realised together with Antonella Dicuonzo, Francesco Giomi, and Ludovico Peroni in December 2020. *Sounds of the Pandemic: International Online Conference* was organised by Università degli Studi di Firenze, Dipartimento SAGAS, together with the centre for music research, production, and education, Tempo Reale, and was held online due to the restrictions imposed by the health crisis. It was part of the activities of our research group *Come suona la Toscana*, which is responsible for a triennial project within the national initiative *Heritage, Festivals, Archives: Music and Performing Practices of Oral Tradition in the 21st Century*, involving the Universities of Roma 'La Sapienza' (leading the project), Cagliari, and Torino. The project is funded by the Italian Ministry for University and Research, in the framework of its Relevant Researches of National Interest (PRIN) funding scheme.

We would like to extend our heartfelt thanks to Antonella, Francesco, Ludovico, and all those who contributed to the conference, especially the Tempo Reale and SAGAS staff, and all the scholars and artists who delivered their papers there, making it a great experience of scholarly and human communion in very hard times. We would also like to thank Zachary Androus, for his immense help in revising and formatting the manuscript; our editorial assistants Adam Woods and Emily Tagg at Routledge, and commissioning editor Hannah Rowe, for approaching us with the wonderful proposal for this volume.

Finally, we would like to dedicate this book to Flavio Maurizio, born on 16 February 2022, the first grandson of Maurizio Agamennone.

This volume is the result of research carried out at Dipartimento SAGAS of Università degli Studi di Firenze, and benefits from research funds awarded to Maurizio Agamennone (PRIN 2017 project B54I18010540001).

Introduction

Understanding a Pandemic through Sound

Maurizio Agamennone, Daniele Palma, and Giulia Sarno[1]

On 28 January 2021, with the world suffering the effects of the second COVID-19 wave, around 130 live music clubs in Italy simultaneously posted pictures on social media of their lowered shutters and locked front doors. All of the images featured the same graphics: a big question mark, a set of dates (the year of the club's founding – 2021, followed by a *?*), and the hashtag #ultimoconcerto (i.e., last concert).[2] Hinting at the fact the 2021 might be the year those venues would shut down for ever, this action marked the beginning of a campaign – *L'ultimo concerto?* – intended to raise awareness of the critical conditions facing the live popular music industry (specifically its venues and workers) one year into the pandemic, a period in which the restrictions enforced by the government to stem the spread of the virus had prevented clubs from programming any concerts. Fear of the total collapse of a crucial sector 'for people's cultural growth, and supporting social cohesion' was articulated in a manifesto on the campaign's website www.ultimoconcerto.it, clamouring for recognition, political intervention, and ultimately reform.[3]

A couple of weeks after the launch, a sensational announcement was made: on 27 February 2021, at 9:00 p.m., all the clubs involved in *L'ultimo concerto?* would live stream a full concert from their premises, to be seen free of charge on the campaign's website. The line-up of this peculiar festival, both dispersed (physically) and convergent (virtually), featured prominent names in the Italian popular music scene, along with lesser-known local bands. On the night of the event, an audience of more than 100,000 people connected to the website from their homes, eager to get what they were promised – a live music feast to compensate for all the cancelled or unprogrammed shows of the past year, and the drastic compression of sociality and cultural life. However, those who were able to access the website (and many could not because of the volume of traffic) soon realised the festival was not so much a musical occasion as a political web mob: instead of live-streamed concerts, the many little squares on the website's homepage broadcasted pre-recorded videos, lasting one to ten minutes, that were shot at each of the venues participating in the campaign.

The videos were produced and edited with varying degrees of professionalism. Some would just exhibit an empty stage for some time, with show lights paradoxically illuminating the scene to the rhythm of silence; some were more complex, featuring backstage moments with the announced artists dressing, chatting casually, and getting ready (not) to perform. Most involved sound engineers, technicians, stage managers, roadies, and all kinds of club workers actually doing what they would do on a normal concert night. Many showed performers walking on stage only to remain standing silently and awkwardly, or making dejected statements. Some would rely on words, some on wordless narrations; some used crowd noise to produce an alienating contrast with the visual lack of an audience; others would make the empty spaces resound meaningfully; some resorted to muting the audio altogether. Most videos focused on the bleak present, while a few intertwined it with nostalgic shots from pre-pandemic concerts with bands playing in front of full crowds. Despite

DOI: 10.4324/9781003200369-1

2 *Maurizio Agamennone, Daniele Palma, and Giulia Sarno*

this variety of solutions, every video ended the same way: during the final 40 seconds, before the campaign credits, a text rolled down, revealing the sense of this action – no concerts would be offered, in order to highlight beyond any possible doubt how bad the situation had become for people professionally involved in live music, and the risk of an entire cultural and economic sector's demise.

As social media comments from the audience show, the initial excitement soon turned into confusion and bewilderment, which then gave way to critical reactions comprising a wide array of feelings: if many understood and approved, albeit sadly, of the operation, a very vocal part of the public was hurt, outraged, and even enraged.[4] They felt fooled, betrayed by their favourite artists, and they questioned the efficacy of a campaign that affected, and was mainly visible to, people already convinced of the cultural and social importance of live popular music, thus failing to hit a more meaningful target (that is, politicians). The day after, 28 February, a three-and-a-half-minute long video was published by the campaign's organisers (a montage of all the broadcasted videos);[5] this was accompanied by a short text further explaining the reasons behind what could have been (and in some cases was) mistaken for a cruel joke, with an invitation to the public to support the campaign through its next steps, exemplified by a new manifesto articulating very concrete requests, and political perspectives for saving the industry.[6] The campaign resonated considerably in the national press; nevertheless, at the time of our writing, one more year has passed, and 'the next concert' has not yet happened. This February 2022, a new press release was published on the campaign's website, significantly rebranded as *Nessun concerto!* (i.e., no concerts!), as Italian live clubs are still closed, unable to program any sustainable activity under the ongoing restrictions associated with the fourth wave of COVID-19.

There are many reasons why we chose to discuss this example as a way to open this edited collection, the aim of which is to address the multiple relations between sound and the pandemic. First of all, *L'ultimo concerto?* allows us to situate ourselves – both as human beings experiencing the pandemic, and as music scholars observing it – with respect to the phenomenological complexity of an event that was unprecedented for many reasons. Indeed, as has been widely noted, when compared with the previous great crises of the 21st century – the terrorist attack on the twin towers in 2001 and the 2007–2008 financial crisis – the pandemic presents some singular and exceptional traits, which were particularly evident in spring 2020. First of all, it was both an extra-systemic and systemic shock, in that it originated from the random action of a non-human pathogen whose progression nevertheless depended on the enormous capacity for circulation and interconnectedness at the heart of neoliberal capitalism. In this, perhaps the most significant element is that the pandemic has been a truly global crisis: it has directly affected the lives of a large portion of the world's population within a short time and in homologous forms, by virtue of an unprecedented overlap between the speed of contagion, the spread of information through mass media channels, and the political, technological, and social responses (Teti 2020, 13).

Moreover, the powerful impact of the pandemic in its early stages relied on its intrinsic capacity to engender an *indifferentiating* process, that is, a sudden and dramatic suspension of any kind of hierarchical or functional difference between the self and others that, for better or worse, regulates organised societies.[7] As highlighted by anthropologist René Girard, with his theories of mimetic desire, violence, and the sacred ([1972] 2013), and by Mary Douglas in her work on the conventions of hygiene and the dangers of contamination ([1966] 2002), the anthropological process of *indifferentiation* is one that epidemics share with other types of natural or anthropic catastrophes (like earthquakes, famines, and wars) and with certain ritual forms. This process has fundamental bearings on culture and the symbolic order. Indeed, as a counterpart of the very same mechanism, the pandemic imposed

Introduction: Understanding a Pandemic through Sound   3

itself both as a multiplier of the differences intrinsic to the Capitalocene,[8] highlighting and amplifying a wide range of divides (including technological, economic, racial, and gender), and as a catalyst for further processes of differentiation, particularly evident in the construction of new categories of heroes and scapegoats, and in the progressive polarisation of public opinion into dichotomous positions, especially after the advent of vaccines.

Against the backdrop of these general features, it is quite likely that our initial interest in the sounds of the pandemic was prompted by some of the specificities of how the rapid spread of coronavirus was faced in Italy, and also that our view was influenced by its local repercussions, both short and long term.[9] Drastic measures to fight the disease's outbreak were adopted in Italy as soon as 21 February 2020, when public spaces were closed in several villages in the north of the country; on 9 March, two days before WHO declared the pandemic status, a nationwide lockdown was announced, the first to be experienced in a Western country. All non-essential activities were suddenly shut; our urban soundscape fell under a silent spell.[10] By 19 March, the country's death toll surpassed that of China: Italy was put under the spotlight as the epicentre of the virus outbreak outside of Asia. Only a few days after the lockdown was enforced, international press was reporting on balcony singing in Italian towns under lockdown (e.g., Thorpe 2020). At that point, the multifaceted aural dimension of the pandemic had already emerged: its entanglement of new silences and sounds; the wide array of values, meanings, and affective hues they carried, and how they crucially contributed to marking these unprecedented conditions.

Because of its unfortunate primacy, Italy was a privileged observatory for these new phenomena: by September 2020, when we launched the call for papers for what became our *Sounds of the Pandemic: International Online Conference* (from which this volume originates),[11] many scholars were starting to investigate them all over the globe, as the overwhelming response we received in terms of submissions showed us. Most of the chapters in this book were developed from papers given at said conference – for this reason, they focus mainly on the first months of the pandemic, analysing the processes and issues that sound and music embodied during that period. In this respect, one of the main elements at the heart of this volume is the value and strength of testimony in and of itself, as a driving force that precedes and orients any subsequent operation devoted to the elaboration of a given experience – in this case, that of the pandemic. Therefore, many of the chapters gathered herein attest to and pursue the need for documenting phenomena and practices that, at the time of this writing, have already mostly disappeared and are apparently destined to not be repeated, as well as processes that are still in progress, but not yet fully consolidated.

The second goal that this volume seeks to achieve is to detect how these different phenomena, practices, and processes – their conditions of appearance and modes of existence over time – were implicated in the broader social and cultural mechanisms underlying the collective construction of the pandemic experience – that is, in the complex and mobile set of perceptions, affective projections, and cultural representations in-between individual reflexivity and the reconstruction of collectiveness. In other words, if the multifarious sounds of the pandemic played a pivotal role in the processes of sense-making and reconfiguring individual or group agency, then the authors gathered in the following pages attempt to read them in light of both local epiphenomena and the overall, progressive transformation of the pandemic experience. They strive to grasp the asymmetries, stratifications, continuities, and discontinuities otherwise hidden under the veil of apparent homogeneity that the virus itself seems to have imposed. In our opinion, this ultimately allows for the problematisation of radical narratives of the pandemic as a *rupture* or an *exception*, introducing instead a critical gaze that underlines its extreme mobility, and simultaneously recognises links with the pre-pandemic past, and the potential implications for the post-pandemic future.

The volume is divided into three main sections. The first two are thematically oriented: a series of case studies dedicated to the unparalleled disruption of acoustic environments, and the emergence of peculiar sound phenomena during the spring of 2020; and another on new forms of musicking (Small 1998) fostered by social restrictions and the shift to digital life. The third section, on the other hand, is intentionally transversal, presenting selections that contextualise the pandemic in longer-lasting processes. Our goal is not to give a comprehensive account of the several surveys and research paths undertaken in the last two years, many of which are still ongoing – as testified, for instance, by the activities of the international network MUSICOVID, among the most visible of these initiatives.[12] Rather, we chose to circumscribe a limited array of topics within the subject, isolating certain phenomena that we believe to be particularly significant, also by virtue of their relevance beyond the temporal boundaries of the health emergency. In this sense, the *L'ultimo concerto?* campaign makes a good starting point, as it encapsulates a number of elements characterising the relationship between sound and the pandemic, allowing us to introduce the contents of this volume as they emerge at the intersection of major thematic axes.

Listening to the Pandemic Silence

A fundamental trait of the communication strategy at the core of *L'ultimo concerto?* was its insistence on silence, expressed both aurally in the videos, and verbally in the texts associated with the campaign (such as press releases and social media posts). Here, silence became a complex metaphor, simultaneously signalling the interruption of social and cultural life, and the impossibility of the live music community to make itself heard in the political arena. The effectiveness of this metaphor can be measured against a fact: as many of the chapters in this book show, silence operated as a potent signifier of life under lockdown during the pandemic, revealing itself as a multifaceted aggregate of perceptive acts and symbolic encoding. At the immediate grasp of anyone's ears, silence was the aural counterpart to the images of stillness, emptiness, and disruption widespread throughout media. It became a site for certain fundamental binary oppositions to emerge and be negotiated: reflexivity and inner dialogue versus relationality and outer stimuli, indoor/private spaces versus outdoor/public ones, survival versus life, nature versus culture, and human versus non-human. Given these features, we find it fruitful to follow Michael Bull's proposal of a 'bodily approach to sound and silence': if we recognise that 'emotion and fear … play their role in the phenomenology of sonic experience, then a range of "different" silences might be discovered that challenge us methodologically' (Bull 2021, 27).

Several chapters of this book address this challenge by analysing the new pandemic soundscapes, and their multifarious reception, through diverse approaches. In their inquiry into people's responses in Wrocław, Renata Tańczuk, Robert Losiak, and Sławomir Wieczorek clearly show how the sudden acoustic void brought a radical and suprising rediscovery of sound communication, even if for others this same experience triggered discomfort and, in some cases, severe anxiety. A deep fascination with the hi-fi environment freed from noise, largely relegated to the margins by a graphic-video-centric culture, fuelled a real *praise* of silence. Changes in acoustic ambiances are also at core of Dominika Moravčíková's chapter on the sonic phenomena which emerged in post-communist Czech Republic and Slovakia. A personal account of the author's confinement is the point of departure for her reflection on how sound and silence mediated the porous relationship between the private and the public in culturally related ways, ultimately challenging the apparent homogeneity of the sonic experience of the pandemic. A direct personal experience is also the starting point for Ludovico Peroni, who explores the practice of making ambient sound recordings during lockdown through the lens of recent literature about atmospheres and audiotactile music.

Introduction: Understanding a Pandemic through Sound   5

In his chapter, Peroni shows how what he calls *pandemic soundscaping* allowed for the emergence of a new aura that presents peculiar aesthetic and poietic features.

The theme of silence as a *space of possibility* is further developed by Theodoros Lotis, who focuses on the act of listening as the primary interface between the individual and the environment. Drawing on recent research in cognitive sciences, Lotis hypothesises that the sudden abatement of low-information background sounds occasioned the emergence of ear-driven processes of *latent learning*, highlighting the necessity of acoustic transparency as a mean to develop a better sense of oneself in space. Pandemic silence can thus be celebrated, and seen as an opportunity to shed light on the unsustainable dynamics of the Capitalocene. Given its irreducible plurality, however, pandemic silence must also be deconstructed, as it often embodies forms of *silencing* that go well beyond the *presentness* of sonic materiality to imply processes of power, structures of privilege, and marginalisation. To this end, James Deaville analyses the soundscapes of disaster in Western audio-visual news coverage from China and Italy at the first stages of respective lockdowns. While Wuhanese citizens were invisibilised through footage of empty and silent cities, media representations of quarantined Italians relied heavily on balcony musicking, thus highlighting how media used sound, or its absence, to evoke issues of race, ethnicity, and nationality deeply rooted in the neoliberal capitalist imaginary.

Mediatising Reality

One of the most peculiar features of the pandemic transformation of sonic domains is undoubtedly the unprecedented extent to which social media (and mediatised environments and products more generally) constituted the key, and almost sole, site and paradigm of musicking. This was especially so during the first and second waves of the pandemic, when confinement measures were enforced, and social life in physical spaces was interrupted in most countries. Indeed, the postulate for the success of *L'ultimo concerto?*, or better for its conception, was the enormous spread of live-streamed music performances we witnessed since spring 2020. The event promised to be the utmost online concert experience (for an Italian audience interested in popular music) in the time of the pandemic. Its peculiarity though – an announced live event turning out to be a pre-recorded video broadcast, in a way a parody of the concert streaming format – engages a major critical issue of this volume, that is, the concept of *liveness* and its key role in reflections on mediatisation processes.

Stemming from Philip Auslander's pioneering work (1999), liveness in a mediatised world has been intensely debated well beyond the field of musicology, involving at a minimum theatre-performance studies, and media studies (see Dixon 2007; Salter 2010; Hjarvard 2013; Hadley 2017; Bratus 2019). In line with this multifaceted and cross-cutting field of reflection, the *L'ultimo concerto?* campaign elicits some questions: how was the live experience reconfigured in light of its required transfer to virtual, technologically mediated environments? Did the pandemic foster new ways of conceiving liveness? Or did it just make some features of life in the post-digital era more significant, ubiquitous, and evident than ever? A number of authors in this volume attempt to answer these questions, often through (digital) ethnography, by focussing on specific issues of mediatisation as an operational meta-process that encompasses (1) the formal adaptations of given practices (or texts) to the functioning of specific platforms; (2) the experience of liveness in and of itself, through new forms of simultaneity and engagement aiming at (re)constructing a shared feeling of places and of *being there*; and (3) eventually, the (new) modalities of participation and sociability made possible by social media networks.[13]

In their chapter, Daniel Margolies and J.A. Strub take into account the diverse responses of Mexican musicians to the pandemic, with a major focus on how YouTube

6   *Maurizio Agamennone, Daniele Palma, and Giulia Sarno*

and Do-It-Yourself media channels. By mediating the action of *son huasteco, jaroco,* and *calentano* musicians, and participation by their publics, these channels helped establish an early communitarian response to the crisis. YouTube and Mexico City are also under the lens of Michaël Spanu, who analyses a number of *cumbia sonidera* videos produced before, as well as during, the pandemic, stressing how they facilitated a reshaping of *sonidero* culture, and assessing their heritage value in the domain of digital platforms. Ignazio Macchiarella and Marco Lutzu bring their readers to Sardinia, one of the richest Italian regions for traditional music making: their chapter clearly shows how some Sardinian musicians took advantage of the forced digital transfer to restate the social values and functions of *canto a concordu* (a local form of religious multipart singing) and of *cantzonis* (monodic narrative songs widespread in the south of the island). In the same light, Antonella Dicuonzo analyses the pandemic repertoire and performing practices of Roma and Sinti musicians settled in Italy, highlighting how new communication channels made viable by the health crisis simultaneously facilitated a multifaceted move towards pride in their cultural origins, and a process of gaining visibility within social contexts that usually marginalise them.

The issues of music production through digital means and social networks are further explored in Mark Thorley's chapter, which provides an overview of the knowledge, background, situation, skills, and behaviours necessarily implied by remote forms of music making before and during the pandemic. In doing so, he highlights the transformations that the sociological status of musicians has undergone throughout the last decades, especially in the field of popular music. The topic of musical performance, instead, is at the core of the chapter by Alessandro Bratus, Alessandro Caliandro, Fulvia Caruso, Flavio Antonio Ceravolo, and Michela Garda; they combine different methodological standpoints to discuss a series of case studies on traditional religious rituals, operatic performance, discourses about live clubs, and the circulation of musical performance through social media. Finally, Luca Befera brings us to the Elbphilharmonie in Hamburg, where Alexander Schubert's *Genesis* premiered in spring 2020. The author analyses how the composer readapted the piece – a virtual reality video game implying avatar-musicians controlled by the public from home – and his original conception about its performance, according to the necessities raised by the pandemic, noting how this unexpected state of uncertainty ultimately enhanced the expressive potential of *Genesis*.

Uncertainty is a key word here. As one of the most recognisable features associated with the outbreak of coronavirus (Rettie and Daniels 2021), uncertainty has many connotations. Its individual psychological implications (anxiety, depression, and all kinds of mental stress) emerge several times throughout the pages of this book, just as they emerged in the audience's response to *L'ultimo concerto?*. On the one hand, then, uncertainty was perceived and suffered at an individual level. On the other, it had meaning from a social, collective point of view, which was powerfully symbolised by the big question mark that appeared on the campaign's name and graphics: a dramatic synecdoche of the economic uncertainty in which musicians and industry workers all around the world found themselves. Just as the psychological effects of uncertainty hit the most vulnerable individuals harder, so did the economic effects, causing thousands of people to lose their livelihood in Italy alone. As Suzel Ana Reily and Regina Machado make clear in their chapter, devoted to the Aldir Blanc Emergency Bill in Brazil, political implications come into play: the financial support and policy actions taken by governments reflect the diverse levels of recognition and legitimacy granted to the various fields of music making.

Musicking as a Technology of the Self

The strong emotional responses that the *L'ultimo concerto?* campaign evoked – notably, a shared feeling of betrayal, inducing dismay, grief, and rage that might sound somewhat

Introduction: Understanding a Pandemic through Sound   7

disproportional – shed light onto the affective values that many people projected onto sound and music during the pandemic. Audience members either praised or condemned the campaign precisely for its emotional impact, which was undeniable: they were promised a significant moment of relief, and they were given more anguish. Whether or not they thought it was fair for the organisers to do so in order to make a political point, their reactions explicitly or implicitly highlighted how participating in online music events was considered a precious resource to alleviate the psychological distress caused by the pandemic. It helped people shape their days in a meaningful way, giving them a feeling of activity and a sense of community, or just making them forget their troubles for a while. Generally speaking, by executing functions of 'mood or affect regulation' and 'social cohesion,' which 'appear particularly valuable in times of medical disaster and isolation' (Chiu 2020, 2), musicking in all its forms played a significant role in the strategies that people implemented to cope with these uncertain times.

Perhaps more than any other practices, at least during the spring 2020 lockdowns, the *epic* of the balcony as a new stage for performing the self and the collective represented the perception and expression of a formidable loneliness that suddenly, and with an impetuous crescendo, overwhelmed millions of people. Many, but especially the youngest, felt induced to exorcise marginalisation and relieve themselves of this *anxiety of sociality*. So they suddenly projected themselves outside and beyond their places of imprisonment (apartment buildings, above all), jutting out over acoustically accessible streets and squares, in order to show themselves present through the most easily achievable procedures. They drew on the most disparate repertoires: from song writing to memories of dance songs, from simple sonic gestures (such as clapping hands or striking kitchen utensils) to brief episodes of solo performance on very different instruments, played with very varied skills.

For example, Nakshatra Chatterjee and Srijita Biswas account in their chapter for the Janata Curfew, a voluntary lockdown that took place in India on 22 March 2020: at the suggestion of the Prime Minister Narendra Modi, people celebrated healthcare workers through diverse gestures of sonic euphoria, in ways homologous to what happened in other countries all over the world – such as with the Clap for Carers, a weekly flash mob introduced in the UK as early as 26 March 2020. Hence, *making noise together* was a key element in the participatory expression of solidarity, and in the establishment of new forms of rituality devoted to the pandemic heroes – a category that, as we said before, emerged as a consequence of the differentiation processes fostered by the pandemic. Nevertheless, as Alessandro Greppi and Diane Schuh highlight in their chapter, these very same sonic practices could be repurposed as a mean of protesting against government. This was particularly the case in France, where balconies and windows became sites of socio-political dissent, hosting campaigns such as the #CortègeDeFenêtres, and *casserolade* concerts that parodied the local applause ritual for caregivers.

The multifarious forms of pandemic musicking can be read as a *technology of the self*, a set of real-life musical activities and experiences (from practice to consumption) 'that actors use to elaborate, to fill out and fill in, to themselves and to others, modes of aesthetic agency and with it subjective stances and identities' (DeNora 1999, 54). Hence, pandemic musicking can be properly appreciated as a *mediator* of experience (Hennion 2017; Born 2005; Born and Barry 2018): a site of transformation that serves a wide array of functions (care, well-being, sociability, objectification, and symbolisation) and, ultimately, compensates for the loss of unmediated interactions.[14] Both Melanie Wald-Fuhrmann and Niels Chr. Hansen point towards this direction, in light of their research within the aforementioned MUSICOVID network. In her chapter, Wald-Fuhrmann proposes a comprehensive survey on the relevant types and repertoires of what she calls *coronamusicking*; she analyses them in light of psychological theories on music's functions to highlight how these novel practices can offer precious insights into music-related ways of coping with crises. The

8 *Maurizio Agamennone, Daniele Palma, and Giulia Sarno*

coping capacity of music is also at the core of Hansen's Afterword, which raises questions about how thematically tailored repertoires and practices may emerge (and be studied) in relation to future crises, from natural disasters and climate change to new wars.

Nevertheless, the very same capacity of musicking to foster consolatory processes and moods may also be strategically repurposed within extra-musical power structures and technologies, implying ambiguous issues of nostalgia, identity, and exclusion. Such a Foucaultian approach informs Deaville's chapter, and also clearly emerges in the chapters by Esteban Buch, and Makis Solomos. Buch interrogates the idea of resilience and social cohesion through music by focussing on the destabilisation of consolidated spatial configurations of music making, and on the porous borders between self-expression and social bonds. Solomos is even more radical: he explicitly states that during the first lockdowns, art and music were mostly subjugated to the logics of digital capitalism, their consolatory power becoming a Trojan horse which facilitated a temporary suspension – or worse, erasure – of any other social or political function. On the contrary, the developments of techno-capitalism forced the arts and artists to find new modes of production and consumption, in line with ideas of degrowth and social justice.

Time and Memory

Finally, the unfolding of *L'ultimo concerto?* over these past two years allows us to grapple with the complexity of the temporal perspective with which we are dealing. Indeed, one of the effects of the pandemic has been to alter people's perception of time (Irons 2020; Ogden 2021), and of themselves in the movement and succession of the days, weeks, and months of the usual civic and religious calendars. The possibility of being connected from home, even at a great distance, can disorient and unbalance the habits inherent to movement in space, such as those related to sociability and commuting. This condition also interferes with the memories and habits of circadian processes, freeing for oneself large portions of time otherwise destined to routine daily activities, or to night rest. Thus, the availability of conspicuous fragments of social experience in mediatised, repeatable environments and products creates forms of asynchronousness: for instance, lectures, conferences, seminars, even entertainment offerings, can all be enjoyed at night, thus interfering further with social processes and circadian rhythms.[15] The result is a very mobile perception of oneself in the movement of time, and a continuous rearrangement of temporal processes. This asynchronousness has then a particular significance in the case of solitary participation in activities designed as social and shared experiences while closed in one's own house, where asynchronous timelines experienced by physical bodies under lockdown overlap in the same, synchronous, virtual environment, with multiple impacts on the production of meaning, sense of being there, and values.[16] In this light, it is worthwhile to reflect on the construction of significant timelines through the pandemic's 'giant blur' (Williams 2020) that sound, in many ways, can foster.

To go back to *L'ultimo concerto?*, the time span from 2020–2022 (associated with the #nessunconcerto hashtag on which the campaign is insisting at the time of this writing in February 2022), signifies a tragic void in the life of music workers that must be filled as soon as possible. Expanding on this, the perceptual disruption and sense of suspension that the pandemic triggered might be one of the reasons behind individual and collective narrative and self-representational strategies often relying on the performance of acts of memory (Bal, Crewe, and Spitzer 1999), through nostalgic re-activation of archival materials on social media accounts and personal pages during this time of social isolation and uncertainty. Research on cultural recall has shown

Introduction: Understanding a Pandemic through Sound 9

the importance and the elusiveness of the future in all acts of memory ... when despair and uncertainty about the future cast their shadow on the present, only a selective, debris-free, past remained as a potential anchor for personal and group stability and identity.

(Spitzer 1999, 101)

In this cultural framework, infused by nostalgia for a more meaningful (often idealised) past, and fear for an uncertain future, sound can operate as a powerful memory device anchoring people to their social and collective identities. This clearly emerges from several chapters in this volume: from the sonic experience of the 'virus trains' in India, recalling the 'ghost trains' that travelled across the country during the 1947 Partition (Chatterjee and Biswas), to the suspended pilgrimage to the Holy Trinity shrine in Vallepietra, Italy (Caruso); from the appropriation and rewriting of fandom musical practices within virtual audio strategies in televised football (Sarno), to the construction of #*otherbeats*, an archive of participatory location rhythm performances illustrating the effects of sonic leakage, mediation, temporal and spatial distortion (Sagesser).

Nothing shows the importance of these acts of memory better than the repurposing of traditional and folk music in many balcony performances in Italy during lockdown: videos of *tammurriate* in the southern regions of the country, or the Sienese 'Canto della Verbena,' circulated virally over the web and attracted significant attention in the worldwide press (see Deaville, this volume), as well as in scholarly observations (e.g., Chiu 2020). As ethnomusicologists, our duty is to underline the critical implications of such processes; for instance, how narrations often relied on long-standing stereotypes (Italians sing almost by definition), or how these heartwarming sonic reflections of the pandemic could contribute to a celebration of resilience that risks diverting attention from the reality of political failure. However, we simply cannot be insensitive towards these powerful accounts attesting to the relevance of participatory music making (Turino 2008) in keeping people together through hard times.

Notes

1 The three editors are responsible for writing the first section of this Introduction; Giulia Sarno authored the second and third sections ('Listening to the Pandemic Silence,' 'Mediatising Reality'); Daniele Palma authored the fourth and fifth ('Musicking as a Technology of the Self,' 'Time and Memory').

2 See for instance www.facebook.com/LOCOMOTIVCLUB/photos/a.263612243695143/3833878353335163/.

3 The original manifesto, which was replaced several times, can be retrieved here: https://web.archive.org/web/20210224124445/https://www.ultimoconcerto.it/manifesto/.

4 See the over 900 comments on the Facebook post from the night of the event (www.facebook.com/ultimoconcerto/posts/117087550368109).

5 The video (www.facebook.com/ultimoconcerto/videos/956070555151490/) got over 9,300 reactions, more than 1,100 comments, and was shared over 5,000 times on Facebook.

6 The text can be retrieved here: https://web.archive.org/web/20210227214011/https://www.ultimoconcerto.it/manifesto/.

7 Put very simply, if the invisible virus was (and is) potentially everywhere, then anyone could be a source of contagion, and therefore anyone was everyone's enemy, even the dead, like those in Bergamo that were transported at night on military trucks with the impossibility of guaranteeing them proper funeral honours.

8 The concept of Capitalocene was coined by Andreas Malm as a possible alternative to ecological narratives drawing on the idea of the Anthropocene as a geological era, which was introduced by Paul Crutzen and Eugene Stoermer (2000). In particular, debates about Capitalocene try to move

10 *Maurizio Agamennone, Daniele Palma, and Giulia Sarno*

beyond the dualism intrinsic to the 'Green Arithmetic,' that is, 'the idea that our histories may be considered and narrated by adding up Humanity (or Society) and Nature, or even Capitalism plus Nature' (Moore 2016, 2). Instead, Capitalocene seeks to point out the geo-historical dimension of capitalism as an ecological regime in itself, that is, as a system based on the 'mutually constitutive transformation of ideas, environments, and organisations that co-produce relations of production and reproduction' (Moore 2017, 33).

9 The same case of *L'ultimo concerto?* clearly represents this state of affairs: although the campaign was inspired by a Spanish initiative, once it was carried out in Italy it addressed specific national conditions, such as the extreme fragility of the local system for producing independent music. The original Spanish campaign was called *El último concierto?* (www.elultimoconcierto.com).

10 Many contributors to this volume refer to R. Murray Schafer's well-known proposals. His most influential book, *The Tuning of the World*, was published in 1977 by Knopf, New York; it was reprinted under a different title (*The Soundscape: Our Sonic Environment and the Tuning of the World*), but with almost identical content (minor paratextual adjustments only), in 1994 by Destiny Books, Rochester. Our editorial choice was to indicate the 1977 edition when contributors mention the book in general; instead, citations of specific passages refer to the current 1994 imprint.

11 For further information, see https://soundsofthepandemic.wordpress.com.

12 The network was started as early as 19 May 2020 by Niels Chr. Hansen and Melanie Wald-Fuhrmann at the Max Planck Institute for Empirical Aesthetics (Germany), and provides an indispensable tool for anyone interested in music in times of coronavirus. For an overview of the more than 80 ongoing research projects, as well as information on conference meetings and scholarly outcomes, see www.aesthetics.mpg.de/en/research/department-of-music/musicovid-an-intern ational-research-network.html.

13 This tripartite theoretical and methodological framework was established and further developed by Italian media sociologists Laura Gemini, Stefano Brilli, and Francesca Giuliani (2021), in a chapter devoted to the theatrical experience during the pandemic. We find borrowing this framework useful, as it helps in summarising the array of topics that authors engage throughout this volume.

14 Drawing on anthropology and sociology, Georgina Born in particular has developed the topic of musical mediation, with the aim of mobilising ontological discourses on music and pointing to a thorough understanding of musical experience as a complex of embodied, affective, and cultural acts (2010, 8–10). In this, her reflections find a productive resonance within recent theorical work in the field of media studies: see Grusin (2015), and Couldry and Hepp (2017).

15 A paradigmatic expression of these trends can be found in so-called *remote learning*, especially in universities, where new users are now emerging who study and prepare almost exclusively on recorded lessons left in online repositories available to those who live their own and different times, elsewhere and asynchronously with the teaching offered in the classroom.

16 All of these features also pertain to (social) media networks, before and beyond the pandemic, and can be described with the concept of 'context collapse': a situation that 'occurs when people are forced to grapple simultaneously with otherwise unrelated social contexts that are rooted in different norms and seemingly demand different social responses' (boyd 2014: 31).

References

Auslander, Philip. 1999. *Liveness: Performance in a Mediatized Culture.* London: Routledge.

Bal, Mieke, Jonathan Crewe, and Leo Spitzer, eds. 1999. *Acts of Memory: Cultural Recall in the Present.* Hanover: Dartmouth College Press.

Born, Georgina. 2005. 'On Musical Mediation: Ontology, Technology and Creativity.' *Twentieth-Century Music* 2 (1) (March): 7–36. https://doi.org/10.1017/S147857220500023X.

Born, Georgina and Andrew Barry. 2018. 'Music, Mediation Theories and Actor-Network Theory.' *Contemporary Music Review* 37 (5–6): 443–487. https://doi.org/10.1080/07494467.2018.1578107.

boyd, danah. 2014. *It's Complicated: The Social Life of Networked Teens.* New Haven: Yale University Press.

Bratus, Alessandro. 2019. *Mediatization in Popular Music Recorded Artifacts: Performance on Record and on Screen.* Lanham: Lexington Books.

Bull, Michael. 2021. 'Sounds Inscribed onto the Face: Rethinking Sonic Connections through Time, Space, and Cognition.' In *The Bloomsbury Handbook of Sonic Methodologies*, edited by Michael Bull and Marcel Cobussen, 17–34. New York: Bloomsbury.

Chiu, Remi. 2020. 'Functions of Music Making under Lockdown: A Trans-Historical Perspective across Two Pandemics.' *Frontiers in Psychology* 11. https://doi.org/10.3389/fpsyg.2020.616499.

Couldry, Nick and Andreas Hepp, eds. 2017. *The Mediated Construction of Reality*. Cambridge: Polity Press.

Crutzen, Paul J. and Eugene F. Stoermer. 2000. 'The Anthropocene.' *IGBP Newsletter* 41 (May): 17–18. www.igbp.net/download/18.316f18321323470177580001401/1376383088452/NL41.pdf

DeNora, Tia. 1999. 'Music as a Technology of the Self.' *Poetics* 27 (1) (October): 31–56. https://doi.org/10.1016/S0304-422X(99)00017-0.

Dixon, Steve. 2007. *Digital Performance: A History of New Media in Theater, Dance, Performance Art, and Installation*. Cambridge: MIT Press.

Douglas, Mary. (1966) 2002. *Purity and Danger: An Analysis of Concepts of Pollution and Taboo*. London: Routledge.

Gemini, Laura, Stefano Brilli, and Francesca Giuliani. 2021. 'Liveness e pandemia. Percorsi di ricerca sulla mediatizzazione del teatro a teatri chiusi.' In *Shockdown: Media, cultura, comunicazione e ricerca nella pandemia*, edited by Giovanni Boccia Artieri and Manolo Farci, 141–160. Milano: Meltemi.

Girard, René. (1972) 2013. *Violence and the Sacred*. London: Bloomsbury Academic.

Grusin, Richard. 2015. 'Radical Mediation.' *Critical Inquiry* 42 (1) (Autumn): 124–148. https://doi.org/10.1086/682998.

Hadley, Bree. 2017. *Theatre, Social Media, and Meaning Making*. Cham: Palgrave Macmillan.

Hennion, Antoine. 2017. *The Passion for Music: A Sociology of Mediation*. London: Routledge.

Hjarvard, Stig. 2013. *The Mediatization of Culture and Society*. London: Routledge.

Irons, Rebecca. 2020. 'Quarantime: Lockdown and the Global Disruption of Intimacies with Routine, Clock Time, and the Intensification of Time-Space Compression.' *Anthropology in Action* 27 (3): 87–92. https://doi.org/10.3167/aia.2020.270318.

Moore, Jason W., ed. 2016. *Anthropocene or Capitalocene? Nature, History, and the Crisis of Capitalism*. Oakland: PM Press.

Moore, Jason W. 2017. *Antropocene o Capitalocene? Scenari di ecologia-mondo nell'era della crisi planetaria*. Translated by Alessandro Barbero and Emanuele Leonardi. Verona: Ombre Corte.

Ogden, Ruth S. 2021. 'Why Covid-19 Might Be Making Us Lose Our Sense of Time …' *The Cognitive Psychology Bulletin* 6 (1). https://researchonline.ljmu.ac.uk/id/eprint/14300.

Rettie, Hannah and Jo Daniels. 2021. 'Coping and Tolerance of Uncertainty: Predictors and Mediators of Mental Health during the COVID-19 Pandemic.' *American Psychologist* 76 (3): 427–437. http://dx.doi.org/10.1037/amp0000710.

Salter, Chris. 2010. *Entangled: Technology and the Transformation of Performance*. Cambridge: MIT Press.

Small, Christopher. 1998. *Musicking: The Meanings of Performing and Listening*. Middletown: Wesleyan University Press.

Spitzer, Leo. 1999. 'Back through the Future: Nostalgic Memory and Critical Memory in a Refuge from Nazism.' In *Acts of Memory: Cultural Recall in the Present*, edited by Mieke Bal, Jonathan Crewe, and Leo Spitzer, 87–104. Hanover: Dartmouth College Press.

Teti, Vito. 2020. *Prevedere l'imprevedibile. Presente, passato e futuro in tempo di coronavirus*. Roma: Donzelli.

Thorpe, Vanessa. 2020. 'Balcony Singing in Solidarity Spreads across Italy during Lockdown.' *Guardian*, 14 March 2020. www.theguardian.com/world/2020/mar/14/solidarity-balcony-singing-spreads-across-italy-during-lockdown.

Turino, Thomas. 2008. *Music as Social Life: The Politics of Participation*. Chicago: University of Chicago Press.

Williams, Alex. 2020. 'The Year of Blur: How Isolation, Monotony and Chronic Stress Are Destroying Our Sense of Time.' *New York Times*, 31 October 2020. www.nytimes.com/2020/10/31/style/the-year-of-blur.html.

Part I

Accounts

Sounds from a World under Lockdown

1 Listening to the First Lockdown

The Auditory Experience of Wrocław's Inhabitants[1]

Robert Losiak, Renata Tańczuk, and Sławomir Wieczorek

Dedicated to the memory of Dorota Wolska

The reality of everyday life, taken for granted as self-evident, is a sphere in which we follow instructions derived from common-sense knowledge (see Berger and Luckmann [1966] 1991, 38–39). These instructions tell us, among others, what we should do in case our routine ways of acting are disturbed. Problems that undermine our routines make us question, to a greater or lesser extent, the obviousness of everyday life. Some of them we can reintegrate into the sphere of the self-evident without much difficulty; others will require more in-depth exploration of common-sense knowledge in search of interpretations that make such reintegration possible. Profound disturbances of everyday reality call for new models of behaviour, which will be internalised with time and will become just as routine and habitualised as those that regulated our actions before the problem occurred. The COVID-19 pandemic has undermined the self-evident nature of everyday life. Some procedures for acting in the thus altered reality and its interpretations were already known to us before, though most of us had not previously been forced to follow them or refer to them. The pandemic reality, and the lockdown in particular, overstepped the boundaries of everyday life and manifested itself as a separate enclave leaving specific meanings, ways of acting, and different experiences in its trail. The pandemic broke the routine; even everyday conversations lost the sense of ease and coherence that they had enjoyed before its onset. The situation we found ourselves in when the first lockdown was announced may be compared to that which Peter Ludwig Berger and Thomas Luckmann cite as an example of the loss of casualness in exchanges as a result of a break in routines, and consequently the appearance of a potential threat to 'the taken-for-granted reality' ([1966] 1991, 34). They suggest that we imagine the effect on casualness of an exchange like this: 'Well, it's time for me to get to the station,' 'Fine, darling, don't forget to take along your gun' (ibid., 172). The lockdown was introduced in Poland on 16 March 2020, five days after the WHO announced a global pandemic. As a consequence, our everyday conversations lost their casualness, and we began to engage in such rather unsettling exchanges as: 'It's time for me to go shopping,' 'Fine, darling, don't forget to take your mask and gloves, and remember not to come close to your acquaintances in case you meet them.' The exceptional character of this situation, the sense of danger, and the restrictions meant that the time of the lockdown was a very intense and emotionally charged, and thus highly memorable, experience.

The breakdown of everyday routines and interactions was reflected in the auditory experience, which underwent radical transformations as a result of the restrictions and limitations. The soundscape,[2] which, like our exchanges, supports the fundamental reality, changed as well. Our sound environment has its easily recognisable spatial and temporal dimensions. We take its usual sounds so much for granted that they become largely imperceptible to us.

DOI: 10.4324/9781003200369-3

16 *Robert Losiak, Renata Tańczuk, and Sławomir Wieczorek*

We normally do not reflect attentively on our everyday auditory experience. On principle, we harken to and analyse the soundscape of our everyday reality only when it begins to bring sounds which we interpret as warning signals, or ones that are out of place in a given spot and disturb the ambiance to which we have been accustomed. We also listen attentively when we switch from a pragmatic to an aesthetic mode of perceiving reality.

The lockdown, or rather our changed forms of daily life in cities, workplaces, homes, and public venues associated with rest and entertainment, sensitised us to the auditory dimension of those loci and disturbed our routine ways of experiencing them. Reduced road and air traffic; limits imposed on our ability to move and gather in public and even private space; professional work and schoolwork now being done from home – all of this radically altered the soundscapes of the places in which we live. Since the change was sudden and profound, it was *heard* by everyone. We attempted to domesticate the reality of the lockdown and its new sound environment, and to work out suitable models for their interpretation and ways of acting within their confines, which would allow us to relate it to the well-known sphere of everyday life.

Assumptions, Categories, and Methodology

We decided to study the reception of Wrocław's soundscape during the first lockdown (16 March to 24 April 2020) while it was still ongoing. As we had learned from our previous research (Losiak and Tańczuk 2014), extensive in-depth interviews would have been the best technique for collecting data about the emotional and axiosemiotic aspects of urban auditory experience. However, since official restrictions and fear of contracting the disease precluded the use of this method, we opted for questionnaires containing both closed and open questions, which were dispatched to our respondents by electronic mail. We also asked internauts visiting the Facebook profiles of the University of Wrocław Institutes of Cultural Studies and of Musicology to complete our questionnaire. Its pilot version was later slightly modified; we nevertheless decided to include and analyse the data collected from those original forms as well.

Our research considered the ways in which respondents interpreted, evaluated, and emotionally reacted to sound phenomena. The questions we asked concerned changes in the soundscapes of interviewees' domestic space, workplace, and the city at large. We inquired about the aesthetic qualities of those soundscapes and their new elements. Part of the questionnaire dealt with respondents' reactions to the sounds of the city, their sense of living comfort in the changed sound environment, and whether they missed some particular types of sound. One of the questions concerned the dominant mood in the city during the pandemic. Under the lockdown we also conducted participant observation and archived comments on the experience of that period that were published online, in the press, and in books by the inhabitants of Wrocław and other Polish cities. We collected our students' diaries and notes in which they reflected on the soundscape of the lockdown. Together, this material constitutes a substantial supplement to the questionnaire responses, which, as we are well aware, largely came from academic circles. The basic body of data that we submitted to analysis comprises 43 completed questionnaire forms, 12 student reports, and a set of diaries with entries by 24 authors (see Łątkowska 2020). Some of our respondents and authors spent the first spring of the pandemic outside Wrocław. We could nevertheless use their accounts to some extent as supplementary and comparative material.

Our questions were related to the assumption that sound plays a major role both in everyday social interactions and in relations between human and non-human beings. We focused our attention on the auditory experience of urban dwellers, that is, on the ways in which they experience the changing sound environment in which they live. The crucial components of this experience are sensory and affective sensations, emotions, and axiosemiotic acts (defined as those which bestow value and meanings on the auditory phenomena we perceive). This proposed approach to the auditory experience is related to two other categories, namely, those of *soundscape* and *ambiance*. As mentioned in note 2, we interpret the former, following Schafer (1994), as a sound environment viewed from the perspective of an individual or a community.[3] A soundscape is thus a dynamic, changeable sound environment perceived by and reacted to by individuals. As an 'axiosemiotic object' (Pietraszko 1980, 53–73), a soundscape is the result of valuation and semiotisation performed by those individuals.[4] The other category, that of 'ambiance,' comes from Jean-Paul Thibaud (2011), who presented its basic meaning in his article 'A Sonic Paradigm of Urban Ambiances.' Thibaud defines ambiance as 'a space-time qualified from a sensory point of view' (ibid.). The notion concerns the feeling and sensing of a given place, with an emphasis on the somatic-sensory dimension of urban experience. As a quality specific to a particular situation, ambiance is always related to a 'specific mood expressed in the material presence of things and embodied in the way of being city dwellers' (ibid.). Ambiance is subjective and objective at the same time since it is related to the experience of people and to the place which they co-create. Thibaud also observes that urban ambiances 'create a subtle interweaving of synaesthesia and kinaesthesia, a complex mixture of percepts and affects, a close relationship between sensations and expressions' (ibid.).

Thibaud further remarks that 'sound can help us to record, document and describe the dynamics of an ambiance' (ibid.) and reminds us that, by being surrounded by sound, as by ambiance, we become immersed in our environment. Sound situates us in the world. It co-creates any given place and, when experiencing sound, we cannot separate it from its place. Most significantly, as the French sociologist claims,

> sound can mark the urban character of an ambiance. At a very basic level, the nature of the acoustic signals is quite significant: ongoing traffic noise, compact voices in a crowd, electronic device beeps or background music in public spaces operate as indexes of an urban ambiance. But more fundamentally, the density of micro-events, the loss of intervals and pauses, the reverberated sound of enclosed space and the fast pace of street life plunge us into the urban world. No need to say that sound is a very powerful medium to express the sensory ecology of urban spaces.
>
> (ibid.)

The correlations between sound and urban ambiance, as observed by Thibaud, may also be expressed by referring to the notion of soundscape. The experience of ambiance is invariably situated in a definite space, similarly to that of soundscape. In our analysis, soundscape is a key dimension of urban ambiance; it influences its aesthetic qualities and emotional colouring. Soundscape, as we read in Thibaud, is what makes it possible to distinguish individual ambiances from one another: 'Because sound is context-sensitive, it can help us to clarify the situatedness of each singular ambiance. In other words, the sense of audition is sufficiently accurate to properly qualify and precisely distinguish one ambiance from another' (ibid.). Our research on the experience of the soundscape of lockdown aimed to discover the ways in which the city's inhabitants experienced and perceived the sonic environment in which they found themselves, and how they evaluated and interpreted that environment. At the same time, on the basis of the relationship between soundscape and ambiance indicated above, we set out to define the characteristic qualities of ambiance in

18   *Robert Losiak, Renata Tańczuk, and Sławomir Wieczorek*

places where Wrocław's inhabitants stayed or that they visited. That ambiance is also shaped by other sensory stimuli apart from acoustic factors.

The lockdown restrictions changed the sound environment, which in turn significantly affected the urban ambiance of the places where our respondents spent their time. Another process we witnessed was what we call the internalisation of the city. This consisted in limiting the experience of the city and life in it to private space and the closest vicinity of the respondents' places of residence. Domestic space and the immediate neighbourhood became their urban microcosm. Confined to their flats, the urban dwellers were forced to participate in the urban reality from the perspective of their own window or balcony, to harken to that life through the walls. Notably, such attentive listening became perforce more attractive and engaging. As equally significant as the changes in the city's soundscape and accessibility were the changes in the manner of listening to the sound environment, and the increased importance of listening itself.

This chapter discusses the results of our analysis of material that we collected. We have limited ourselves to presenting the experience of the sound environments related to home and the neighbourhood; to the valuation of silence, which transformed the urban soundscape; and to the reception of characteristic sounds of the pandemic which bestowed a particular affective tonality (see Thibaud 2015) on the urban ambiance: all kinds of public announcements about the threat of the epidemic, the sirens of ambulances, and the sounds of human voices distorted by masks. In the last section we put forward hypotheses and present conclusions from our research.

Multiplied Schizophonia

In this period of confinement, everyone's place of residence became a hub for auditory involvement. Home (the house or flat) was not only the space in which most of us stayed and worked nearly all the time, but also a vantage point for monitoring the reality outside. Home mapped out the boundaries of perception which, owing to the restrictions, shrank to the soundscape and landscape accessible from the window, balcony, or the home garden. This may have been the reason for the increased auditory sensitivity that led our respondents to make numerous interesting observations. These comments, which also come from the authors of the notes and diaries we have collected, concern the sounds produced or prompted by other household members, neighbours, passers-by, as well as those emitted by household devices or appliances, and the building's technical infrastructure. Though the respondents quite competently located the sounds in the more immediate or more remote space, they largely put both types of sound on the same level, which blurred the distinctions between domestic noise in their closest proximity and sounds coming from neighbours or the housing estate at large. The pandemic confinement made the latter more recognisable than previously, and thus also more familiar or even intentionally domesticated.

Respondents' attention was frequently attracted by human voices. They stressed first and foremost their increased and far-from-obvious presence at various times of the day, as well as their expressive variety, from conventional conversation to laughter, shouting, and children's squeals, to the sound of domestic rows.[5] Descriptions of human voices emphasise their acoustically oppressive character (loud piercing squeals, shouts, and laughter), as well as their social significance (neighbours conversing with one another from balconies, domestic arguments, and reprimanding children). Notably, some types of noisy behaviour such as partying and carousing were perceived not only as bothersome but also as a source of potential epidemic risk ('the high level of noise irritated me since it revealed they were disregarding the authorities' orders', R–19).[6] In our respondents' minds, then, the sounds of neighbours' activity transformed into signals about possible sources of infection.

The Auditory Experience of Wrocław's Inhabitants   19

The collected interviews make it possible to discuss the ways in which the lockdown's domestic soundscape was experienced. Respondents described it as previously unrecognised and only now discovered anew. Becoming familiar with the sounds of one's place of residence is not tantamount to accepting them. Neighbourhood sounds were frequently perceived as oppressive, interfering, unpleasant auditory phenomena of the confinement period. One of our respondents wrote: 'I began to feel tired of living "together" with my neighbours whom I do not know' (R–18, Warsaw). It was that incessant exposure to sounds coming from neighbours, in the situation of being in a way compelled to listen to them attentively, that created the impression of the (previously unrealised) closeness of other people and their constant presence in the interviewees' personal space. The walls that separated them from others proved not only far from soundproof but also, in a sense, illusory. Their permeability to sounds need not have constituted a problem in times when people did not stay at home nearly all the time. In the period of confinement, however, it became harder to accept. The respondents found themselves in an acoustically shared space which they had already experienced before, but not in such an intense fashion. Some of them also realised that they were actively contributing to that space, which entailed some neighbourly duties as well. One of our respondents wrote:

> During the pandemic, the sounds of the external environment were significantly reduced due to the restrictions, which made those coming from the neighbours more audible. ... This was not such a great discomfort to me as the awareness that I was now more easily heard by others, too.
>
> (R–30)

The pandemic-time auditory experience of domestic space facilitated the emergence of what Schafer (1994, 214–225) calls an acoustic community, in this case grounded more in the sense of acoustic spatiotemporal proximity than in the communicative and symbolic dimension of the sounds being heard. Our interviewees discovered they were able perfectly to identify members of that community, not only members of their own household, but also neighbours, whose individual sonic expressions they could easily identify. This is aptly reflected in the following response: 'I began to be able to recognise what sounds were generated by which of my flatmates; I was able to tell who was entering the flat or preparing meals in the kitchen by the sounds they were making' (R–22).

The prospect of being isolated and confined has an alternative in the form of media, of which the inhabitants take full advantage. The computer and television screen, the radio, and the phone become the acoustically dominant devices of the spaces we inhabit. This introduces a new aspect into the analysis of domestic soundscape experience during the pandemic. Radio and television broadcasts, social media, streaming services, and the like allow us to participate in social (including political, cultural-artistic, even religious) life from home to a much greater extent than before. We could speak here of a kind of *internalisation* of public social life, and the transformation of domestic space into an urban microcosm. In their descriptions of this new to them new-to-them private-zone soundscape, the respondents focused on the sense of acoustic overload and chaos. In the context of working from home, one of them observed that 'the home became more noisy, abuzz with the talk of online meetings' (R–15). This sensation of noise was enhanced by the fact that several persons staying together in one flat or even room participated in online work, school classes, or entertainment at the same time: 'Everyone wanted to see something else on the computer or the TV. Sounds from one or another kind of receiver came from every room' (R–11). This effect became even more bothersome when household members had to speak loudly into microphones. As we can see from the responses, incessant attentive listening to news and announcements ('at the

20 *Robert Losiak, Renata Tańczuk, and Sławomir Wieczorek*

start of the pandemic, television and radio news was listened to at my home round the clock'
[R–50, Serby]), together with virtual participation in the rich offering of attractive online
concerts, theatrical spectacles, and film shows, soon created a sense of being overwhelmed
and overloaded by media content. An excellent example of this process can be found in a
statement in which a respondent admits that, after the enormous initial enthusiasm with
which she greeted the successive events streamed online and open-source materials, day by
day she felt more and more discouraged by their influx. The excess of online information
and content, requiring time spent almost exclusively in front of the computer led, as she puts
it, to a sense of being, in a way, 'detached from reality,' weary of the multitude of sounds
and events.

Analysis of the collected materials legitimises the hypothesis that what our respondents
experienced in domestic space was a form of multiplied schizophonia. The category of
schizophonia introduced by Schafer (1994, 273), refers to the situation of an acoustic split.[7]
People admitted that they felt suspended between the reality of their home space and the
virtuality of the sound environments mediated by the radio, television, the internet, and
recordings. The pandemic schizophonia of domestic space was the result of obligatory
online classes or work, as well as the strongly felt need to communicate with the outside
world during the confinement period, and to build an internal virtual zone of auditory
comfort and intimacy in a situation where private space was shared with others. That latter
tendency was related to the practice (frequently signalled by the respondents) of listening
through headphones and of constantly enclosing themselves in the space of their favourite
music, which becomes a kind of personal 'sonic cocoon' (Flügge 2011). The domestic
soundscape of the pandemic time was thus experienced as schizophonic in a multiplied
form, since the respondents found themselves in many mediated sound spaces at once,
as, for instance, when staying in the virtual sonic environment of their workplace while at
the same time receiving echoes of other sonic spaces shared by their online interlocutors
(such as sounds coming from their colleagues' private flats during online meetings). Such a
case is directly commented on by a respondent who taught an online lesson, which, as she
explains, was

> an acoustically demanding experience because of … the additional sounds that interfered
> with the lesson's sonosphere, coming from family members, various equipment, etc.
> This was precisely the situation I had with the child of one of my students, who was
> crying incessantly. I do not use headphones during my lessons lest I should go crazy. I
> thus let these lessons resound in all the parts of my flat (fortunately I live alone).
>
> (R–2, Warsaw)

In the case of several household members being simultaneously immersed in online work,
however, headphones proved the only solution to this difficult situation of perceptive clash:

> The first week of being confined to our home was hard for us first and foremost due to
> the sounds we produced, with which we mutually disturbed one another. The only choice
> (though far from ideal) was to transfer part of our sound reality into the headphones.
> From the moment of making this decision, each of us has enclosed themselves in their
> own separate sound world for several hours a day. I principally use them most of the
> time, first of all because they are wireless, so I can cook and clean the house in them, at
> the same time listening to a lecture or an audiobook, or to CDs on Spotify. … Unreal
> sounds thus make up the majority of my present life; sounds which do not come from
> the here-and-now of my environment, but are mediated by the headphones, without

which I cannot imagine how I could possibly function in this period of being confined at home.

(R–53)

The author is aware not only of the internally conflicting character of acoustic reality within the limited boundaries of domestic life, in which 'head-space' (Schafer 1994, 119)[8] proves to be the only auditory sanctuary and chance for isolating oneself, but also, as she mentions, of the 'unreal' nature of the sounds that surrounded her.

In many respondents, living among unreal sounds and interpersonal contacts being mediated by the media generated a longing for the real and the non-digital, including non-mediated sounds. The phonic yearnings they describe best reflect the shortcomings of the auditory situation they experienced during the period of strict lockdown. These longings concerned everyday situations and events that involved the voices of persons close to them (for instance, 'dad slurping his coffee' [R–39]), as well as the microsounds of human presence: 'moving chairs back, clothes rustling, the sound of taking notes by hand' (R–22) during academic lectures. These are sounds which, as one of the respondents observed, 'are not audible during distance learning' (R–22). People longed to hear human voices in the streets, in cafés, lecture halls, and during neighbourly conversations.

This longing for contact with live sound environment, expressed by our respondents, was not merely a reflection of their desire to return to the normality of pre-pandemic life. The need for contact with digitally unmediated sound can be considered as autonomous and unrelated to the state of looking forward to the end of restrictions. It is quite likely that the situation in which the respondents found themselves only revealed the existence of this need. The importance of contact with unmediated sounds is evident, among others, in those statements from our respondents which express their disillusion with, possibly even a rebellion against, music accessible in recorded or streamed form. 'Before the pandemic I enjoyed recordings and live concerts equally,' wrote a respondent, 'but now nothing can replace live contact with sound; not even the best recording can fully reproduce it.' 'Currently I do not watch any streamed material and do not listen to recordings. ... During walks I stopped using headphones, which had previously been an indispensable element whenever I left home' (R–50, Serby).[9]

Ambivalent Silence

While flats during lockdown filled, according to our respondents, with sounds previously not heard there at all, or heard with much less intensity – the city fell silent. The sense of silence became the dominant auditory experience and the primary difference between the soundscape of the lockdown and the one that Wrocław city dwellers had been accustomed to. The following statement is emblematic of this situation: 'The city certainly grew simply quieter, by which I mean fewer sounds of people and traffic, as well as overall less noise pollution' (R–30). Under the first lockdown, TV news viewers, and internet users were treated to films and photos of deserted cities. Those new images had a powerful effect. The sight of empty streets and squares, free from the everyday hustle and bustle, of movement and traffic noise, certainly became associated in the audience's heads with silence and soundless space. It would be worthwhile to ask to what extent the change in urban soundscapes, the hushing-up pointed out in the questionnaires, resulted from direct auditory experience, and to what extent it was suggested for instance by visual stimuli and media coverage.

Be it as it may, this silence in the city was experienced and assessed in two different ways. Most responses were positive; the respondents perceived the changed urban soundscape as comforting, relaxing, soothing, and blissful. It helped them to rest and they felt very well in it. Some described it as a sense of long-awaited relief, of rest from the previously experienced urban noise and excess of acoustic stimuli. Such opinions may probably be related to individual acoustic sensitivity: 'I enjoyed the silence' (R–3); 'I like peace' (R–31); 'my hypersensitive ears were treated to soothing silence, and it was genuine rest' (R–12, Warsaw). There were also some respondents who presented a negative assessment of this change in urban soundscape, indicating, first and foremost, that the silence intensified the sense of tension and anxiety evoked by media news ('something sinister could be felt in the air' [R–36, Białogard]). These repliers interpreted the silence as a confirmation of the epidemic threat and an obstacle to finding their place in the changed environment. It generated a feeling of alienation in them ('I can't recognise my own city' [R–29]). Familiar spots lost their sonic identity in their perception, which created a sense of unease, anxiety, and terror. The pandemic silence brought to their minds associations with death and disaster; it signified 'lack of life' (R–41). These respondents described it as 'horrifying', and the metropolis enveloped in silence as 'dead' (R–11), 'a ghost city' (R–32). 'Silence in the city,' wrote another person, 'is a reason to worry' (R–20). Yet another interviewee interpreted the city as 'hushed' by some kind of external force, or even violence. The silent city was perceived as strange and uncanny: 'what was most astonishing was the weird silence all around' (R–32). This silence formed a peculiar urban ambiance for which respondents used the metaphor of a dream to render the experience of urban life having become unreal, suspended, of a surprising break in the city's life similar to the inactivity of nightly repose: 'Strange ... as if from a dream' (R–16); 'all this disappeared, as though the inhabitants had fallen into a sleep' (R–42, Warsaw). There were also comparisons to an apocalypse (R–12, Warsaw) and a surreal film location (R–8, Warsaw), which the quiet and deserted streets supposedly had come to resemble.

Positive valuation of silence was clearly related to the axiological-interpretative framework that one group of respondents applied to the lockdown transformations, according to which environmental values are a priority.[10] For this group, silence in the city is an *improvement* in the sound environment since it eliminates sound pollution and transforms a lo-fi urban soundscape into a hi-fi one (see Schafer 1994, 43). Silence, interpreted as a reduction of urban noise (especially related to transport), brought a chance for a more in-depth, attentive form of listening to the immediate sound environment. Subtle sounds that had previously been harder to notice in that soundscape were now revealed. This concerned, first and foremost, the sounds of nature ('birdsongs could be heard more clearly when road traffic got smaller' [R–4, Warsaw]; 'fewer engines and reduced traffic means that there are more birds singing and the swoosh of trees behind the window can be heard more easily' [R–19]). Embracing this framework does not preclude an awareness of the negative consequences of the observed soundscape change for the city's identity. Negative valuations, instead, resulted from adopting an interpretative framework that primarily emphasises the negative social and cultural consequences of the change, including the sense of alienation and isolation. Within this framework, silence is interpreted as a message carrying a threat to the community and to individuals. *Opponents* of the pandemic-time silence stressed, first and foremost, the lack of human voices in urban space ('It was quieter; the absence of people's voices made it an unpleasant experience' [R–33]), the fact that there was no evidence of human activity, and the much-awaited freedom of behaviour, which manifests itself, among others, in laughter ('No sounds of people working could be heard outdoors ... only rarely could one hear people talking or laughing heartily' [R–39]). This group of respondents also expressed their longing for, among other things, the sounds coming from the nearby school, the voices of children playing in the yard, and of people talking in front of the shop. 'The moderate noise ... related to the habitual behaviour of the city's inhabitants' (R–36, Białogard) was identified with the sounds of life, while the silence that now fell over the city was identified with 'dying' (R–36, Białogard). From this perspective, silence becomes an unwelcome aspect of the urban ambiance that should be eliminated. Some of the respondents recalled the strategies of coping with silence which they adopted, such as listening to the perfectly familiar soundtracks of their favourite films: 'I tried to dispel [the silence] by watching, for the hundredth time, films which I already knew well, which also helps me concentrate, because these sounds are familiar and, in a way, reassuring' (R–39). Also worth noting is the fact that the negative appraisal of pandemic-time silence could change, and people could become habituated to this phenomenon. We recorded two instances of such a change. In both cases, silence, originally viewed as unpleasant or strange, was domesticated with the passage of time and became a source of comfort and pleasure, including even some degree of aesthetic satisfaction.[11]

Conversely, as mentioned above, overall positive valuation of silence may likewise be coupled with some negative emotions or appraisals. It is this kind of duality in the experience of the pandemic hush that some of our respondents pointed out: 'My feelings are ambivalent. On the one hand, I feel better because the city has become less noisy; but I also feel worse since I cannot recognise "my" city' (R–29). This last statement demonstrates the difficulty of unequivocally assessing the new urban soundscape. The confusion is caused by mutually contradictory convictions: one related to silence's environmental value and the subjective sense of auditory comfort, the other concerning the destructive impact of silence on the city's unique ambiance. A similar ambivalence is evident in the following comment: 'I felt better since the place grew quieter, though I missed the hubbub of the streets and the thought of such a teeming city becoming so desolate was sometimes depressing' (R–43).

The Lockdown's Soundmark

What contributed to the atmosphere of anxiety and terror was, according to our respondents, not only silence, but also the megaphone announcements concerning the state of epidemic emergency, the necessity of staying at home in order to avoid infection, and the recommended behaviour. These announcements became a kind of negative soundmark for the lockdown.[12] They were megaphoned in public space, in local streets and lanes as well as along main roads, at crossroads, in parks and green areas, and from vehicles belonging to various uniformed services (mainly the police, but also municipal guards and fire brigades).[13] The announcements were loud enough to be heard inside flats and were mostly listened to in domestic space. The respondents listed them very frequently among the sounds they recalled from the lockdown period, those that were particularly well audible, as well as among sound phenomena previously unknown in the city, and among sounds perceived as unpleasant. The repertory of unique sounds of the pandemic also includes pre-recorded announcements constantly played back in shops (such as those appealing to customers to be quick with their shopping and use disinfectants) and on public transport.

The pandemic-time announcements aroused strong emotions and were negatively assessed as *unpleasant*. In one of the responses, we read: '[these messages] were played back at every bus or tram stop ... so that listening to them over and over again became very exhausting' (R–14). Another person compared the various types of announcements heard under the lockdown and added that 'those made from radio cars were the worst' (R–33). The negative qualities and shades of fear associated by our respondents with those messages represented an entire spectrum, from extreme responses which interpreted them as acoustic evidence of a catastrophe in progress (described as 'hellish,' 'apocalyptic,' 'appalling,' 'oppressive,' 'depressing,' 'evoking very bad associations,' and 'the most bothersome thing heard during the pandemic' [R–20; R–14]),[14] to more moderate ones (in which these announcements

24   *Robert Losiak, Renata Tańczuk, and Sławomir Wieczorek*

were labelled 'grave,' 'rather threatening,' 'provoking anxiety,' or merely 'unpleasant' [R–30; R–31]).[15] Not without significance for such emotional reactions to the announcements was their content and the way they were worded. They exhorted people to feel responsible for others ('the health and lives of ourselves and our loved ones: parents, grandparents, children, friends, and neighbours ... depend on each person's responsible conduct' [Policja.pl 2020]) and informed about legal sanctions ('those who break the regulations concerning public order will be liable to prosecution' [Bielsko-biała.policja.gov.pl, n.d.]).

Such announcements disturbed the familiar ambiance of the city and were therefore negatively appraised. The announcements concerning the lockdown, sent from what the respondents described as 'roaring' radio cars which 'brutally cut into the silence' and spoilt the pastoral atmosphere of the hushed city, were thus perceived as 'irritating' (R–31). This negative assessment of the pandemic soundmark was expressed in responses to questionnaires and in diaries also in terms of contrasts, such as the clash of springtime sounds in the city with the unpleasant police announcements. One inhabitant of Warsaw wrote in his pandemic-time diary: 'an open window, wonderful sunlight, I can even smell the Vistula River'; music is playing (W.A. Mozart's Piano Sonata No. 12 in F major, K. 332), but it is drowned out 'now and again by the gibberish from the loudhailers' (Bień 2020, 346) placed on the radio cars.[16]

In descriptions of these megaphoned messages, we can also find a reflection of cognitive helplessness in the face of the pandemic reality. The respondents seem to have proved unable to incorporate the experience they were facing into the existing categories of their knowledge about everyday life. The announcements were perceived as 'unreal,' 'creating a sense of unease,' or 'slightly surreal' (R–4, Warsaw). Many commented that they had never heard such announcements from uniformed services made live or in public space. One person stressed that the only related previous experience was that of consumer credit being advertised from loudspeakers mounted on a car. Another pointed out that such phenomena had previously only been known 'from films or the internet' (R–31). The respondents also dwelt on the acoustic aspects and the articulation of the recorded messages, which they claimed were 'unnaturally loud' and 'piercing' (R–30), reaching their flats suddenly and unexpectedly, coming from a car that was cruising the neighbourhood. In the responses we find references to the tone of the disembodied voice (or rather, to quote Barry Truax, that of a voice 're-embodied via the loudspeaker' [2012, 62]) and the (male) gender of the speaker. The person who read out the warning was said to have sounded 'serious,' 'rather threatening,' and 'unemotional,' which altogether created an 'unpleasant impression' (R–2, Warsaw).

The megaphoned messages frequently triggered general associations with war, and sometimes specifically with one historical event, such as World War II or the martial law introduced in December 1981 by Poland's communist authorities. Such references were mediated not by respondents' individual memories, but by 'communicative memory' (see Assmann 2011, 34–41) – namely, the recollections of family members and other witnesses of these events, as well as 'cultural memory' (see Assmann 2002) derived from films, literary works, and memoirs. An example of this latter type of acoustic memory transmission and its reflection in language can be seen in the references to police megaphones. In respondents' statements (and in many other sources we have analysed)[17] they were quite frequently described as *szczekaczki*, a coinage from the period of World War II (derived from the Polish verb *szczekać*, 'to bark'), originally used as a contemptuous word for the loudhailers installed by the German invaders in the streets of Polish cities in order to broadcast announcements about military campaigns and repressive measures imposed on the Polish population (including lists of those executed by firing squad).[18] Also the (German) language of the occupying forces was commonly referred to as 'barking,' as in wartime poems where we read about 'their [the Germans'] barking speech' (Gelbard 1946, 18) and 'barking loud in German like dogs' (Borowski 1972, 116). According to linguist Piotr Nowak, however, this term only has 'an unclear and obscured meaning' in present-day Polish, 'if any at all' (2007, 101). Nevertheless, the animal connotations render this word inherently negative, and Nowak claims it is still used to describe phenomena viewed as hostile and alien. This may be the reason why none of the questionnaires assessed the megaphoned messages as useful and conducive to improving the epidemic situation in Poland.[19]

Notably, video recordings of police cars with mounted loudhailers cruising the deserted streets were almost immediately (in late March 2020) incorporated by internet users into their own messages and commentaries on the pandemic. New soundtracks were added, made up of remixed statements full of obscene words, vulgar anti-clerical songs, and football fans chanting what they think about rival teams.[20] Only one such a remake was of a different, more humorous character (the added sound signal was that of food vans selling frozen products). Such remakes modify the above-presented, quite uniform image of the announcements' emotional perception, but they do not invalidate the thesis that the phenomenon was appraised negatively as hostile and alien. The videos may be interpreted as a way of releasing the tension caused by the pandemic anxiety and frustrations. On the other hand, it was a rather obscene form of ridiculing the police, and the overall message was anti-hierarchical and anti-establishment. The negative responses to the pandemic's soundmark may have resulted from interpreting them first as signs of threat and destruction of the hitherto safe everyday reality, and second as tools of an oppressive system.

Can this negative attitude to pandemic soundmarks change? Can they become a neutral and acceptable element of soundscape? We have only recorded one case of a person whose attitude to them changed with time. She claimed that 'originally they fuelled fears (by bringing to mind my parents' memories of the martial law), but after a week I managed to get used to them' (R–48, Ostrów Wielkopolski). Much more frequently, however, respondents admitted that the police warnings were among those sounds they would rather never hear again in their lives. Their negative connotations and content prevent, or at least slow down, the habituation of this element of soundscape; it never turned into a neutral backdrop for everyday activities.

Conclusions

Our analysis of the interview material on soundscape experience in private space has led us to formulate two hypotheses: first, that an auditory community emerged, in which individuals shared the same spatial and temporal experience of sonic presence; and second, that multiplied schizophonia was experienced in the form of the perception of, and participation in, several auditory spaces at once during online meetings. The former hypothesis was already to some extent confirmed by our earlier research into the reception of Wrocław's soundscape, which pointed to the role of sounds heard through the walls of respondents' flats in building neighbourly communities. Sounds from the neighbourhood play a major role in cementing social bonds, as Aleksandra Kil (2014) argues on the basis of Bruno Latour's Actor-network theory (ANT). The second of our hypotheses needs, in our view, to be researched in more detail before it is confirmed. Generated by the new digital media, multiplied schizophonia is the experience of simultaneous participation not only in different sound environments, but also in the sonic dimension of different social interactions. Though the perceiving subject is not an intentional agent of these interactions, he or she nevertheless becomes their witness, and sometimes also their involuntary and unwelcome participant.

Significantly, the respondents pointed to the need for contact with sounds of human activity and with live music unmediated by the media. Under the lockdown, they longed

26 *Robert Losiak, Renata Tańczuk, and Sławomir Wieczorek*

for the microsounds of actual human presence in public and private space and for the live experience of music-making. This need for unmediated contact with real-life, non-media generated sounds may prove relevant in acoustic design, as well as in the fields of aesthetics and ambiance. There are many possible conclusions with respect to designing sound environments in situations of social emergency such as the pandemic confinement. One concerns the necessity properly to construct the public audio announcements megaphoned in the streets, played back on public transport, and in shops and offices. Both their linguistic form and content, which must be persuasive but balanced, and their acoustic-sonic setting and quality, need to be carefully thought out.

Wrocław's soundscape, changed by the lockdown, is one of the elements that constitute the city's unique ambiance. Thibaud (2015, 43) writes that ambiance may not be reduced to just one aspect such as the sound environment. Ambiance is jointly created by such different correlated constituents as climate, architecture, human activity, light, types of public transport, and so on. Though all of them are important, what many of them have in common is that they are either a source or a medium of sound. The deserted streets, inhabitants' reduced activity in public space, distance maintained during social interactions, queues in front of shops (due to the limits imposed on the number of customers inside), the smaller number of commuters on public transport, closed cafés and restaurants, cars cruising the streets with megaphoned announcements, human voices muffled by the protective masks – these are only some of the most obvious elements that constitute the ambiance of a city under the lockdown, influencing the visual, auditory, haptic, and olfactory experience.

The city's soundscape was perceived as hushed and appraised positively or negatively depending on the adopted axiosemiotic interpretative frameworks. The calm-bringing silence was, however, frequently ambivalent. The pleasant experience of the calming down of urban life was sometimes accompanied by fear, anxiety, or sadness. This affective bias of the city's ambiance was also caused by other, non-auditory and, as we have pointed out, in many cases non-sensory factors. Media images of cities under the lockdown; broadcasts informing about the COVID-19 epidemic and its consequences; data concerning the number of new cases, patients put on ventilators, and deaths, but also grassroots civic initiatives such as facemask sewing, making meals for medical personnel, and helping senior citizens with their shopping; the war metaphors used by journalists; the constant references to the 1918 Great Influenza epidemic as an event to which the current situation could be compared, as well as to such literary texts as Albert Camus' novel *The Plague*, and films such as *Contagion*[21] – all of these certainly exerted an impact on how the urban ambiance was perceived and how this experience was articulated. The language used to convey wartime experiences, still remembered thanks to intergenerational transmission and cultural texts, was now applied to describe the city's soundscape and atmosphere. The memory resources mentioned above and analogous experiences objectified in language and in cultural artefacts, as well as meanings related to them, constituted a store of knowledge upon which our respondents drew.

(Translator: Tomasz Zymer)

Notes

1 This chapter has been published with financial support from the Excellence Initiative – Research University (IDUB) programme for the University of Wrocław.
2 We will use the terms soundscape and sound environment synonymously. They refer to the full spectrum of sonic phenomena taking place within a given space, as viewed from the perspective of an individual or collective subject. We adopt such an understanding of soundscape from Truax (2001).

The Auditory Experience of Wrocław's Inhabitants 27

3 R. Murray Schafer's concept of soundscape has been criticised, among other reasons, for its focus on eyesight, which leads to ignoring the unique character of the auditory experience, the processual nature and dynamism of sound phenomena, and the listening subject's active participation (Rodaway 2002, 82–114); the normative character of that concept (Kelman 2010, 212–234; Stern 2013, 181–193); and the reduction of multisensory landscape perception to its auditory dimension alone (Ingold 2011, 136–139). Mindful of these critical opinions, we have related our analysis of soundscape to Jean-Paul Thibaud's (2011, 2015) idea of ambiance, which takes into account the multisensory character of our experience of the environment. We do not pass value judgments on any existing soundscape; instead, we have left its valuation to our respondents.

4 The category of axiosemiotic object (Pietraszko 1980, 53–73) concerns an object to which a meaning and value have been assigned, and through which values therefore manifest themselves.

5 Respondents also listed many other sounds related to household members' and neighbours' daily activity, such as preparing meals, hoovering, renovating, watching television, playing music instruments and singing, children playing in the flats, moving furniture, etc., as well as sounds emitted by household devices and appliances.

6 Quotations from the collected source material have been marked in our article as follows: R (for respondent) plus the number of the response. Place names are quoted only for those quotations from interviews, accounts, and diaries, which do not concern the city of Wrocław. All the questionnaires, students' works, and reports have been stored at the Library of the Institute of Cultural Studies and Musicology, University of Wrocław.

7 The concept of schizophonia 'refer[s] to the split between an original sound and its electroacoustic reproduction. Original sounds are tied to the mechanisms that produce them. Electroacoustically reproduced sounds are copies and they may be restated at other times or places. I employ this "nervous" word in order to dramatize the aberrational effect of this twentieth-century development' (Schafer 1994, 273).

8 Schafer introduced the notion of 'head-space' (1994, 119), used in colloquial youth discourse for the sound space of music experienced through headphones or earphones.

9 Such statements suggest a wider question: is it possible that the pandemic confinement may result in a revival of more traditional forms of participation in musical culture?

10 Interpreting silence within this framework could also have been influenced by reports that diminished traffic would improve climate conditions and would help alleviate the environmental crisis. TV news and websites presented images of animals venturing into the cities in the absence of humans. Nature, it was claimed, could now take a rest from humankind.

11 To the question concerning the perception of Wrocław in the pandemic period, one of our respondents replied as follows: in the beginning there was 'a sense of anxiety and expectation, surprise, fear, but at times also moments of satisfaction derived from the new calm in the city and the smaller number of people seen at every place' (R–3).

12 We have revised the meaning of Schafer's original term by assigning to it a negative rather than positive connotation. The thus conceived soundmark is no longer a 'protected' and 'specially regarded' sound signal, but it still remains 'unique' and functions within a community. See Schafer 1994, 10.

13 Such announcements appeared already a few days after the Minister of Health's regulation of 20 March 2020 announcing the state of epidemic in Poland. They were made for a period of up to two to three weeks associated with the most rigorous restrictions concerning the freedom of movement (binding until 20 April 2020).

14 This is also how the media coverage was constructed. One of the tabloids wrote about a 'horrifying' announcement heard by the inhabitants of Cracow. The extent to which the individual experience of the lockdown was indebted to media discourse calls for further research. We may hypothesise, however, that this discourse did have an impact on the ways emotions and experiences were articulated, and on how the urban ambiance was valuated.

15 Only in two cases can a slightly different attitude to these announcements be detected, associated with a certain degree of curiosity about their sound. Both respondents write about 'alarming' and 'unpleasant' messages, but one of them also called them 'intriguing' because they brought

reminiscences of 'wartime,' while the other person's attention was attracted to their acoustic aspect and the various intonations of the speaker reading out the announcements.

16 The author of the diary attempted an interesting transcription of the distorted announcement heard from a passing car: 'Avrav-sh-rbamsh-do-do-urshmba-departure-nonono-every-opportunity-brshmrr-assembly-yyyyyrtd-yyyyytsrt' (Bień 2020, 346).

17 For instance, in one of the pandemic-time diaries, its author noted a change in the soundscape of Cracow, commenting in an entry for 23 April 2020 that 'the *szczekaczki* have been silent for several days' (Sambor 2020, 409).

18 In the memories of Anna Czocher (2007, 22) we read of 'hundreds of people with their heads bowed down' and how the collective listening to these announcements was accompanied by silence and tears.

19 We have met with a relatively favourable opinion once only, in an online commentary on a film recording of an announcement from a radio car: 'A ghastly impression. *Black Mirror* through and through, it sends shivers down my spine. But if it can help to make the passers-by aware [of the threat], then it's necessary.' This commentary interestingly associates the pandemic situation and police announcements with a popular dystopian TV series (Wyborcza.pl 2020).

20 For instance, in: www.youtube.com/watch?v=S8WQXF2sCDM; www.youtube.com/watch?v=Vk1yxU7TFAY.

21 Immediately after the lockdown was announced, websites nominating the most important books and films about epidemics appeared on the internet. They were even recommended to be read and watched while under quarantine. To quote one of the authors' personal experiences, in the early days of the lockdown a neighbour asked Renata Tańczuk to lend her Gabriel García Márquez's *Love in the Time of Cholera*. She explained that everyone claimed that book was now a must-read.

References

Assmann, Aleida. 2002. 'Vier Formen des Gedächtnisses.' *Erwägen, Wissen, Ethik* 13 (2): 183–190.

Assmann, Jan. 2011. *Cultural Memory and Early Civilization: Writing, Remembrance, and Political Imagination*. Cambridge: Cambridge University Press.

Berger, Peter Ludwig and Thomas Luckmann. (1966) 1991. *The Social Construction of Reality: A Treatise in the Sociology of Knowledge*. Harmondsworth: Penguin Books.

Bielsko-biała.policja.gov.pl. n.d. https://bielsko-biala.policja.gov.pl/ka2/informacje/wiadomosci/284 633,Policjanci-przestrzegaja-przed-lamaniem-przepisow-o-zwalczaniu-epidemii.html.

Bień, Paweł. 2020. 'Silva rerum.' In *Wiosnę odwołano. Antologia dzienników pandemicznych*, edited by Mirosława Łątkowska, 337–358. Kraków-Warszawa: Instytut Literatury.

Borowski, Tadeusz. 1972. 'Odrodzenia człowieka.' In *Poezje*, edited by Tadeusz Borowski, 116. Warszawa: Państwowy Instytut Wydawniczy.

Camus, Albert. 2021. *The Plague*. Translated by Laura Marris. New York: Alfred A. Knopf.

Czocher, Anna. 2007. 'Gadzinówki i Szczekaczki. Ze wspomnień o okupowanym Krakowie.' *Biuletyn Instytutu Pamięci Narodowej* 12 (83): 14–22.

Flügge, Elen. 2011. 'The Consideration of Personal Sound Space: Toward a Practical Perspective on Individualized Auditory Experience.' *Journal of Sonic Studies* 1 (1). www.researchcatalogue.net/view/223095/223096.

Gelbard (Czajka), Izabela. 1946. 'Pieśń o kupcu żelaznym Abramie Gepnerze.' In *Pieśni żałobne getta*, 18. Katowice: Wydawnictwo Julian Wyderka.

Ingold, Tim. 2011. 'Four Objections to the Concept of Soundscape.' In *Being Alive: Essays on Movement, Knowledge and Description*, 136–139. London: Routledge.

Kelman, Y. Ari. 2010. 'Rethinking the Soundscape: A Critical Genealogy of a Key Term in Sound Studies.' *The Senses and Society* 5 (2): 212–234. https://doi.org/10.2752/174589210X1266838 1452845.

Kil, Aleksandra. 2014. '(Od)głosy zza ściany. Audiosfera wrocławskich sąsiedztw w perspektywie teorii aktora-sieci.' In *Audiosfera Wrocławia*, edited by Robert Losiak and Renata Tańczuk, 283–297. Wrocław: Wydawnictwo Uniwersytetu Wrocławskiego.

Losiak, Robert and Renata Tańczuk, eds. 2014. *Audiosfera Wrocławia*. Wrocław: Wydawnictwo Uniwersytetu Wrocławskiego.

Łątkowska, Mirosława, ed. 2020. *Wiosnę odwołano. Antologia dzienników pandemicznych*. Kraków-Warszawa: Instytut Literatury.

Márquez, Gabriel García. 1988. *Love in the Time of Cholera*. Translated by Edith Grossman. New York: Alfred A. Knopf.

Nowak, Piotr. 2007. 'Opozycja swój/obcy a skuteczność komunikacji – wybrane zagadnienia.' *Etnolingwistyka* 17 : 99–108.

Pietraszko, Stanisław. 1980. 'O sferze aksjosemiotycznej.' In *Problemy teoretyczne i metodologiczne badań stylu życia*, edited by Andrzej Siciński, 57–73. Warszawa: IFiS PAN.

Policja.pl. 2020. 30 March 2020. https://policja.pl/pol/aktualnosci/186748,Krakowska-Policja-wykorzystujac-system-naglasniajac-apeluje-o-pozostanie-w-domac.html.

Rodaway, Paul. 2002. *Sensuous Geographies: Body, Sense and Place*. London: Routledge.

Sambor, Mariusz Paweł. 2020. 'Mistrz ukradkowej fotografii.' In *Wiosnę odwołano. Antologia dzienników pandemicznych*, edited by Mirosława Łątkowska, 397–423. Kraków-Warszawa: Instytut Literatury.

Schafer, R. Murray. 1994. *The Soundscape: Our Sonic Environment and the Tuning of the World*. Rochester: Destiny Books.

Stern, Jonathan. 2013. 'Soundscape, Landscape, Escape.' In *Soundscapes of the Urban Past: Staged Sound as Media Cultural Heritage*, edited by Karin Bijsterveld, 181–193. Bielefeld: transcript Verlag.

Soderbergh, Steven, director. *Contagion*. Warner Bros. Pictures, 2011. (1hr, 45 min.)

Thibaud, Jean-Paul. 2011. 'A Sonic Paradigm of Urban Ambiances.' *Journal of Sonic Studies* 1. www.researchcatalogue.net/view/220589/220590.

Thibaud, Jean-Paul. 2015. 'The Backstage of Urban Ambiances: When Atmospheres Pervade Everyday Experience.' *Emotion, Space and Society* 15 (May): 39–46. https://doi.org/10.1016/j.emospa.2014.07.001.

Truax, Barry. 2001. *Acoustic Communication*. Westport: Ablex.

Truax, Barry. 2012. 'Voices in the Soundscape: From Cellphones to Soundscape Composition.' In *Electrified Voices: Medial, Socio-Historical and Cultural Aspects of Voice Transfer*, edited by Dmitri Zakhari and Nils Meise, 61–79. Göttingen: VetR Unipress.

Wyborcza.pl. 2020. Facebook, 25 March 2020. www.facebook.com/krakowwyborcza/posts/10158299004753912.

2 Together in Discipline and Turmoil

Remembering Public Sounds during the COVID-19 Pandemic in the Czech Republic and Slovakia

Dominika Moravčíková

Introduction

On 15 May 2020, Slovak Roma activist and member of the European Parliament Peter Pollák walked up the slightly worn-out asphalt road to the Roma settlement in the town of Žehra, Slovakia, along with dozens of officials, journalists, TV crews, and casual observers.[1] Even from far away, whistling, screaming, and clapping could be heard as the group slowly approached the crowd of masked, quarantined citizens that had assembled behind a blockade of police tape. Each of the groups escorting Pollák was following him to witness the strange affair that was about to take place – one unlike anything seen before in Žehra, or Slovakia for that matter, and that had been utterly unimaginable only months ago. Once the group had reached the police tape, and the crowd hushed itself, Peter Pollák began his statement, facing the restless listeners on the other side of the barricade. He started by expressing his thanks to the local community for being so patient and liable during the time that they were involuntarily locked in the village territory. He then moved on to the important message: 'I want to announce joyful news. Today, at half past three pm, you can go wherever you want.' Seconds after his last word, a cacophony of excited yelling, victory screams, and hand clapping broke out among the men, women, and children. And, after just a moment, the boisterous, uncontrolled turbulence mutated smoothly into the chanting of the word 'starosta,' which means mayor – in this case, the mayor of Žehra, who was responsible for negotiating their rights during the time that their entire village was sealed by police tape.

The residents of Žehra were kept inside a designated territory from 9 April until 15 May 2020. This state-imposed regulation, in response to the rapid spread of COVID-19 infections in the local Roma settlement, was implemented through militarised surveillance, with the help of the police and national army. Unfortunately, Žehra was also not the only case. In April 2020, a total of 6,000 people in five Roma villages were put in a militarised blanket quarantine. The term 'closed village' was coined in Slovak media, referring specifically to the militarised quarantines of Roma settlements. In her reportage from the field, support assistant from the organisation *Healthy Regions*[2] who worked inside the quarantined territory in a protective suit, wrote about the atmosphere on that day, when Peter Pollák arrived in Žehra to deliver his statement: 'We were thrilled, and there was dance. And I joined them. I didn't even need that [protective] suit anymore' (Mižigárová 2021, 127).

This account of the morale in Žehra after Peter Pollák's visit, together with the videos from the moment of his address, reveal something noteworthy about the importance of the sonic affective markers (scream, clapping, dance music) in the expression of the community's victory against both the virus and the punitive measures of the state. But what remains unexplored is how the people involved in this situation (and the people who watched it on

DOI: 10.4324/9781003200369-4

Remembering Public Sounds during the Pandemic 31

screens) sonically structured their experience and social engagement throughout the stages from being disciplined to regaining a state of normalcy.

In this chapter, I discuss the auditory remembrance of the public pandemic experience in the Czech Republic and Slovakia, specifically in the case of the collective emotional sound interventions and their possible legacy – not as an absolute, material archive, but rather as an imaginary 'system of significations' (Castoriadis 2005, 146) that constitutes the selection, construction, and perhaps also invention of acoustic memories from the time of the pandemic, whether it is the practice of claiming space for aural privacy, creating a sense of solidarity and belonging in a crisis, or proliferating the pre-existing and, during times of uncertainty, politically important collective identities that, through music performances, represent social actions cloaked as purely cultural affairs (Herzfeld 2004, 2).

By discussing a personal account of acoustic changes in domestic space, and the publicly disseminated sound practices with concrete social roles, I will comment on the private versus public division of sensory experience, scream as a political expression, and variations of sound-specific social engagement through the newly invented outlets for cultural intimacy during the national lockdowns. Ultimately, I am interested in the question of how can all these categories of sound practices be remembered once the initial affective responses to the pandemic anxieties are gone, and how can the aural remembrance of the pandemic deepen the knowledge about the larger social responses to this global emergency.

Remembering Sound?

In the introduction to *Sound Souvenirs: Audio Technologies, Memory and Cultural Practices*, Karin Bijsterveld and José van Dijck state that although sound as an area of memory production has gained increasing recognition during the recent decades, books on cultural memory have often neglected sound 'as constitutive of remembrance' (2009, 12). Cultural studies scholar Michele Hilmes goes so far as to say that studies of sound will always be academically marginal due to the secondary position of sound in the way we sensorially construct our knowledge about the world (2005, 249). Sound reproduction technology does not help the case of sound-as-memory either, because recordings can never be characterised (and trusted) as solid memories in an ontological sense. Specifically, sound historian Jonathan Sterne argues that 'the recording is less a memory and more a mnemonic' (2003, 320). The reason for this designation is that the mediated sound event is always only imagined, since recording inherently involves a transformation in an abstract object of the performance *before* it is reproduced (ibid.).

Other sound studies scholars have explored the parallels between the visual and the sonic realm of experience. For instance, John Mowitt points to the lack of terminology in English vocabulary around sound that would constitute an equivalent to 'gaze' – the distinctive term in Lacanian analysis of visual culture (2015, 6). To emulate the theoretical intricacy of the *gaze*, Americanist Jennifer Lynn Stoever devises a similar term for the politically charged perception of sound – 'the listening ear.' In her book *The Sonic Color Line: Race and the Cultural Politics of Listening*, focused on exploring the racially defined practices of sound and listening throughout the history of the United States, Stoever introduces the listening ear as 'an aural complement to and interlocutor of the gaze' and 'a socially constructed ideological system producing but also regulating cultural ideas about sound' (2016, 13). Consequently, the listening ear functions as a sensory mechanic for effecting hegemonic ideas and significations that occur 'at the intersection of class, sexuality, gender, and race' (ibid., 15), while it also imposes a standardising measure for what is perceived as 'natural, normal, and desirable' (ibid., 14).

32 Dominika Moravčíková

In the language category of sound, it is worthwhile to explore parallels to the *image* in the sense of a general ideation of space and time. Predictably, *image* can be partially translated to the vocabulary of sound perception as *soundscape* – a term that was introduced to the field of music studies and acoustic ecology by composer and music teacher R. Murray Schafer to describe a totality of various concurrent sound events (1969, 51).[3] However, in more recent literature, soundscape describes a 'perceptual construct' of the acoustic environment, heard or remembered (Brown, Gjestland, and Dubois 2016, 5). But *image* also comes with a larger array of meanings. Philosopher Cornelius Castoriadis argues that *image* – in the meaning of the way something is thought of – is involved in the construction of a particular historical period and ontological condition in which its inhabitants imagine to dwell:

> every society defines and develops an image of the natural world, of the universe in which it lives, attempting in every instance to make of it a signifying whole, in which a place has to be made not only for the natural objects and beings important for the life of the collectivity, but also for the collectivity itself, establishing, finally, a certain 'world-order.'
>
> (2005, 149)

It is debatable whether an ideation of the world and historic time through the domain of soundscape can constitute a concept grounded in real sensory memory. By referring to historian of psychology Douwe Draaisma, who studies how the ability of people to recognise voices from the past declines over time, Karin Bijsterveld and José van Dijck point attention to the fact that there is a difference between *recognising* and '*mentally imaging*' sounds, specifically human voices from the past (2009, 13). This issue of memory decline is not limited to voice recognition but extends to the fundamental question of sound's potential to be memorable. Aural historian Emily Thompson, who writes about the sonic dimensions of modernity, points attention to the tendency of the specific historical soundscapes – which she explains as both natural phenomena *and* the culture through which these phenomena are constructed and perceived (2012, 117) – 'to melt into air' and delete themselves 'from the historical record' (ibid., 125). But it appears that sounds can be very memorable when it comes to personal perception with traumatising impact. In her essay titled *Earwitnessing: Sound Memories of the Nazi Period*, Carolyn Birdsall explores 'the role of sound within personal and social contexts of remembering' (2009, 169), and pursues the theorisation of traumatic sound memories (as opposed to those invoking nostalgia and positive sentiment) described in oral history interviews. Birdsall's inquiry into acoustic memories of traumatic events challenges the 'visually biased notion of witnessing' (ibid., 179) through the concept of 'earwitnessing' that refers to a memory and testimony 'informed by auditory experience' (ibid., 170). In my account of auditory remembrance, I seek to observe how particular, and perhaps isolated, sound practices can build up the remembered soundscapes of the COVID-19 pandemic era in a geographically specific political and cultural condition.

Mutual Trajectories

When the first lockdown measures went into effect in spring 2020 in both Slovakia and the Czech Republic at roughly the same time,[4] the territory of the Czechoslovak region – during the three decades following the Velvet Revolution in 1989, divided into two separate invented nation-states that remain deeply connected by the flow of cultural and economic activity, and by the shared identity of the past – was cut in half by border closures and the halt of international public transportation. Unable to travel back home to Slovakia, and confined to my tiny student room in Prague, I became immersed in the acoustics of my anxiously maintained privacy, varied by brief exposure to the public space of supermarkets and parks. During the lockdown, the room behind my wall in the neighbouring apartment – a room which I never saw and only heard – suddenly ceased to be a kid's room and was transformed into an adult's newly established home office. The soundscape of children's games and rhymes, xylophone, and rattle were replaced by a soundscape of business phone calls and online meetings. The high-pitched child's voice was replaced by an adult's baritone, and excitement and play were replaced by the father's stoicism and routine. After the child had been relocated to some other room, and I had lost the initial object of my acoustic voyeurism, my working days started to be structured according to the business hours of the neighbour (which means that I would wake up to the sounds of him working). The material properties of sound allowed me to acoustically map a given portion of the neighbouring household and, in some sense, become a part of it. As Brandon LaBelle argues in his study *Acoustic Spatiality*, the emplacement of sound generates a structure that connects otherwise unrelated subjectivities, functioning as 'an emergent community, stitching together bodies that do not necessarily search for each other, and forcing them into proximity, for a moment, or longer' (2012a, 1). Our lockdown units were not only porous to sound, but our bodies were also connected through abstractions, associations, and triggers of acoustic proximity that disrupted the previously taken for granted acoustic habitat, in my case through the cultural markers of space that identify it as domestic and intimate or as working and official.

The relational nature of sound poses multiple challenges in compartmentalising the public and private aural experience during the pandemic. I argue that the relationality of sound parallels the limits of social privacy in the sense of a space protected from public connections and interference since the very possibility of privacy is unequally distributed among populations. Therefore privacy is always confined to and restricted by politics. In his article, *The Performance of Secrecy: Domesticity and Privacy in Public Spaces*, Michael Herzfeld goes so far as to claim that privacy belongs to the domain of the public because secrecy and privacy 'as form[s] of social agency' (2009, 136) have performative elements because they must 'be performed in a public fashion in order to be understood to exist' (ibid., 135), and by definition, the very 'requirement of performativity' is that it can only be done in public (ibid., 142). Consequently, every privacy stems from public negotiations and habits that are performative. This binary is even more intricate in the area of acoustic privacy. Sound can get around architecture, walls, and social constructs of property – in the words of Brandon LaBelle, sound 'pushes against architecture; it escapes rooms, vibrates walls, disrupts conversation' (2012b, 470). But at the same time, technology and behaviour exist that may reinforce the private/public disjuncture of sound. In his article *Quiet Comfort: Noise, Otherness, and the Mobile Production of Personal Space*, sound studies scholar Mack Hagood introduces 'soundscaping' – a method of creating aural privacy through noise-cancelling headphones – a technology that travellers use to 'suppress the perceived presence of others' by creating their own sonic spaces (2011, 575). Hagood renders the term soundscaping with a complexity that does not centre privacy as something entirely psychological and unrelated to politics. Instead, he argues that the sensorially erased presence of the other represents a sanitation of sounds which are recognised as 'noise,' and that could be, according to him, any sound that is subjectively othered. The very perception of something as *other* is 'socially constructed and situated in hierarchies of race, class, age, and gender' (ibid., 574). This social production of otherness is theoretically related to the aforementioned aural-gaze generating listening ear, and its split between acoustic normalcy and aberrance (Stoever 2016, 14). Soundscaping is, therefore, very much a political tool. The sensory technology of 'shutting down' the undesired sensory surroundings can also metaphorically operate similarly on a larger scale of political and affective relations across individuals, social groups, and media discourses. The historically and spatially situated usage of

34 Dominika Moravčíková

'suppression of the other' could be observed in the contested space of the imagined national community[5] of Slovaks that could or could not include the Roma minority, depending on the optics of deciding who can be granted the complete package of citizen rights and who may be partially excluded from this category.

In the opening of this chapter, I depicted the historical fragment of Peter Pollák's address to the quarantined citizens in Žehra, who delivered a strong and loud emotional reaction to the information about regaining their freedom of movement. It is noteworthy that this multigroup performance of power, surveillance, media presence, decision-making in public health, and political rhetoric was delivered by politician Peter Pollák as an emancipated member of the Roma minority. Such characteristics situate this event in a complicated position regarding the distribution of agency, control, and political achievement. But it is also challenging to define the publicness of this sound-distinctive collective act. I argue that the screams and clapping of Roma people celebrating the end of militarised quarantine were private, in the sense of performing a mutual affective capacity to undergo and withstand the hardship of surveillance and denial of mobility rights from inside an intimate community in which they shared this frightening experience. The people from Žehra were not screaming primarily towards the group behind the police tape, and in fact, these sounds were not meant for them at all. Instead, they performed the robust joy of victory to *each other*, signalling to one another the competence for the shared affective condition in which they could, through bodily sounds, synchronously celebrate the persistence of their populace. On the other hand, their screams were also public, in the sense of 'social, political, and cultural articulation' (Hofman 2015, 43) of the affect that has become socialised as an emotion. This affect, which Ana Hofman describes as 'raw material' (in this case located in the bodily force of the screams and handclapping), represents a 'pre-personal' ingredient for emotion – a term that refers to an affect that is already ideologically articulated (ibid.). The shouting sounds produced by the voices that, in the words of Ulrike Sowodniok, enable materialisation of 'interpersonal relations, situated expectations, and concepts of the self' (2021, 119), could therefore be read as an affective exercise of community belonging on the intersection of the visceral and the social, the voices functioning as 'a corporeal matter, and a resonating memory-house of personal and societal existence and its processes' (ibid., 118). I propose to accept the undivided double-layeredness of sound significations produced by these bodies – in other words, to appreciate that it *is* political to designate belonging to a group via a specific vocal expression, while it is *also intimate* to use these sounds as building blocks for a collective entity organised around a specific affect (almost as in the theoretical exploration of erotic sounds by Holger Schulze, *'they recognize each other'* in a 'fundamentally liberating and mutual intimate experience' [2021, 158]). Put another way, the crowd in Žehra built up its performance from the 'raw material' of affect that had developed into a social organisation of emotion – which represents an affect that had already been collectively affirmed and reflected.

The sound performance of people from Žehra could also be theorised within the politics of emotion framework developed by Sarah Ahmed, who studies the meaning of love as a fundamental element of group identity formation. By referring to Freud, Ahmed argues that 'groups are formed through their shared orientation towards an object' (2003, 130), which could be one that is missing, while such a condition of absence generates pain and closeness among those who share this love that cannot be reciprocated. This absent object could be, for instance, 'the nation' (ibid.). I propose that in the case of a smaller community in Žehra, this love object is materialised very selectively by a group that shares a tangible experience of suffering inside the quarantine zone. It is also worth mentioning that while members of the local Roma community in Žehra mark their divergence from the outsiders on the other side of the police tape through their disproportionate loudness (especially in comparison

Remembering Public Sounds during the Pandemic 35

to the volume of the politician's speech), their shouting also plays into racial stereotypes of being noisy and untameable. As Gábor Fleck and Cosima Rughiniş have pointed out, Roma are often perceived as loud through the stereotype of their inherent cultural dedication to entertainment (2008, 37). They emphasise that these 'loud' entertainment practices ('looking at TV, listening to music, hanging out, playing games or sports' [ibid.]) are necessary for maintaining a sense of contentment through their daily hardships (ibid., 38). This was doubly true for the residents of Žehra after weeks in militarised quarantine.

It also must be pointed out that this ethnic signification of particular audible expressions has always been present in the perception of Roma voices through the ears of the Slovak ethnic majority. This racialised logic of perception can be explained via the cultural mechanics of 'the sonic colour line.' This term, introduced by Jennifer Lynn Stoever, refers to the ways through which African Americans are subject to racial distinctions through the perception of resonant events and stereotypes, like particular modes of speech, musicking, and listening, by encoding their vocal expression as 'dangerous noise, outsized aggression, and a threatening strength' (2016, 3). Consequently, the perception of the victorious screams of the residents of Žehra, like any other sound practices related to the pandemic, depend on pre-existing affinities of emotion, media discourse, politics, and racial distinction, and the remembrance of these embodied soundscapes is confined to these larger social mechanisms of power, agency, and processes of selection for imprinting in the collective memory. My concern is that these larger processes may result in erasure of the minority experience, and the people of Žehra might thus fail to be considered in the national history of dealing with COVID-19 because, to put it simply, they are too *loud* and too much *not us*. Instead, it is conceivable that their story will be allocated to the memory compartments of difference; the narrow terrain of discourse and archive dealing with *the other*. But while I doubt that this historic moment will escape othering or downplaying of the patience and courage of the people from Žehra in the major historical records of pandemic mitigation in Slovakia, I want to emphasise that the shouting performance of the quarantined residents nevertheless had a very real political articulation that unified the community 'as a shared object of love' (Ahmed 2003, 135). In contrast to this localised group performance of agency and belonging that was openly politically contoured, I want now to shift attention to the group performances that are covert in their politicalness and civic ambition. In the following part of the chapter, I will discuss public sound practices that aspire to influence the larger public sphere through the subtle logic and discourse of cultural intimacy.

Sounds of the Privatised Public

On 25 March 2020, at half past noon, an initiative called *Pianos on the Street* organised a music performance that was broadcast on state-owned Czech television and radio. The goal was to unite the Czech nation in singing and dancing in their windows, on their balconies, and on the streets to 'Není nutno' ('No need') – a widely popular feel-good folk song written by Jaroslav Uhlíř in 1983, which offers a lesson about not taking anything for granted. The project received its share of headlines, and many public figures joined the imagined collective music performance in their households. This event, like many others of the same nature that could be seen and heard all over the world during the time of national lockdowns, could be characterised by a vital motivation involved in the production of public sound practices during the time of uncertainty – a will to reinforce the already existing and consistent collective identities and use them for the protection of social stability. As Georgina Born writes, 'music and sound can be enrolled as means of social regulation and control – through the production of subjectivities, the enactment of power, the organisation of spatial boundaries and the affirmation of identities' (2013, 22). An interesting spin on this goal of solidifying

36 *Dominika Moravčíková*

hegemonic identities could be seen in the group performance of the international Roma anthem 'Gelem, gelem' that was delivered by Roma musicians of various nationalities and streamed online as a part of *Festival Khamoro*, dedicated to the International Roma Day on 8 April 2020. In this case, the sense of national belonging and comforting message of solidarity and identity-based companionship during a crisis intersected with the emancipatory discourse of a minority that also declares its affiliation to the imagined global Roma nation.

The cultural significance of each of these performances lies in the calls for social solidarity at a time when the emotional infrastructures of living in the COVID-19 pandemic era were not yet built. It is nonetheless necessary to recognise this specific performative solidarity as always attuned to particular social groups with the goal to collectively attain good pandemic citizenship. In the words of Adeela Arshad-Ayaz and M. Ayaz Naseem, 'a complementary partnership between the state and the citizenry' is imperative for dealing with the public health challenges of the pandemic (2021, 1). By invoking carefreedom and soothing sentiment, the messages of solidarity, patience, and obedience channelled through sound performances like 'Gelem, gelem' or Není nutno' in physical or online public spaces carry serious roles in maintaining the general image of a stable society[6] – as Michael Herzfeld puts it, a cultural form is, in this case, used 'as a cover for social action' (2004, 2). I propose that the said public sound practices and their involvement in the mentioned state-citizenry marriage resides in the manoeuvres of cultural intimacy – a powerful phenomenon explained by Herzfeld as a social poetics that implements 'the familiar building blocks of body, family, and kinship' (ibid., 5) to nurture a space for 'common sociality' while producing a sense of 'the familiarity with the bases of power' (ibid., 3), ultimately functioning as 'an antidote to the formalism of cultural nationalism' (ibid., 14). Ethnomusicologist Martin Stokes further explains cultural intimacy as a cultural logic based on 'an intimate, as opposed to official, idea of the nation' that is built upon 'a sustained and consequential imagination of public life in affectionate terms' (2010, 16, 193). I believe that this specific affectionate framework for imagining public life can be used as a fruitful approach for theorising the acoustic remembrance of the pandemic experience. As I noted in the previous account of the screams in Žehra, sound is a great carrier of messages related to intimate group belonging thanks to its capacity for bodily engagement, which can be tuned in an acoustic organisation that is culturally moulded and accessible only for a specific social group. But in the case of music performances that were offered to the vast publics as reminders of the little pleasures one should be thankful for even during a public health crisis (*Není nutno*), or reminders of the validity of a marginalised group identity (*Gelem, gelem*), the goal was more far-reaching and less openly political than the performance of victorious screams in Žehra. Whereas the *culturally intimate* music performances were seemingly structured around a sense of homely togetherness and mutual experience of living through unpredictable times, in their core they actually contained a social call – to be patient, stay at home, and follow the advice and commands from the national leadership, perhaps also without worrying too much about the issues of unequal surveillance of marginalised groups, or the socioeconomic inequalities that during the national lockdowns have been nothing but amplified. Since these performances presented themselves as enacted by a purely cultural impulse, while being in their core organised as a social action, they represent examples of how cultural intimacy 'may erupt into public life' (Herzfeld 2004, 3). This public life has been, however, heavily privatised during the pandemic. Still, new media enabled the public's capacity to perceive and latch onto the culturally intimate message-making. As Kate Lacey puts it in her book *Listening Publics: The Politics and Experience of Listening in the Media Age*, 'the re-sounding of the public sphere through new media technologies further entrenched the privatised modern public, characteristically encountering public life within domestic space' (2013, 114). And whereas the privatisation of social transactions intersects with the

Remembering Public Sounds during the Pandemic 37

demands of cultural intimacy to support social cohesion and large social entities (like the nation), these hegemonic goals also cross paths with 'the fragmentation of the public space, and the proliferation of alternative formations ("counter-publics")' (Stokes 2010, 4). Taking into consideration that it is unrealistic to assume any amount of coherence and completeness in the imagined love object of the lockdowned nation, it appears that smaller formations like the crowd in Žehra are by definition more tangible and less performative than the imagined national crowd. Also, their emotional soundscapes seemingly stem from impulses that are too raw and untamed to contain themselves within the intricate game of cultural intimacy. It might be tempting to assume that the performance of belonging in Žehra is put together only by affect and, therefore, not socially coordinated. But even intimate groups like the one in Žehra rely on the abstraction of a shared social body and its deliberate organisation into a collective entity that synchronises itself acoustically. Ultimately, both types of audible public events discussed – the 'community' shouting and the 'national' music performances – are shaped by strategies and contexts that we should keep in mind when listening to them and recollecting how they sounded.

Conclusion

In this chapter, I inquired into the ways that acoustic experiences of the COVID-19 pandemic were developed in the specific geographic and cultural context of the post-Communist Czech Republic and Slovakia. Because the subject of sound memory is so elusive, it is not possible to conceive a definitive recipe for listening to and remembering the acoustic changes and social actions effected through sound during the pandemic. I am instead offering a simple reminder that a society is always tuning in to an aural space that is selective – whether through technology, cultural tactics, or social distinctions – and that we should listen to the publicly memorised acoustics of the pandemic with a critical ear. In order to establish a counter-practice to the hegemonic instincts of collective memory, our studies and mappings of the acoustic textures specific to the pandemic times should aim to critique and ultimately break the appearance of homogeneity in the experience of the public and private soundscapes, and also question the forming narratives that may favour experiences that only come from particular types of bodies and places. Through a detailed contextualisation of the individual capsules and also collectively curated reservoirs of memory, we might imagine and ultimately work out an unbiased sound heritage for the post-pandemic publics.

Notes

1 Peter Pollák's team streamed his statement on Facebook (Pollák 2020).
2 Healthy Regions is a state-subsidized organization focused on developing temporary improvement measures in public health.
3 Schafer uses the term *soundscape* for the first time in his handbook of music pedagogy titled *The New Soundscape: A Handbook for the Modern Music Teacher* (1969). His later work, *The Tuning of the World* (1977), provides a comprehensive account of his soundscape theory that has been thoroughly discussed and criticized by sound studies scholars. See, for instance, Schmidt (2019).
4 On 12 March 2020, the Slovak government closed all international airports and ordered a quarantine of 14 days for all travellers returning from abroad. In the Czech Republic, the border closure to all foreigners was effected a few days later, on 16 March 2020.
5 As Benedict Anderson stated, a nation 'is imagined because the members of even the smallest nation will never know most of their fellow-members, meet them, or even hear of them, yet in the minds of each lives the image of their communion' (2006, 6).
6 I would argue that this stability of the Czech society promoted by public performances like 'Není nutno' is tied to the contemporary condition of liberal democracy in the Czech Republic, under

38 Dominika Moravčíková

current leadership that has promised a quick return to previous freedoms and economic stability. To use the words of Michael Herzfeld: 'the nation-state is ideologically committed to ontological self perpetuation for all eternity' (2004, 21).

References

Anderson, Benedict. (1983) 2006. *Imagined Communities: Reflections on the Origin and Spread of Nationalism*. London: Verso.

Arshad-Ayaz, Adeela and M. Ayaz Naseem. 2021. 'Politics of Citizenship during the COVID-19 Pandemic: What Can Educators Do?' *Journal of International Humanitarian Action* 6 (3): 1–4. https://doi.org/10.1186/s41018-020-00089-x.

Ahmed, Sara. 2003. *The Cultural Politics of Emotion*. Edinburgh: Edinburgh University Press.

Bijsterveld, Karin and José van Dijck. 2009. 'Introduction.' In *Sound Souvenirs: Audio Technologies, Memory and Cultural Practices*, edited by Karin Bijsterveld and José van Dijck, 11–22. Amsterdam: Amsterdam University Press.

Birdsall, Carolyn. 2009. 'Earwitnessing: Sound Memories of the Nazi Period.' In *Sound Souvenirs: Audio Technologies, Memory, and Cultural Practices*, edited by Karin Bijsterveld and José van Dijck, 169–181. Amsterdam: Amsterdam University Press.

Brown, A. Lex, Truls Gjestland, and Danièle Dubois. 2016. 'Acoustic Encounters and Soundscapes.' In *Soundscape and the Built Environment*, edited by Jian Kang and Brigitte Schulte-Fortkamp, 1–16. Boca Raton: CRC Press.

Born, Georgina. 2013. *Music, Sound and Space: Transformations of Public and Private Experience*. Cambridge: Cambridge University Press.

Castoriadis, Cornelius. 2005. *The Imaginary Institution of Society*. Cambridge: Polity Press.

Fleck, Gábor and Cosima Rughiniş. 2008. *Come Closer. Inclusion and Exclusion of Roma in Present-Day Romanian Society.* Bucharest: Human Dynamics.

Hagood, Mack. 2011. 'Quiet Comfort: Noise, Otherness, and the Mobile Production of Personal Space.' *American Quarterly* 63 (3): 573–589. https://doi.org/10.1353/aq.2011.0036.

Herzfeld, Michael. 2004. *Cultural Intimacy: Social Poetics in the Nation-State*. London: Routledge.

Herzfeld, Michael. 2009. 'The Performance of Secrecy: Domesticity and Privacy in Public Spaces.' *Semiotica* 175 (January): 135–162. https://doi.org/10.1515/semi.2009.044.

Hilmes, Michele. 2005. 'Is There a Field Called Sound Culture Studies? And Does It Matter?' *American Quarterly* 57 (1): 249–259. https://doi.org/10.1353/aq.2005.0006.

Hofman, Ana. 2015. 'The Affective Turn in Ethnomusicology.' *Muzikologija* 18: 35–55. https://doi.org/10.2298/MUZ1518035H.

LaBelle, Brandon. 2012a. 'Acoustic Spatiality.' *SIC – Journal of Literature, Culture and Literary Translation* 2 (2): 1–14. https://doi.org/10.15291/sic/2.2.lc.1.

LaBelle, Brandon. 2012b. 'Auditory Relations.' In *The Sound Studies Reader*, edited by Jonathan Sterne, 468–474. London: Routledge.

Lacey, Kate. 2013. *Listening Publics: The Politics and Experience of Listening in the Media Age*. Cambridge: Polity Press.

Mižigárová, Alžbeta. 2021. 'Žehra odrezaná od sveta.' In *Celkom (ne)obyčajné veci izolovanej doby*, edited by Mirka Ábelová et al., 124–130. Bratislava: Lindeni.

Mowitt, John. 2015. *Sounds: The Ambient Humanities*. Oakland: University of California Press.

Peter Pollák. 'Aktuálne zo Žehry. Ďakujem Vám všetkým za podporu. PODARILO SA.' Facebook, 15 May 2020. www.facebook.com/307556876112136/videos/253175149260308.

Schafer, R. Murray. 1969. *The New Soundscape: A Handbook for the Modern Music Teacher*. Scarborough: Berandol Music Limited.

Schafer, R. Murray. 1977. *The Tuning of the World*. New York: Knopf.

Schmidt, Ulrik. 2019. 'Sound as Environmental Presence: Toward an Aesthetics of Sonic Atmospheres.' In *Oxford Handbook of Sound and Imagination*, edited by Mark Grimshaw-Aagaard, Mads Walther-Hansen, and Martin Knakkergaard, 517–534. Oxford: Oxford University Press.

Schulze, Holger. 2021. 'The Intimate.' In *The Bloomsbury Handbook of the Anthropology of Sound*, edited by Holger Schulze, 147–161. New York: Bloomsbury Publishing Inc.

Sterne, Jonathan. 2003. *The Audible Past: Cultural Origins of Sound Reproduction*. Durham: Duke University Press.

Stoever, Jennifer Lynn. 2016. *The Sonic Color Line: Race and the Cultural Politics of Listening*. New York: New York University Press.

Stokes, Martin. 2010. *The Republic of Love: Cultural Intimacy in Turkish Popular Music*. Chicago: University of Chicago Press.

Sowodniok, Ulrike, Angela Ankner, Anna Weißenfels, BelindaDuschek, Holger Schulze, and Jonny Labrada Ramirez. 2021. 'The Voice.' In *The Bloomsbury Handbook of the Anthropology of Sound*, edited by Holger Schulze, 111–128. New York: Bloomsbury Publishing Inc.

Thompson, Emily. 2012. 'Sound, Modernity, and History.' In *The Sound Studies Reader*, edited by Jonathan Sterne, 117–129. London: Routledge.

3 Listening to the Hustle and the Hush

Sound, City, and the Pandemic

Nakshatra Chatterjee and Srijita Biswas

Introduction

To think of an Indian cityscape through sound, one must consider the differences in terms of physical space compared to other major cities of the world. The population is a primary factor that defines the human exchanges of the Indian everyday.[1] The hustling aliveness of Indian cities is reflected through life lived on the streets. In times before the pandemic, the steady rhythm of urban life hinged on the quotidian. A typical day would begin with the tinkling bells of the milkman, newspaperman, or garbage gatherer in the morning, then a casual stroll through busy roadside markets with imperceptible hawker cries as the day progresses, followed by the regular share of *adda*[2] in assembly points like tea shops or cafes, and the frequent fare of slogans from crowds bustling in rallies and roadshows or dispersing after a day's football or cricket match. The soundscape is incomplete without the regular traffic in its incessant honking. Yet, to much surprise, traffic signals can be an interesting space for perceiving intersectional sound elements that seem to have a crucial bearing in our everyday lives. For instance, in the city of Kolkata, when the traffic stops at the signals this can be a moment to relish for a passionate listener. Just when the vehicles stop, usually a pleasant musical composition by Rabindranath Tagore[3] is played on speakers adjacent to the signal post. This endeavour is meant to relieve people from the mundane everyday experience of being stuck in the sea of traffic. The auditory dimension of the street is mediated through all these sonic events. This noisy and vibrant cityspace functions as a platform for a sensory experience that arises from the activity of everyday life. The resonance of the living and the non-living in the physical space should be conceived as a frame shared by two phenomena: the energetic wave of sound cutting across the urban space and the patterns of listening employed to perceive it. This pervasive resonance renders the features of how individuals make sense of the place in terms of what sounds they hear. This aural understanding associates the individual with a larger social fabric, where they feel a sense of belonging and sustenance. The impact of the COVID-19 outbreak, mounting to the condition of a pandemic, has overwhelmingly altered the deep-rooted world order in every possible way. Like every other city in the world, it has had an overall impact on the pre-existing conditions of Indian cities. Dialogues are exchanged on how the pandemic has influenced the social, political, economic, and environmental spheres. People are still trying to grapple with the new circumstances by gaining a new rhythm in acceptance of the new normal. The situation certainly is a setback for urban life in several ways. It leaves an inescapable void to trace the repercussions lodged by the pandemic.

In this chapter, we will illustrate a perspective, through a cultural lens, that reflects on how people's relationship with the city has changed in terms of making sound and listening. We rely heavily upon digital records, as our mobility was restricted by the ongoing lockdown. We clubbed our own respective experiences with sources drawn from YouTube videos, newspaper reports, and social media posts on influential platforms such as Twitter. For

DOI: 10.4324/9781003200369-5

the purpose of this study, we took into account the sonic events that occurred specifically with the onset and during the initial days of the COVID-19 pandemic, which significantly changed the *aural topography* of Indian cities. The idea of aural topography is employed in the sense of how sound production or performance and its perception in the physical space stimulates the emotional landscape through the affective impulses of the auditory, catering to personal and collective memory. We focus on the Janata Curfew event, which we identify as the day when a significant activity of collective noise-making was undertaken on a large scale and, as a result, a plethora of characteristic sonic events was produced. We argue that the collective sound production in the Janata Curfew caused a shift in the soundscape. We consider this day as a milestone marking India's entry into its official lockdown phase, an event that redefined the soundscape as well as was instrumental in distinguishing the time of the pandemic from the time preceding it. The absence of familiar sounds following the imposition of a nationwide lockdown caused an unprecedented and uncertain situation. In an attempt to identify the prominent sonic responses following the day of the Janata Curfew, we have mapped the events that are fraught with sonic relevance, that occurred due to the pandemic and due to the necessary lockdown. In this respect, we have chosen to highlight the migrant crisis that arose due to the lockdown and the consequential loss of jobs, disrupted communication, and immobility. We have critically read the literal chaos of the situation being caught in intra- and inter-state transit, and have picked up cues from the aural cacophony produced by the desperation of migrants at terminals and on roads, by the honking of the trains on which they travelled, and also the emergency announcements delineating the COVID-19 protocols. Taking these observations as a premise, we address the fundamental questions of collective resilience, agency, and citizenship, as shaped by select sonic events during the pandemic, by applying the theoretical concerns of *sonic citizenship* (Brueck, Smith, and Verma 2020).

Sounding South Asia

A comprehensive understanding of Indian sound studies can be instrumental in situating this chapter along the axes of the disciplinary purview. Conveying a 'sonic turn,' the past decades have generated a handful of seminal texts concerning the historical (Thompson 2002), cultural (Corbin 1998), technological (Sterne 2003), and literary (Picker 2003) implications of sound. These publications provide an impetus to the thoughts and beliefs about sound discourses that consolidates the discipline of sound studies. These developments contribute to the re-evaluation of some traditional questions pertaining to modernity, knowledge, and experience. However, there is one obvious objection to the prospect of sound studies that is rightly pointed out by Jonathan Sterne: 'the West is still the epistemic centre for much work in sound studies' (2005, 71). Of course, it is inarguably a fact that while early scholars of sound studies imagined the East as sensuous and closer to hearing, they did not care to include a single essay or a write-up that pioneered the way of 'sound' thinking in any anthologies or critical readers.[4] The Eurocentric view comes under critical investigation with the emergence of recent studies signifying a *sonic turn* beyond the ideological demarcations of so-called Global North or West, as expressed by Radano and Olaniyan (2016), and Steingo and Sykes (2019). In order to widen the possibility of the investigation, Steingo and Sykes assume that the discourses of the South and sound are 'conjoined concepts lying at the heart of modernity,' and suggest the deconstructive thinking of sound 'not as the South (or as analogous with the South) but, rather, in and from the South' (2019, 4). In India, sound studies is still in the stage of emerging as a discipline, and most of the earlier works come from musicological scholarship with greater emphasis on Indian classical, folk, and popular song traditions. In the last decade a number

42 *Nakshatra Chatterjee and Srijita Biswas*

of publications have emerged from media studies, film studies, anthropology, music studies, and cultural studies that exhibit new prospects. A noteworthy work, *Indian Sound Cultures, Indian Sound Citizenship,* edited by Laura Brueck, Jacob Smith, and Neil Verma (2020), proposes a new outlook on the question of citizenship while specifically situating the field of sound studies in Indian contexts.

We identify the notion of the soundscape as the primary object of study. R. Murray Schafer's *The Tuning of the World* (1977) is one of the foundational texts that reflects on the radical study of the soundscape. For him, the idea of 'soundscape' refers to specific sounds and sonic environments for consideration. However, unlike his concern with ecological grounds, in this chapter we exploit the cultural aspects of the soundscape.[5] If the urban soundscape is a conjugation of the whole sonic materiality of the environment, it also signifies the ways of perceiving the environment. In this regard, two obvious considerations for the development of the chapter need to be discussed. First, the nature of the urban soundscape that the pandemic has drastically changed, and second, the manners of listening that have shaped the subjective experience of the urban space. In the class-based society of an Indian city, we consider the location of the listeners in the socio-economic structure because it reveals how their status in the society determines their participation in, and reception of, the sound making. This chapter encompasses how the privileged have responded to the presence and absence of sound in the lockdown. It also digs into the lives of those residing along the margins of the class divide. An analysis of the sonic events during the lockdown reflects on how the performance of sound and the performance of politics has formulated a complicated aural projection of the postcolonial city. During the lockdown, people are forced to stay in isolated spaces. It implies that the chances of moving out and participating in daily life are curtailed. Hence, visual faculties are put to rest. This passivity of visual sense makes way for other senses to come into play, specifically listening. Therefore, the practice of listening demands special attention by the ear that can capture a wide range of sounds and their absence. It goes without saying that this chapter cannot elaborate on all the nuanced elements of soundscape and detailed listening practices. It focuses on particular cases of specific sound production and sonic exchange that illustrated the pandemic in broad detail within the spectrum of human experience both historically and culturally. Taking into account the events and records of significant sound production, this chapter situates the pandemic in the Indian cities through a study of its soundscape, and also documents the socio-political capital of sound in building narratives that extend from the personal to the political, and even the nationalistic.

Janata Curfew: A Euphoria in Serious Hours

The Janata Curfew[6] was a voluntarily self-imposed curfew by the people, lasting for 14 hours – from 7:00 am to 9:00 pm – on 22 March 2020. It was observed nationwide responding to an appeal from Prime Minister Narendra Modi on 19 March. The curfew was a test run to check the country's preparedness for a longer period of shutdown. The Prime Minister also requested that people make noise for five minutes at 5:00 pm on the day of the curfew by clapping, *taali bajao* (*taali* means clap and *bajao* is the act of clapping), or by banging vessels, *thaali bajao* (*thaali* is a plate, often a metallic one), or by ringing bells, *ghanti bajao* (*ghanti* means bells). The directions were to make noise to convey gratitude to the COVID warriors, who are primarily doctors, nurses, health workers, and other essential service providers.[7] Millions of Indians came out on their balconies and rooftops to bang pots and pans in a show of support for the frontline workers. The Prime Minister's appeal was broadcast through the country's leading media outlets, a programme that he uses to establish a direct medium of communication with the common people. The broadcast acted like a sonic cue that circulated at warp speed across diverse publics through the media platforms. Analysing the Prime Minister's influence through his speech, Praseeda Gopinath, in her essay 'Narendra Modi Speaks the Nation: Masculinity Radio and Voice,' writes about how 'his strategic vocal practice in and through his massively popular radio broadcast, *Mann Ki Baat*, crafts a gentle everyman national saviour even as it constructs a new national public' (2020, 152). Citizenship is negotiated by responding to the Prime Minister's call and by actively participating in the collective act of making noise in this particular case of the Janata Curfew, which was an attempt to integrate the masses on a national level.

Narendra Modi's act was both appreciated and criticised. Clapping as a gesture of thankfulness turned into a festive euphoria with people going overboard and making noise with acts such as ululating, bursting crackers, and so on.[8] Since the mandate was to remain within the premises of one's home, some citizens criticised the act as they highlighted the constricted space in their area. For instance, in an array of tweets, people from Mumbai reported the unavailability of balconies in the densely populated city which is the financial capital of India (NEWS18 2020). On the other hand, people brought out huge processions and rallies in an almost carnivalesque fervour. Bizarre videos of women frantically banging brass plates (Times of India 2020c) and people dancing to chants like 'Go Corona Go' surfaced all over digital media. As a matter of fact, since the outbreak of the pandemic, several theories were put forward that the vibrations produced from the clamour of the vessels would kill the virus. For instance, on 11 March 2020, two YouTube videos showing how musical performances combined with religious practices were to ensure the departure of the virus from India had already circulated widely. Accompanied by a group that had gathered to spread awareness about COVID-19, Union minister Ramdas Athawale chanted slogans of 'Go Corona Go! Go Corona Go!' (Times of India 2020a). It seemed that they were chanting slogans as if to turn the virus back from India. The video went viral and even made it to a DJ mix (Anup K. R. 2020). In another video shared over the internet, a group of women was seen singing *bhajan*[9] invoking the coronavirus to leave us, 'corona, *bhaag jaa*' (Crux 2020). These two videos supported widely circulated misinformation that vibrations from banging vessels would kill the virus. A few days later, in a Twitter post, the Indian superstar Amitabh Bachhan claimed in a now-deleted tweet:

> AN OPINION GIVEN: 5 PM; 22nd Mar, '*amavasya*,' darkest day of month; virus, bacteria, evil force at max potential & power! Clapping *shankh* vibrations reduce/ destroy virus potency Moon passing to new '*nakshatra*' Revati. cumulative vibration betters blood circulation.'

(Economic Times 2020)

The study of these phenomena brings forth local and cultural contexts that are particular to India, or at least more relevant to South Asian contexts. Dipesh Chakrabarty laid forth the necessity of the task to 'provincialize' Europe as a disciplinary renewal, because European thought is at once 'indispensable' to and 'inadequate' for thinking through historical consciousness and political modernity in non-European contexts (2012, 6). In the context of sound production during the initial days of the COVID-19 pandemic, this was a common observation on digital platforms, where the floating videos of musical practice on balconies in European countries, namely Italy and France, were happening. The performance in the respective isolated spaces stemmed from the need for social isolation, yet turned into an exhibit of social interaction during the pandemic, when people were advised to stay isolated. In India, the appeal of expressing gratitude to the frontline workers was welcomed by standing in doorways or on balconies where clapping and banging vessels were performed along with religious rituals. The correlation between religion and sound has been meaningfully established in Indian culture. Religious symbolism and beliefs are attached to certain sounds, and it was best illustrated on the occasion of the Janata Curfew when the use of conch shells, hand-held gongs, cymbals, and so on incorporated in the *thaali/taali bajao* act can also be perceived as expressing religious materiality. Sukanya Sarbadhikari writes, 'the conch is studied as a direct material embodiment of the sacred domestic. Its materiality and sound-ontology evoke a religious experience fused with this-worldly wellbeing (mongol) and afterlife stillness' (2019). The religious practices in Hinduism are necessarily sonic and reverberate with chants and musical instruments. This oral-aural orientation is a social practice that comes from the Vedic Age.[10] Vedic concepts of *shruti*, or the act of listening, had been a pre-modern South Asian practice of documenting and transferring religious knowledge. Therefore, listening has been a culturally significant practice when seen through a South Asian religious ethos such as Hindu practices. Thus, it is duly established that sound and music has been culturally crucial in invoking the auspicious, even warding off evil.

This was a comprehensive scenario on the day of Janata Curfew that officially marked the beginning of an uncertain lockdown, or the onset of the new normal in India due to the COVID-19 pandemic that changed the functioning of daily life for everyone. The Janata Curfew was a day marked by its characteristic features of sound production, which ranged from being random, rhythmic, melodious, explosive, to being on the whole mostly *noisy*. The days following it were heralding the suspension of usual human life during the lockdown, a life without the presence of familiar sounds, one of unusual silence and noiselessness.

Noise: The Necessary Evil

Two days after the Janata Curfew, a nationwide lockdown of 21 days was announced. The lockdown caused a complete shutdown of mills and temporary labour jobs. It affected the hand-to-mouth daily wage earners who mostly migrate to the larger cities for their livelihoods. It robbed the migrant workers of their jobs, and the disconnectivity in public transport caused by the lockdown put them in a situation of 'stuckedness' as introduced in a situation of immobility (Hage 2009).

The COVID-19 virus is a contagion whose origin is foreign to India. Migration has been one of the ways the virus has spread rapidly. Thus, any form of migration whether, intra- or inter-state, raised red flags and the word 'migrant' became an alarming term, floating as a keyword for the times of the pandemic. Migrant workers were the worst affected by the lockdown imposed with very short notice. They lost livelihoods and shelter. Unable to survive the expenses of the metropolis, and having lost their jobs as factories were shut, the cash-strapped migrant labourers were homeward bound. The desperation at bus terminals like that of Delhi's Anand Vihar or stations like Surat exhibited the dreary reality of the people who could not be privileged to have afforded to stay within the comfort of their homes. It was starkly evident that when the nationwide lockdown kicked in, hundreds of thousands of migrants found themselves without any means of sustenance.

The aural topography of such a situation can be characterised as cacophonous, fearful, and definitely uncertain. The desperation at bus terminals, railway stations, and roads was not only visible but also audible as ambulance sirens, police sirens, and announcements to maintain social distancing and emergency directions were broadcast to the migrating crowds. The *New York Times* reported from Surat that 'the crowds surged through the gates, fought their way up the stairs of the 160-year-old station, poured across the platforms and engulfed the trains' (Gettleman et al. 2020). The encompassing cacophony consisting of the announcements from enquiry points, from the police ensuring distance, and the general chaos of the entire activity of desperation to travel at the points of the exodus, which itself had an aural identity of alertness. It was rooted in the production of noise in a situation of utter distress and hopelessness facing a journey that they were about to embark upon due to the recently unleashed lockdown. According to David Novak, noise stands for 'subjectivities of difference that break from normative social contexts. It interpellates marginal subjects into circulation, giving a name to their unintelligible discourses even as it holds apart unfamiliar ways of being' (2015, 130).

Noise has often been denounced because of its meaninglessness and obscurity for interpretation. Noise, however, has played a role in portraying human expression in the discursive context of the COVID-19 pandemic. Noise, as a means of communication, is crucial paraphernalia for disseminating information and connecting people to each other. Listening to noise can also trigger affective responses and can create situations prompting involvement in the action of making noise. The migrants' cries for help, coupled with encompassing sounds like horns, emergency directives, and train honks, are subversive when seen in the light of the spatial reclamation of state-assigned directives to maintain order and social distancing. This phenomenon of reverse migration was visible when it was about migrants returning to their native homes, all the while robbing the city of its usual hustle and leaving it to the loneliness of soundlessness. The cacophony encompassing the announcements, sirens, car and bus horns, along with the migrants in desperate enquiry is itself an identity alike and everywhere when heard together. The honks of COVID-19 trains, arranged by the Indian Railway to take people back home from the metropolises, were perceived as a harbinger of the threat of contagion in the places from which people embarked. Some of these trains were labelled as *Shramik Specials* (*shramik* means labourer in Hindi) as they were used specifically for taking migrants back home from the cities. 'There was a very direct correlation between the active COVID cases and the trains … It was obvious that the returnees brought the virus,' said Keerthi Vasan V., a district-level civil servant in Ganjam (quoted in Gettleman et al. 2020). The collective sound, creating a sonic territory that engulfed the silence of the lockdown, is a negation of the silence of the lockdown, almost as a protest by migrants against the conditions that put a stop to their livelihoods and added a host of difficulties to their lives, compelling them to migrate and submitting them to a state of both economic and existential uncertainty. Thus, this phenomenon in itself can be seen as both a metaphorical and real manifestation of the *cry of the migrants* that was not adequately addressed by the authorities (here by authorities we mean both the people and/or the private owners they worked for, in particular, and the state in general). This act of intra-state reverse migration can be interpreted as robbing the agency of citizens in the metropolis during the pandemic, when the literal and figurative voices of the migrants are raised in forms of desperate cacophony, questioning the aspect of sonic citizenship performed in the Janata Curfew.

The Rhapsody of Silence

The large metropolis, a milieu of clamour and chaos, took an oath of silence with the imposition of lockdown. The tremendous congestion of urban crowding, due to the arbitrary co-habitation of people and their cultures, faced an unusual threat. Social distance, once unimaginable, all of a sudden became an imminent reality. This unavoidable transformation brought a massive lifestyle change by forcing people out of their normal ways of life. A sudden rupture in the rapidly increasing noise of the modern cities leaves a scathing impression on the formative elements of the ecology of urban sound. This newly built soundscape, which prompted unconditional indulgence to silence, leaves space to contemplate its strange functioning. Our everyday world is so saturated with sounds that sometimes silence seems to be a relieving break. In the days of lockdown, silence became a state of absence of the familiar sounds that we identify within our day to day lives. While the days also marked the reappearance of long-unheard ones, the absence of the regular began to have various effects on the lives of people and their surrounding soundscape. For instance, the missing mechanical sound of looms rendered the workers in Surat sleepless, as if to the workers the familiar sound of the looms worked as a white noise lulling them to sleep. A report in the *Times of India* reads:

> Migrant workers in India's man-made textile hub – Surat – are forced to stay home due to lockdown, they are woefully missing the high decibel clatter of the powerloom machines that pound their ears during the 12-hour shift. The workers are so used to this that many are spending sleepless nights these days.
>
> (*Times of India* 2020b)

The silence on the streets was frequently disrupted at night and was breached through occasional emergency announcements and ambulance sirens imparting an untimely fear. On the other hand, there are pleasant developments in the soundscape, with the birds being again heard at daybreak and dusk, as the usual cacophony of the city had come to a halt. Silence seemed to be haunting the suddenly hushed-up cityspace, stimulating it in more ways than one. It provided people with the quiet to be creative of their own accord. Indian ambient musicians took this moment of tranquillity to compose soothing music. Uddipan Sarmah from the band Aswekeepsearching shared his experience of the anxious days of lockdown:

> But I later found peace. This pandemic has given us the moment to pause and look inward, reevaluate things and focus on everything that we have forever ignored.' He continues, 'our sound has a lot of soundscapes and ambient layers which take it to calmer space,' emphasizing that 'ambient leaves a lot of space to think, meditate and get better with mindfulness.'
>
> (New Indian Express 2020)

The prevailing stretch of silence that persisted during the initial days of COVID-19 was replaced rightfully and necessarily with diverse sound productions that arose observing the Janata Curfew to mark the official commencement of nationwide lockdown in India.

Conclusion

Urban sound is an inseparable part of the city's everyday life. The function of these sounds is realised more than ever after the COVID-19 pandemic changed the ways of our lives. We have illustrated the relevant sonic activities and alterations that are contextual to both cities and human life, from a South Asian perspective. Narrativizing the discreet yet attributive events that took place in the urban soundscape as a result of the pandemic is meaningful and compelling when it comes to understanding how the soundscape of our everyday life was altered as we started living a life in bubbles. To trace certain transitions in the sound world heralded by the pandemic, it is necessary to imagine how the urban aural culture of twenty-first-century India has been shaped. The modern history of the nation plays a significant role in shaping its everyday lives. For instance, the special trains (*Shramik Special*) that transported migrants during the COVID-19 pandemic were apprehensively called 'virus trains' (Gettleman et al. 2020). It symbolises a grim human tragedy. These trains are reminiscent of the *ghost trains*, which used to travel across the borders during the 1947 Partition of British India. These trains would arrive at their destination ablaze, their passengers burned to death for having fallen prey to the Partition brutalities comprised primarily of communal violence. The modern nation-state of India was born in violence. The historic association between trains (and their sounds) and atrocities specific to Indian modern culture probably added to the way the virus trains, and the sounds associated with them, were perceived in India during the pandemic. One such association is expressed through the creative rendition of the audiovisual composition *Different Trains 1947* (2017),[11] which incorporated train sounds as perceived and recollected from the memories of those who lived through the tumultuous 1947 Partition. This piece rekindled the collective trauma associated with the event that rendered people helpless in the face of the sudden need for migration and movement. The honking of the trains heralded gory horrors as they often carried mutilated bodies of people from across the borders. Trains have been a major symbol of migration and memory in the country's history of Partition.

The city is the chosen site of inquiry for this chapter and the city emerges repetitively in the political and aesthetic imagination as the powerhouse of nationalistic thought and the locus of nation-building in post-Independence India. However, instead of a homogenised body of existence, the contour of the city appears fragmented due to uneven urbanisation (see Sandten and Bauer 2016). Hence, the palimpsestic developments of the cityspace, bifurcating the modern and the non-modern spaces, create a contested infrastructure that destabilised the 'right to the city' (Harvey 2008) for everyone belonging to the socio-economic structure. Architect and social activist Jai Sen propagates the term 'unintended city' when he writes about the 'other' society which has never been a part of modernising postcolonial India, but always remains under state supervision (1996, 2978). This huge mass of officially castaway 'obsolete' city dwellers relentlessly 'provide(s) the energy – literally the cheap labour – that propels both the engine of civic life in a Third World society and the ambitions of its modernizing elite' (Nandy 1998, 2). The time of lockdown, when privileged people sought resilience and passed their time being creative from within the comfort of their homes, was a probation of sustenance for the 'unintended' living in deprivation. The instances of discriminatory political treatment, when migrant workers were forced to migrate back to their homes, highlight the questions of agency to live in the city, redefining the attributes attached to the idea of a citizen. The soundscape, similarly, is a discursive space that imparts the idea of a citizen. By employing sonic perception and recognition of the produced sound, the producer-listener citizen justifies the ideas of citizenship.

Notes

1 Architect and urbanist Lakhsmi Rajendran writes:

> in Mumbai there is merely 1.28 sq. metres of public space per person, compared to 31.68 sq. metres in London or 26.4 sq. metres in New York. In this context, public spaces in India such as railway stations, bus stops, or local parks serve multiple functions. They often contain informal market places, space for spontaneous social gatherings or interactions, and act as a matrix where everyday life can occur.
>
> (Rajendran 2020)

2 As Dipesh Chakrabarty says, 'it is the practice of friends getting together for long, informal, and unrigorous conversations' (2012, 181).

3 The state government installed loudspeakers at traffic terminals to play *Rabindrasangeet* (songs composed by Rabindranath Tagore) whenever the traffic came to a stop at the intersections. This was explained as an attempt to emphasise the rich Bengali culture and acknowledge the great poet and the first Indian Nobel Laureate in Literature (1913) Rabindranath Tagore.

4 One can refer to what Radano and Olaniyan said, 'from Edmund Carpenter, Walter Ong, and R. Murray Schafer all the way (arguably) to Claude Lévi-Strauss and Marshall McLuhan, non-Western peoples are positioned as closer to sound and hearing than their European counterparts' (2019, 3). For more see McLuhan ([1962] 2011, 15–16).

5 For more details, see Corbin 1999, Truax 2001, and Thompson 2002.

48 Nakshatra Chatterjee and Srijita Biswas

6 The Hindi word *janata* means 'the populace' or 'the people,' so a collective act was being implied through a curfew that was so named.
7 For audiovisual reference and brief report, see BBC News 2020.
8 *Janata Curfew* was criticised as a frivolous act because it turned out to be a carnival which was unsought for mitigating a pandemic demanding social isolation as a control measure. See NDTV 2020.
9 *Bhajan* is a Hindu song tradition of devotional music that is sung to express reverence for the Divine.
10 The Vedic Age (c. 1500–c. 500 BCE) is when the religious texts, namely Rigveda, Yajurveda, Samaveda, and the Atharvaveda were composed.
11 *Different Trains 1947* was an audiovisual composition, collaborative performance by Actress, Jack Barnett, Sandunes, Jivraj Singh, Priya Purushothaman, Iain Forsyth, and Jane Pollard. It was performed at Barbican Hall London (1 October 2017) and Magnetic Field Festival in Rajasthan, India (17 December 2017). This composition is inspired by the framework of *Different Trains* (1988) by the minimalist composer Steve Reich who made the composition on his childhood memories of trains and connected it to the Jewish experience of train travels in the Holocaust.

References

Anup K. R. 2020. 'GO CORONA, CORONA GO (RAP REMIX).' YouTube, 15 March 2020. www.youtube.com/watch?v=cspF9QK5FlA.

BBC News. 2020. 'Coronavirus: Indians Bang Pots and Pans to Support Fight.' 22 March 2020. www.bbc.com/news/av/world-asia-india-51997699.

Brueck, Laura, Jacob Smith, and Neil Verma, eds. 2020. *Indian Sound Cultures, Indian Sound Citizenship*. Ann Arbor: University of Michigan Press.

Chakrabarty, Dipesh. 2012. *Provincializing Europe: Postcolonial Thought and Historical Difference*. Princeton: Princeton University Press.

Corbin, Alain. 1998. *Village Bells: Sound and Meaning in the 19th Century French Countryside*. Translated by Martin Thom. London: Papermac.

Crux. 2020. 'Coronavirus: This Viral Indian Video of Women Singing "Corona Bhaag Ja" Will Leave You in Splits.' YouTube, 11 March 2020. www.youtube.com/watch?v=PhJ88WJb58A.

Economic Times. 2020. 'Janata Curfew Row: Amitabh Bachchan Claims Vibrations from Clapping Destroy Virus, Twitter Schools Him.' 24 March 2020. https://economictimes.indiatimes.com/magazines/panache/janta-curfew-row-amitabh-bachchan-claims-vibrations-from-clapping-destroy-virus-twitter-schools-him/articleshow/74785682.cms.

Gettleman, Jeffrey, Suhasini Raj, Samir Yasir, and Karan D. Singh. 2020. 'The Virus Trains: How Lockdown Chaos Spread Covid-19 Across India.' *The New York Times*, 15 December 2020. www.nytimes.com/2020/12/15/world/asia/india-coronavirus-shramik-specials.html.

Gopinath, Praseeda. 2020. 'Narendra Modi Speaks the Nation: Masculinity, Radio, and Voice.' In *Indian Sound Cultures, Indian Sound Citizenship*, edited by Laura Brueck, Jacob Smith, and Neil Verma, 152–173. Ann Arbor: University of Michigan Press.

Hage, Ghassan. 2009. 'Waiting Out the Crisis: On Stuckedness and Governmentality.' In *Waiting*, edited by Ghassan Hage, 97–106. Melbourne: Melbourne University Press.

Harvey, David. 2008. 'The Right to the City.' *New Left Review* 53 (September–October). https://newleftreview.org/issues/ii53/articles/david-harvey-the-right-to-the-city.

McLuhan, Marshall. (1962) 2011. *The Gutenberg Galaxy: The Making of Typographic Man*. Toronto: University of Toronto Press.

Nandy, Ashis. 1998. 'Introduction: Indian Popular Cinema as a Slum's Eye View of Politics.' In *The Secret Politics of Our Desires: Innocence, Culpability and Indian Popular Cinema*, edited by Ashis Nandy, 1–18. London: Zed Books.

NDTV. 2020. 'On PM Modi's "Taali, Thali Bajao" Appeal, A Swipe from Shiv Sena.' 24 March 2020. www.ndtv.com/india-news/coronavirus-on-pm-modis-taali-thali-bajao-appeal-a-swipe-from-shiv-sena-2199341.

NEWS18. 2020. 'Mumbaikars without Balconies Want to Know Where They Can Clap during Modi's Janta Curfew.' 20 March 2020. www.news18.com/news/buzz/mumbaikars-without-balconies-want-to-know-where-they-can-clap-during-modis-janta-curfew-2543573.html.

Novak, David. 2015. 'Noise.' In *Keywords in Sound*, edited by David Novak and Matt Sakakeeny, 125–138. Durham: Duke University Press.

Picker, John M. 2003. *Victorian Soundscapes*. Oxford: Oxford University Press.

Radano, Ronald and Tejumola Olaniyan. 2016. 'Introduction: Hearing Empire – Imperial Listening.' In *Audible Empire: Music, Global Politics, Critique*, edited by Ronald Radano and Tejumola Olaniyan, 1–23. Durham: Duke University Press.

Rajendran, Lakshmi Priya. 2020. 'In India's Cities, Life is Lived on the Streets – How Coronavirus Changed That.' *The Conversation*, 17 April 2020. https://theconversation.com/in-indias-cities-life-is-lived-on-the-streets-how-coronavirus-changed-that-135232.

Sandten, Cecile and Annika Bauer, eds. 2016. *Re-Inventing the Postcolonial (in the) Metropolis*. Leiden: Brill Rodopi.

Sarbadhikari, Sukanya. 2019. 'Shankh-er Shongshar, Afterlife Everyday: Religious Experience of the Evening Conch and Goddesses in Bengali Hindu Homes.' *Religions* 10 (53). https://doi.org/10.3390/rel10010053.

Schafer, R. Murray. 1977. *The Tuning of the World*. New York: Knopf.

Sen, Jai. 1996. 'The Left Front and the "Unintended City": Is a Civilised Transition Possible?' *Economic and Political Weekly* 31 (45/46): 2977–2982.

Steingo, Gavin and Jim Sykes. 2019. 'Introduction: Remapping Sound Studies in the Global South.' In *Remapping Sound Studies*, edited by Gavin Steingo and Jim Sykes, 1–37. Durham: Duke University Press.

Sterne, Jonathan. 2003. *The Audible Past: Cultural Origins of Sound Reproduction*. Durham: Duke University Press.

Sterne, Jonathan. 2005. 'Hearing.' In *Keywords in Sound*, edited by David Novak and Matt Sakakeeny, 65–77. Durham: Duke University Press.

New Indian Express. 2020. 'The Pandemic has Contributed to a Rise in Calming Lo-fi and Ambient Releases in Indian Music ... We Explore the Trend.' 2 October 2020. www.indulgexpress.com/culture/music/2020/oct/02/the-pandemic-has-contributed-to-a-rise-in-calming-lo-fi-and-ambient-relea ses-in-indianmusic-we-e-28551.html.

Thompson, Emily Ann. 2002. *The Soundscape of Modernity: Architectural Acoustics and the Culture of Listening in America, 1900–1933*. Cambridge, MA: MIT Press.

Times of India. 2020a. '"Go Corona, Go Corona" Chants Union Minister Ramdas Athawale as if to Send the Virus Back from India.' YouTube, 11 March 2020. www.youtube.com/watch?v=4dPd 708Sk98.

Times of India. 2020b. 'Silence of Loom Makes Workers Go Sleepless in Surat.' 15 April 2020. http://timesofindia.indiatimes.com/articleshow/75149207.cms?utm_source=contentofinterest&utm_med ium=text&utm_campaign=cppst.

Times of India. 2020c. 'Watch: Janata Curfew Brings Out Creativity in People during "Thali Bajao".' 22 March 2020. https://timesofindia.indiatimes.com/videos/toi-original/watch-janata-curfew-bri ngs-out-creativity-in-people-during-thaali-bajao/videoshow/74764074.cms.

Truax, Barry. 2001. *Acoustic Communication*. Westport: Ablex.

4 Applauses and Banners, Horns and Fireworks

Tracing the Sonic Expression of French Social Movements during Lockdown

Alessandro Greppi and Diane Schuh

In France, strikes and demonstrations marked 2019 and the beginning of 2020. Several social movements emerged and generated transport strikes, teacher strikes, and large demonstrations in opposition to several bills, including the pension reform bill. The demonstrations against pension reforms began on 5 December 2019 and grew to include several million citizens.[1] The Yellow Vests – a popular movement protesting for economic justice, initiated by the rising costs of fuel – joined the movement on 7 December, and higher education and research personnel in January 2020. At that time, shortly before the confinement, a wide range of protest songs (including 'Le Chant des partisans' [Girard 2010], 'On est là' [Coquaz 2019]), brass bands, traditional Brazilian percussion ensembles called *batucadas*, choirs, screams, sound systems, and political slogans, all gave rhythm to France's political and social life.[2] The social movement never seemed to stop. However, in the face of the accelerating health crisis related to the spread of COVID-19, the French government decided abruptly to confine the country on 17 March 2020. The question arose as to how to continue to express disagreement with public policies when the very possibility of protesting was effectively prohibited by the lockdown. How can the voices of protest be listened to and heard when the vast majority of its actors are confined in their homes?

In this chapter, we analyse the sonic responses to this question. For this study, we conducted a virtual ethnography by collecting visual and sonic documents online. We followed activist groups on social media and analysed their practices: how did they gather and what codes did they reference while shifting the avenues of the protest from the street to the windows and balconies? We studied how they created alternative venues of expression across visual and sonic dimensions. We also analysed how new and traditional media chose to report and portray these practices. Departing briefly from virtual ethnography, we also conducted daily recordings for 15 days outside our windows during the lockdown, focusing on the 8:00 pm clapping event – a social and sonic practice that was established in France and other countries during the first lockdown. Which alternative avenues of expression did these confined struggles find during the lockdown? Did the lockdown inspire new sonic materials and approaches that facilitated dissident expressions? How did these movements position themselves in relation to the 8:00 pm applause for caregivers' ritual, also initiated during the lockdown?

For clarity, we will use the term soundscape in this chapter to refer to the sonic and acoustic environment of the places and milieux we are analysing. With reference to R. Murray Schafer (1994), we are reading the term with the actualisation of the notion of milieu theorised by Augustin Berque (2000). Each place we analysed has specific sonic characteristics that were induced by sonic events circumscribed in a specific time frame (applause, protests, fireworks, etc.). These sonic events were overlaying the usual sonic environment, which was itself also affected by the lockdown. These events reconfigured the soundscape, creating a sonic bursting in an otherwise calmer soundscape. For the sake

DOI: 10.4324/9781003200369-6

of simplification, we will use the term soundscape as a reference to what could be called a punctuating sound environment overlaying the usual soundscape.

We trace the transition of the protest from the street to the window by focusing mainly on the sonic dimension, but we will start by analysing an artefact that facilitated the shift between visual and sound: the protest banner. To this end we collected banner photos taken by activists and shared on social media with specific hashtags. By noting the musical and sonic aspects of these banners, we observed how windows and balconies singularly replaced public forums during the lockdown. We then consider the places of these sonic events, particularly during what became the daily 8:00 pm applauses. Having recorded the applause on a daily basis, we chose one representative excerpt focused on the socialising aspects of this event, and compared this situated expression with other recordings gathered online. We went on to examine the appropriation of this social practice for purposes of protest. Finally, we analysed how the use of horns, sirens, and fireworks allowed for the appropriation of a counter-space of appearance – framed here in response to Hannah Arendt's 'space of appearance' ([1958] 1998, 199), that comes into being wherever people participate in the realm of the political – by invisibilised and excluded populations. We also analysed a collection of recordings made by the parties concerned and shared on social media for the purpose of reclaiming their own narratives.

#CortègeDeFenêtre: When the Songs of the Protests are Displayed in the Window

After several months of struggle and demonstrations, France found itself confined on 17 March. The demonstrators had to reorganise. The ban on public gatherings made the task difficult. The chants of the union processions stopped. The confinement became then, for some, an opportunity to reuse the tools, sounds, and musical materials of traditional demonstrations in a critical and creative way. It was then a question of rethinking the struggle on another scale. During the lockdown, demonstrations organised in Paris or in other big cities took place on a more local scale. New solidarities were created with neighbours and through political events with a micro-local dimension. On social media, the hashtag #CortègeDeFenêtres tagged banners of the demonstrations that were hung from windows during the lockdown. The *cortège de tête* (leading procession), now confined, expressed itself at the window instead. The phenomenon is socially not insignificant. In Europe, where demonstrations traditionally take place in the streets, windows and balconies have historically taken on the role of substituting the public forum. At the border of the public and private spheres, balconies and windows are full of social significance (see Lefebvre 2019).

The balcony, in particular, has been a military device in the Middle Ages, a stage for authoritarian and fascist powers in the twentieth century, and a promise of social progress and emancipation for the working classes in post-World War II Europe (Nedelec 2016). The windows of the #CortègeDeFenêtres served for their part as political flags for claims that fell within wider anti-government protests. These windows participated in a reinvention of notions of public space during the lockdown.

The impact of these actions was amplified through social media. Indeed, social media is becoming a virtual space of interaction for political expression. During the lockdown, these ways of expressing were amplified by the impossibility of gathering outside. We can argue then that the sociology of social struggle, here displaced to balconies and windows, forms part of a broader digitisation of balconies, leading to a virtual conversion of the formal architectural reality of the balcony. This phenomenon existed prior to the lockdown, and reminds us of the way Julian Assange used balconies to stage public appearances from the

Figure 4.1 The slogan reads 'Money for the public hospital, solidarity with the cashiers.'

Figure 4.2 The banner reads 'The corona will disappear; The commune will bloom again.'

Ecuadorian Embassy in London (France24 2012). Notably, in France this appropriation of a private space for political purposes can lead to prosecution (see Franque 2020).

We were able to observe the content of these messages, consisting of slogans chanted or sung during demonstrations (see Figure 4.1).[3] We recognised that the makers of these banners were either reproducing the same banner slogans as those seen during the protests, or painting the slogans that were only chanted. Thus the slogans, sometimes written in verse, privileged the expressiveness of form. Sound effects were translated by the use of rhymes which facilitate oral and rhythmic transmission. The poetic dimension is marked on canvas.

The songs evoked are from the culture of struggle. The Paris Commune[4] is referenced here by the cherries on the banner shown above (Figure 4.2)[5] and the song 'Le temps des cerises' (The time of cherries). Written in 1868, this song is associated with the Commune because its author, Jean Baptiste Clément, was a Communard who fought during the *Semaine sanglante*. Its lyrics have been interpreted as a metaphor of a lost revolution (Martin 2021).

Tracing the Sonic Expression of French Social Movements 53

In general, most slogans evoke struggles that preceded the lockdown and which were adapted to the medical-lexical field related to the fight against COVID-19. The slogans are critical of government policy. Examples include:

Macron don't leave your poor behind
Pangolins against capitalism
Money for the hospital not for capital!
Let's kill the capitalism that kills us! Let's save the public service that saves us!
Euros for the hospital
Money! Money! For the public hospital!
Broken public service = danger
our lives are worth more than their profits
save lives not the economy
2018: macron cuts 4172 hospital beds
Never forgive never forget.

Other slogans drew parallels with the notion of the police state:

All
Covid-19
Are
Bastards
Support for hospitals[6]

Here, the acrostic form was used. The slogan refers to the acronym All Cops Are Bastards. This anti-police slogan was popularised during the British miners' strike of 1984–1985 (Dupont 2021). Here, the confined slogan can also be read as ACAB, Support for Hospitals. This anti-police slogan tries to camouflage itself under an anti-COVID-19 phrase by playing on semantic ambiguity.

Other slogans assimilate anti-state positions to the lockdown:

ACAB however [7]
Covid-19 Everywhere
Justice nowhere
Solidarity with Caregivers, delivery people, cashiers, construction workers …[8]

The punctuation and the exclamation points remind us that we are in the domain of listening and shouting. On one banner, musical notes have even been drawn (two sixteenth notes and two eighth notes), making explicit the association between the two senses, visual and auditory, which is apparent in this approach (see Figure 4.3).[9] Beyond the solfegic precision, it recalls a fight song. More than reading the banner, it is necessary to sing it.

If, confined, the demonstrators could no longer sing together in the streets, they could write their songs and slogans in capital letters. They revived these songs, in an almost mournful silence, in order to continue the struggle, as if the survival of the expressions of their protest depended on this music. It is enough that the passer-by or the neighbour reads and/or sings them so that the message is disseminated. In this way, despite being confined, the demonstrations continue through the multiplication of individual initiatives, at the window or at the balcony. The appropriation of window and balcony spaces for the political purpose of continuing the demonstrations took place in a context that has provoked other

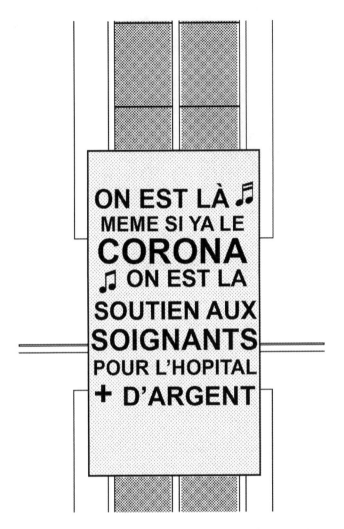

Figure 4.3 The banner reads 'We're here even if there's corona; We are here; Support for caregivers; For the hospital; + money.'

confined collective demonstrations. Thus, these banners were hung in parallel or in opposition to another ritual which was set up from the first day of the confinement: the 8:00 pm applause.

The Applause at 8:00 pm: The Rhythms of a New Confined Socialisation

Appearance of Clapping

Inspired by their Italian neighbours who had already been confined for two weeks, the French began to applaud at 8:00 pm to honour the work of caregivers considered as being on the front lines of the fight against COVID-19. According to Aymeric Renou (2020), a journalist at *Le Parisien*, the first applause was heard in France on Tuesday 17 March, the first day of lockdown. Depending on the neighbourhood and the city, the applause and the wishes of solidarity were extended to workers who continued to go out to pursue their missions essential to the life of the community, in a territory where the virus was active. Invisibilised workers, cashiers, delivery men and women, garbage collectors, and construction workers, found themselves in the media spotlight for a moment. The ritual was well established, and would last throughout the confinement (from 17 March to 11 May 2020). Shortly before 8:00 pm, the French would stand at their window or balcony to applaud the essential workers. The event lasted five, ten, or sometimes 15 minutes. As the month of March progressed, this event, for many the only social moment of the day, transformed into a party.

What is Clapping?

Applause is the sound of two hands clapping together. In his article 'Applause: A Social Behaviour,' sociologist David Victoroff notes that applause is a social mark of satisfaction and approval (1955, 133). He also notes that clapping involves a 'contagious' movement (ibid., 132), and the expression of a 'conventional and institutionalized' gesture (ibid., 134). Finally, he notes that we are here in a gesture of 'communication and language' (ibid.). Thus:

> The manifestations of admiration and approval are thus regulated by the group to which the individual belongs. It is the group that determines the circumstances in which applause is recommended, tolerated or forbidden, and sometimes even regulates its intensity and duration. The applause is thus a social expression of approval, not only as a collective demonstration, but also and especially as a conventional demonstration.
>
> (ibid.)

With these definitions in mind, let's listen to some excerpts from the 8:00 pm applause.

What Clapping Means in Times of Confinement

We can listen to an excerpt we recorded on 10 April 2020, on rue des Hautes Formes in the 13th arrondissement of Paris:

> High Formats recording XIII (6:30): https://soundcloud.com/user-796735567/hautes-formes-xiii.
> From 00:37, we hear a single person clapping quite slowly (almost one beat per second), it is not yet 8:00 pm. Finally, at eight (one minute later), the applause starts in the distance, then a tambourine starts its percussion (at 01:20 approximately), and at 01:37 we hear a yelp siren.[10] At about 03:50, we distinctly hear a rhythmic percussion, probably by means of a stick and a plastic or metallic surface; in any case it seems to be a homemade percussion instrument. We also hear the tambourine trying to keep up with this rhythm by beating along with the beat. At the end, the tambourine takes over and we again hear the slow clapping of the beginning. From 04:53 on, we hear non-musical socializing.

In addition to the rhythmic and sonic dimension of the applause, the confinement highlights the social dimension of the applause. Indeed, as the confinement progressed, the ritual of applause became a moment of resocialisation, especially between neighbours. Every

56 *Alessandro Greppi and Diane Schuh*

evening, after about five minutes of applause, the inhabitants began to converse and to exchange news with one other. In this recording, from 04:53 onwards, we can hear: 'enjoy your weekend,' 'tomorrow, an aperitif!' or 'we'll have some delivered for everyone.' Listening to emblematic recordings of the soundscape of this neighbourhood at 8:00 pm during the confinement[11] allows us to highlight the conventional character of the ritual. Originally conceived as a tribute to the essential workers, this ritual becomes a moment of sharing and musical communication. Indeed, the different applauses answer each other in imitation and develop rhythms whose effect seems to palliate the solitude and the silence of the confinement. The applause is a conventional rite, but it does not make consensus. The hypocrisy that some can detect in the message conveyed, and the social and political manipulations of the gesture, have given rise to numerous criticisms, with some seeing this practice as a way to 'buy a good conscience' (La-bas.org 2020).

Thus, beyond the musical interpretations that one might make, applauding in times of confinement is also a way of situating oneself socially and politically, and of constructing an identity within a movement. These different types of applause can also be compared to the different socio-professional categories that populate the different districts of Paris and in the immediate suburbs (for those that we were able to record). Thus, the music of the applause reveals the social territory in which it is situated (Caenen et al. 2017). Without seeing the architecture, we hear the environment. The ear reveals to us clues on the geographical and political position of the 'applauders.' In this way, certain districts are more receptive to the movements of contestation, and certain activists are going to position themselves in opposition to the conventional applause that they judge to be devoid of sense, or even completely hypocritical. But through criticism, creation and craft is manifested.

Hijacking a Consensual Ritual for the Benefit of Social Struggle

Hijacking the Applause to Make a Connection: The Role of Sound

In the urban environment, the #CortègeDeFenêtres described above, with the banners and the projections (for example, the unauthorised projections warning of the economic consequences of the lockdown for the culture related professions [Vavasseur 2020]) are themselves enough because of the numerous passersby. In rural and suburban areas, where passersby are more scarce, these demonstrations risked going unnoticed. The protesters produced visual objects that they chose to disseminate through other means of communication. They then used social networks in order to reach a wider audience. In both cases, the mobilisation started on social networks and ended at the micro-local level, where it participated in recreating links with neighbours. The mobilisation thus becomes a pretext for forging links on the balcony, at the window, or in the courtyards and other spaces of resocialisation during confinement.

This need to connect is especially evident among the underprivileged, the forgotten of so-called equal opportunities, and the low-income communities in peripheral France excluded from the visual public space. The latter suddenly found themselves in a position to be heard. The desire to be heard by the government became, as in traditional demonstrations, also a call for friendship and togetherness with one's peers. The songs that brought people together in social struggles, before the arrival of COVID-19, reacquired this role by expanding their audience. With the help of loudspeakers, those excluded from the visual space amplified emblematic songs from recent social struggles (see for example the official clip of the Yellow Vests [D1ST1 OFFICIEL RAP 2019]), chanting slogans in the yards of buildings for example, far from the eyes of passersby and media cameras.

The Different Sound Modalities of These Detours: Casserolade, Balcony Concerts for Politically Causes, Slogans at the Window, Sound and Light Shows

The balcony and the window, open to the public space, can thus constitute a space in between, defined here as the threshold between public and private spaces. This space was used as a banner and a place of expression to thank the caregivers. This space has also been used sometimes, through detours, to express feelings of indignation, dissent, and to collectively express a rejection of the government. We will now listen to some emblematic examples of these detours for the benefit of social and political demands:

Shouting, Slogans, Casserolades and DIY Percussion

Some activists opposed the applause that they considered empty of meaning, and even totally hypocritical. One could then hear an appropriation of the eight o'clock applause. People took advantage of this ritual to continue the protests. One of the sound expressions from the protests is called *casserolade*. These are real concerts of pots and pans, for 'to clap with a pan is to protest.'[12] Following the tradition of the charivari, a medieval ritual where noise was used to express a collective dissent when a social rule is broken (see Le Goff and Schmitt 1981), the concert of pans was used to protest against King Louis Philippe in the 1830s. Thus, during the confinement in 2020, we heard again the *casserolade*. This sonic manifestation sometimes provided an addition to the sound environment of the eight o'clock applause.

Let us listen to this first excerpt.[13] A woman at her window with a saucepan:[14] 'Hospitals abused! Macron must pay!'[15] And another example in which we can hear 'Money! Money! For the public hospital!'[16] In this last example, you can hear the rhythmic rendition of the slogan 'Du fric du fric' and then people chanting 'On est là, on est là ...' These are exactly the protest songs that could be heard during the street demonstrations before the lockdown; they are now organised and chanted from windows and balconies.

The Sound and Light Shows and Projections, a Dynamic Extension of the Banners

Other demonstrations involved sound and image. One video, for example, is an anonymous montage set to a soundtrack, found on the Facebook page Nantes Révoltée (2020). We can see and listen to different projections broadcast during the now traditional moment of applause at 8:00 pm. In the first excerpt, these words are projected onto the façade:

> Money for the public hospital
> It's not a war it's an epidemic, less cops more masks
> Money for the public hospital
> THANK YOU to the cashiers, garbage collectors, postal workers
> Money for the public hospital
> Money for public research
> Solidarity is the weapon of the people
> Let's kill the capitalism that kills us, let's save the public services that save us
> Money for the public hospital.

One will notice here the couplet-chorus organisation of the beginning, then a collection of messages that we find on the various fabric banners, and the return of the chorus 'Money for the public hospital.' The second excerpt is an animated image of clapping hands against a background of pan drums. This association could be read as a paradox between the

58 *Alessandro Greppi and Diane Schuh*

symbolism associated with the act of clapping hands – to be viewed as a continuum of the conventional applause – and politically situated pan drums. The third extract is a video of fireworks over a background soundtrack of horns, a foghorn, shouts, and applause (we will come back later to this modality of protest). The fourth extract is a video of a projection of words on a facade, the background sound being applause. We can read:

> Hospitals: Macron – Keep – Your – Tributes – We – Want – Jobs – We – Want – Beds – We – Want – Equipment – We – Want – Real – Salaries – We – Want – Real – Pensions – Neighbours – I – Love – You – Neighbours – We – Make – Noise?

By listening to these examples, we can hear how the soundscape of the protests moved to a new environment between public and private space, creating a kind of political balconism (Dullaart 2014).[17] The confined French cities became noisy again, as the bearers of a protest movement created a new 8:00 pm soundscape: rhythmic, noisy, and political. Unfortunately, windows and balconies were not available to everyone during the confinement. Other sound events took place outside of the window.

Exit the Window: The New Soundscape of Confined Towns and Suburbs in Protest

Diverting the 8:00 pm Applause by Honking Horns

During lockdown, 25 per cent of the French population continued to go to work (Lehut 2020). This segment of the population, considered essential during the crisis, could not always express themselves. Some of these workers, often on the road at eight o'clock, chose to honk rather than applauding. Every evening, during the first lockdown, garbage collectors in downtown Nîmes drove and honked horns to pay tribute to healthcare workers (Corger 2020). Small traders, truck drivers, ambulance drivers, taxi drivers, and Uber drivers followed a similar approach (Lanot 2020). In France, unlike other countries, these sonic expressions quickly took a political turn as medical personnel complained about the lack of government support with a plea to turn these expressions of solidarity into concrete measures to help their cause.

Toulouse cab drivers met every week to applaud, in a sense, by honking repeatedly for the nursing staff. Their tribute was noisy, accompanied by firecrackers and loud applause, and was emblematic of their socio-professional category compared to other groups. Car horns, their 'soundmark' (Schafer 1994, 10), had the effect of increasing the intensity of the demonstration, allowing for the occupation (to use protest vocabulary) of Toulouse's place in the *du Capitole* soundscape (Ouest France 2020). This brings us back to the traditional practices of social movements from the second half of the twentieth and beginning of the twenty-first centuries, which focused on people's occupation of public spaces. During lockdown, the sonic spheres of these places was expanded and magnified to poetically and symbolically circumvent the visual borders of the government decrees banning public assembly. Sound became a platform for protesters to meet and cross the bureaucratic visual borders and limitations of the lockdown. The acoustic properties of sound helped to create a link between neighbours, as this is not constrained by the physical borders of urban spaces. Contact and vision – restricted during lockdown by police controls on public spaces – lost their relevance as instruments of emancipation, leaving room for only the sonic expression of dissent.

Another emblematic example took place when tow truck drivers from Compiegne gathered in their city in April 2020 to thank medical staff. The result of the gathering

Tracing the Sonic Expression of French Social Movements 59

was a concert of horns and sirens superimposed upon one another,[18] an acoustic parade that alternated the repetitive motifs of the sirens with those of the horns, played by the drivers in a counterpoint reminiscent of the 1922 *Symphony of Factory Sirens* by Arseny Avraamov. This concert, so to speak, was all the more interesting in that it unfolded in the historical centre of Compiegne: the streets form sound corridors where the sirens' mechanical and repetitive urgency mixes with frenetic applause, here in the form of horns, resulting from the drivers' annoyance at the government (Oise Hebdo 2020). As far as maritime professions are concerned, at noon on 1 May 2020, ships in commercial service in Corsica paid tribute to the sailors who keep the supply chain going with a foghorn concert, which was not heard by the vast majority of French people, nor recorded, but did in fact take place (Corsica Linea 2020). In the same way, in Marseille, protesters demonstrated from inside their cars in order to circumvent the ban on gatherings of more than ten people.

The common point between all these road and sea demonstrations is that they all thanked and paid tribute to those of France's workers deemed essential during times of crisis. Care workers, transporters, cashiers, sailors, logistics workers, and all the rest had neither the time nor the mental energy to stage protests from home. Their only instrument was the sound of their work vehicles: the car horns and the sirens. Coming from rural and suburban areas, they expressed dissent through vibratory tensions and a sonic dialogue with other working protesters who shared the same struggles. These workers, invisible in normal times, became audible during lockdown.

This particular use of horns echoes the older 'sound referendums' organised on French highways to protest the labour law reform in 2016 where car horns naturally replaced protest songs (B.H. 2016). They also echo the more recent adage of the yellow vests, sung on provincial traffic circles: *un klaxon=un soutien*. Car horns can thus serve as a social and sonic marker for precarious (Le Monde 2019) or 'peripheral' France (Guilluy 2015, 192), and also serve as a critique of the renewed empathy for healthcare workers only a few months after French hospital workers had taken to the streets to protest years of cutbacks across the country. Horns can then serve as a warning, not of incoming traffic, but of the various contradictory political behaviours that the government has been engaged in, often with the support of the wealthier populations who give themselves a clear conscience by applauding at eight o'clock. In comparison, the applause is consistent with the liberal doctrine of small gestures, which now seem derisory when compared with the vastness of the disaster. Finally, through sound and through a socially conscious listening, these horns highlight a two-tier France, prompting us to consider the situation of some workers during lockdown as segregational, with a spatial separation between social classes reinforced by the confinement. It is in fact the poorest and most vulnerable elements of society who were impacted the most, both from medical and social points of view. This trend can be observed virtually everywhere in the world (OHCHR 2020).[19]

Horns and politically mobilised applause position sound in dialogue with social struggles and allow us to rethink the 'space of appearance,' defined by Hannah Arendt as a mere area of visibility ([1958] 1998, 199). Indeed, the space of appearance in which these dissident sonic expressions unfold goes well beyond traditional arenas of visibility by positioning us in relation to invisibilised workers, unrepresented citizens, and all those who are on the margins of the public sphere. Horns and applause can therefore be related to discourses of 'subaltern counterpublics' (Fraser 1992, 123). These discursive arenas develop in parallel with official visual public spheres. They refer to the notion of 'sonic sensibility' (Voegelin 2014, 3) developed by Salome Voegelin, which reorients and upsets the politics of visibility and encourages a critical reflection on government action. Through listening we are given the opportunity to uncover the hidden reality of invisibilised workers, as sensory responses

60 *Alessandro Greppi and Diane Schuh*

encourage us to rise above the taxonomical way we understand the political. We not only listen to the sound of the car horn, we inhabit the life of the dissenting driver if we listen to the sound just for itself, departing the semantic arena of visual politics and exploring non-consensual alternative worlds.

Shedding Light onto an Emerging Counter-space of Appearance Specific to COVID-19

Politics indeed requires a space of appearance that emerges whenever people are together in their manner of speech and action. However, this space does not always exist. In contemporary Western societies, low-income communities are excluded from such a space. This means that part of the population does not appear, it does not emerge into the space of appearance. The lockdown has unsettled this situation. It has allowed the invisibilised and the weak to seize and redefine public spaces emptied by existing powers. The space of appearance is no longer the exclusive domain of a powerful few. It has been deserted, and as such has been opened to the bodies of critical workers and the unemployed, invisible to the rest of society, in rural and suburban areas. Their bodies must no longer enter an inaccessible visible field to exist socially, in so far as this field remained inaccessible for everyone during the lockdown.

The health restrictions and measures put in place during the lockdown effectively outlawed outdoor activities. The act of visual appearance, ontologically fixed and segregational, lost its purpose, leaving dissenting bodies with nothing but the dynamic and common power of the audible and the sonic (Barbanti 2013). New modalities of common interest were thus created. In this context, the traditional space of appearance shifted online onto social media through videos, videoconference calls, and virtual events, while the physical world became a counter-space of appearance where the bodies of critical workers disappeared under the weight of their repetitive work routine and neglect from the ruling classes. The clapping at eight o'clock operated to this end, as a door connecting the two spaces – between the visual world of telework submerged by screens, and the daily sonic reality, often undervalued and overlooked, of critical workers. Coming from rural and suburban areas, they expressed dissent through vibratory tensions and a sonic dialogue with other working protesters who shared the same struggles.

Their car horns and their sirens are loud sounds[20] that occupy space in the same manner that bodies would occupy a street or a square. Sound connects, includes, and overcomes visual restrictions. During the pandemic, sounds transcended the Adornian subject-object dialectic because they carry political meaning that goes beyond their physical and acoustic condition. The essence or raison d'etre of these sounds is to promote social change. Finally, these sounds are instruments for the inclusion of other communities. The sounds of these manifestations of dissent, the object, and the critical workers, the subject, are part of an undertaking bigger than the sound itself and bigger than the act of perceiving sound. It refers to a common vibration, which resonates with the situation of other critical workers. As such, it sheds light on the unfair politics of the lockdown.

The Particular Case of Parisian Suburbs

Another group of unlikely protesters in this new counter-space of appearance is the youth living in the so-called sensitive neighbourhoods of Ile-de-France.[21] On 20 April 2020, an accident involving a motorcyclist and a police car in Villeneuve-la-Garenne, in the Hauts-de-Seine, led to clashes between young residents of the suburbs and the police. These conflicts, which occurred as several videos were denouncing police violence since the beginning of the lockdown, took place in Aulnay-sous-Bois, Gennevilliers, Evry, and Clichy. The soundscape of their protest is different from the sounds of the protests discussed above. Contrary to applause and horns, these improbable demonstrators were not intending to fit into a predefined social movement. Their approach was not one of thanking critical workers or expressing subaltern counterpublics' indignation.

The youth living in these urban neighbourhoods, who face economic and social hardships, set off fireworks and firecrackers not so much to express discontent with this situation or that government policy, but to continue living outside overcrowded apartments in social housing (INSEE 2020). Lockdown deprived them of the little freedom that they had. For them, living outside seems like an economic, social, and existential imperative. The sound of these fireworks and firecrackers operated like sonic bombs defying visual monitoring systems. These sounds became an event in itself: the spectacle of a nocturnal struggle, which may be lost in advance, at least in a short-term perspective, but has the benefit of acting as a warning to the rest of the population about these so-called ghettoised, underprivileged, poor, sensitive, and difficult zones, which are unimaginable to the rest of the population because they are absent from dominant media representations. These youth position sound in dialogue with other forgotten struggles.

The soundscape of these protests is not otherworldly, but it proposes alternative points of view on what the world is and how we live in it, showing what else it could be in the register of the unimaginable. Fireworks are historically featured to accompany official celebrations. Here they are endowed with a political agency of their own (LaBelle 2018). They challenge the singularity of celebration, inciting us to unsettle the real and imagine different layers of possibility. The sound of fireworks becomes an event in itself. It may be a lost battle, but it has the benefit of warning the rest of the population of an alternative future that enables these populations to escape the logic of urban (spatial and therefore visual) ethnic and economic segregation that currently prevails in the discourse on these so-called sensitive neighbourhoods. We thus enter into the register of the unimaginable, because neoliberal ideological alienation is such that an alternative seems impossible, even to imagine, which follows our earlier analysis of the invisibility of critical workers. Finally, we would like to mention the silent protest of the weakest in society, those who have no car horn to sound, no fireworks to display, and perhaps even no window or balcony from which to clap: the dispossessed, the homeless, the lonely. While observing the phenomenology of their silenced routines, ontologically there is sound, the sound they would have sounded had their concerns been heard in pre-pandemic times.

Notes

1 On 5 December alone, according to the CGT (General Confederation of Labour in France), 1,500,000 people protested in France (CGT 2019), while the Ministry of Interior registered 806,000 protesters (Ministère de l'Intérieur 2019).
2 Strikes and demonstrations have been regular and sustained since 5 December 2019 and the calendar of actions did not seem to stop. National educational staff strikes took place on 5, 10, 17 December 2019; 9, 14, 16, 24 January 2020. The strike of the SNCF (French National Railway Company) staff is considered as the longest social conflict in history in France, with 37 consecutive days between December and January. The Yellow Vests movement joined the anti-reform of pensions demonstrations on 7 December, it was then at its 56th act; the last act before the confinement, the 65th act, took place on 8 February. The movement against the PBDA had decided to stop the university and research on 5 March 2020, followed by a large number of actions and demonstrations. See https://universiteouverte.org/.
3 Illustration based on a photo from a collaborative collection of protest banner photography collected on Twitter under the hashtag #CortègeDeFenêtre (Twitter "#CortègeDeFenêtre." @H_Ronac, 21 March 2020. https://twitter.com/H_Ronac/status/1241387327706013697?s=20).

4 The Paris Commune ('La Commune de Paris') was a revolutionary government that seized power in Paris from 18 March to 28 May 1871. It is considered a symbol of popular struggle and a model for communist, socialist, and anarchist movements in establishing a political system based on participatory democracy.

5 Illustration based on a banner posted on Twitter '#CortègeDeFenêtre.' @AnneBonnyBonny, 22 March 2020. https://twitter.com/AnneBonnyBonny/status/1241634984441585664.

6 Posted anonymously, 19 March 2020 in Lille (France). Facebook Page 'Cortège de Fenêtres.' www.facebook.com/Cort%C3%A8ge-de-fen%C3%AAtres-103566351286235.

7 Twitter '#CortègeDeFenêtres.' @matlagratte, 23 March 2020. https://twitter.com/matlagratte_/status/1241801528245129218?s=20.

8 Posted anonymously on 22 March 2020. 'Cortège de Fenêtre' Facebook page. www.facebook.com/Cort%C3%A8ge-de-fen%C3%AAtres-103566351286235.

9 Illustration based on a banner posted anonymously, 22 March 2020. 'Cortège de Fenêtre' Facebook page. www.facebook.com/Cort%C3%A8ge-de-fen%C3%AAtres-103566351286235.

10 We could not determine the exact source (child's toy or adult-owned siren amplifier), the sound is close to the yelp siren of the 'Federal Signal PA300 200 W Siren.'

11 One of the authors was able to record the applause several nights in a row and listen to the rhythms and socialization under construction in this new ecosystem particular to the confinement. A selection of the sound recordings can be listened to online: https://lemondeautre.fr/portfolio/hautes-formes-project-field-recording/.

12 A comment made by a user during a livestream debate hosted by Baptiste Penel on www.twitch.tv/usul2000. Twitch.tv. 26 March 2020. www.twitch.tv/videos/576036367. (Twitch videos are automatically deleted 60 days after the publication.)

13 All sonic illustrations are available at this address: https://soundsofthepandemic.wordpress.com/alessandro-greppi-diane-schuh/.

14 Twitter '#CortègeDeFenêtre.' @Agalitzine, 21 March 2020. https://twitter.com/AGalitzine/status/1241447880847773696?s=20.

15 'Hôpitaux malmenés! C'est Macron qui doit payer (ibid.).

16 'Du fric! du fric! Pour l'hôpital public.' @DelsemmeClaire, 20 March 2020. https://twitter.com/DelsemmeClaire/status/1241080682727825408?s=20.

17 The term 'balconism' is born from the practice of observing the uses and practices specific to the spaces of balconies. The balcony being a space that sits in between the private and public, it participates in the public appearance of the city. It is the space of very specific practices that we can describe and analyse (Dullaart 2014).

18 The duration of the *notes* of some sirens are sometimes held to infinity while others produce 'portamentos' from low to higher registers.

19 'With a projected fall in per capita income in more than 170 countries, people without social protection will be worst hit, De Schutter said. Worldwide, about four billion people have no social protection coverage and those in precarious employment, including the 2 billion workers in the informal sector, are often the first to lose their jobs' (OHCHR 2020).

20 The horn is particularly interesting from a sonic point of view, as its sound pressure has remained that of normal times in modern cities. In Paris, for example, according to estimates released by the Paris municipality (Paris 2019), 10 to 15 per cent of the population is normally exposed to a value above the Lden limit of 68 dB(A). The sound of a horn, like that of sirens, is less appropriate in times of lockdown for its alerting function than for demonstrating or applauding caregivers.

21 These circumscribed territories tend to be defined as the receptacle of most of the ills of French society.

References

Arendt, Hannah. (1958) 1998. *The Human Condition*. Second edition, with an Introduction by Margaret Canovan and a Foreword be Danielle Allen. Chicago: University of Chicago Press.

Barbanti, Roberto. 2013. 'Ecouter le monde.' Presented at *La Semaine du Son 2013*, Paris, France, 15 January 2013.

Berque, Agustin. 2000. *Ecoumène: introduction à l'étude des milieux humains*. Paris: Belin.

B.H. 'Paris: les automobilistes invités à faire du bruit contre la loi El Khomri.' *Le Parisien*, 13 September 2016. www.leparisien.fr/paris-75/paris-75020/paris-les-automobilistes-invites-a-faire-du-bruit-contre-la-loi-el-khomri-13-09-2016-6116987.php.

Caenen, Yann, Claire Deconde, Danielle Jabot, Corinne Martinez, Samira Ouardi, Pierre Eloy, and Luca Jouny. 2017. 'Une mosaïque sociale propre à Paris.' *Insee Analyses Île-de-France* 53, 2 February 2017. www.insee.fr/fr/statistiques/2572750.

CGT (Confédération Générale du Travail). 2019. 'Grève du 5 décembre: réussite générale!' *La CGT*, 5 December 2019. www.cgt.fr/actualites/france/retraite/mobilisation/greve-du-5-decembre-reuss ite-generale.

Coquaz, Vincent. 'D'où vient l'hymne "On est là" chanté dans les manifs depuis un an?'. *Libération*, 10 December 2019. www.liberation.fr/checknews/2019/12/10/d-ou-vient-l-hymne-on-est-la-chante-dans-les-manifs-depuis-un-an_1768086.

Corger, Corentin. 'EN VIDÉO Hommage au personnel soignant, les éboueurs klaxonent en centre-ville de Nîmes.' *ObjectifGard*, 31 March 2020. www.objectifgard.com/2020/03/31/en-video-homm age-au-personnel-soignant-les-eboueurs-klaxonnent-en-centre-ville-de-nimes/.

Corsica Linea. 'Les cornes de brume de nos navires ont retenti en hommage à nos marins, à tous les marins & professions maritimes à travers le monde.' Twitter, 1 May 2020. https://twitter.com/Corsi caLinea/status/1256197051110625281?s=20.

D1ST1 OFFICIAL RAP. 'Gilets jaunes clip officiel D1ST1.' 20 January 2019. www.youtube.com/ watch?v=ix0p5Q1937o.

Dullaart, Constant. 2014. 'Balconism: A Manifesto.' *Art Papers* (March/April). www.artpapers.org/ balconism/.

Dupont, Marion. '"ACAB" ou la rage anti-flics.' *Le Monde*, 26 May 2021. www.lemonde.fr/idees/arti cle/2021/05/26/acab-ou-la-rage-anti-flics_6081472_3232.html.

France24. 2012. 'Julian Assange demande aux États-Unis de cesser "la chasse aux sorcières."' *France 24*, 19 August 2012. www.france24.com/fr/20120819-julian-assange-demande-etats-unis-cesser-cha sse-sorcieres-contre-wikileaks-londres-equateur-ambassade.

Franque, Adrien. 'À Toulouse, Montpellier ou Paris, la police débarque pour des banderoles anti-Macron.' *Libération*, 1 May 2020. www.liberation.fr/france/2020/05/01/a-toulouse-montpellier-ou-paris-la-police-debarque-pour-des-banderoles-anti-macron_1786871.

Fraser, Nancy. 1992. 'Rethinking the Public Sphere: A Contribution to the Critique of Actually Existing Democracy.' In *Habermas and the Public Sphere*, edited by Craig Calhoun, 109–142. Cambridge: MIT Press.

Girard, Quentin. 'Ami, entends-tu les chants de la protestation?' *Slate.fr*, 11 October 2010. www.slate. fr/story/28457/chansons-manifestations.

Guilluy, Christophe. 2015. *La France périphérique: comment on a sacrifié les classes populaires*. Paris: Flammarion.

INSEE. 2020. 'La suroccupation des logements en Ile-de-France est quatre fois plus importante qu'en province en 2017,' press release, 19 November 2020. www.insee.fr/fr/information/4965622.

La-bas.org. 2020. 'Le message du jour: faut-il applaudir à 20h?' *Le répondeur de Là-bas si j'y suis*, 21 March 2020. https://la-bas.org/la-bas-magazine/le-repondeur/le-message-du-jour.

LaBelle, Brandon. 2018. *Sonic Agency: Sound and Emergent Forms of Resistance*. London: Goldsmiths Press.

Lanot, Clément (@ClementLamont). 2020. 'Une centaine d'ambulances vient applaudir les soignants de l'hôpital de Melun qui luttent contre le #Covid_19.' Twitter, 6 April 2020.

Lefebvre, Henri. 2019. *Éléments de rythmanalyse et autres essais sur les temporalités*. Paris: Eterotopia France.

Le Goff, Jacques and Jean-Claude Schmitt, eds. 1981. *Le Charivari. Actes de la table ronde organisée à Paris (1977)*. Paris: Mouton.

Lehut, Thibaut. 2020. 'SONDAGE – Les Français et le confinement: dans le monde du travail, le coronavirus accentue les inégalités.' *France bleu*, 9 April 2020. www.francebleu.fr/infos/societe/ sondage-les-francais-et-le-confinement-dans-le-monde-du-travail-le-coronavirus-accentue-les-158 6355082.

Le Monde. 2019. 'Qui sont vraiment les "gilets jaunes"? Les résultats d'une étude sociologique.' *Le Monde*, 26 January 2019. www.lemonde.fr/idees/article/2019/01/26/qui-sont-vraiment-les-gilets-jau nes-les-resultats-d-une-etude-sociologique_5414831_3232.html.

Martin, Denis-Constant. 2021. ' "Le temps des cerises," comment analyser les rapports entre musique et politique?' In *Plus que de la musique*, 167–185. Guichen: Éditions Mélanie Seteun.

Ministère de l'Intérieur. 2019. 'Félicitations aux forces de l'ordre et de sécurité mobilisées lors de la manifestation du 5 décembre 2019.' 5 December 2019. www.interieur.gouv.fr/Archives/Archives-des-communiques-de-presse/2019-Communiques/Felicitations-aux-forces-de-l-ordre-et-de-secur ite-mobilisees-lors-de-la-manifestation-du-5-decembre-2019.

Nantes Révoltée. 2020. 'Mieux que les applaudissements: solidaires et creatifs.' Facebook, 21 March 2020. www.facebook.com/Nantes.Revoltee/videos/638212806744206/UzpfSTEwMzU2NjM1MTI4NjIz NToxMDY3MjcyMjA5NzAxNDg/.

Nedelec, Cyril. 2016. 'Le balcon comme seuil et dispositif environnemental. Architecture, aménagement de l'espace.' Mémoire de séminaire, Ecole Nationale Supérieure d'Architecture de Toulouse. https://dumas.ccsd.cnrs.fr/dumas-01802992/document.

OHCHR (Office of the High Commissioner for Human Rights). 2020. 'COVID-19 Crisis Highlights Urgent Need to Transform Global Economy, Says New UN Poverty Expert,' press release, 1 May 2020. www.ohchr.org/en/press-releases/2020/05/covid-19-crisis-highlights-urgent-need-transform-global-economy-says-new-un.

Oise Hebdo. 2020. 'Compiègne. Les dépanneurs offrent un concert de honaxons pour remercier les personnels soignants.' *Oise Hebdo*, 29 April 2020. www.oisehebdo.fr/2020/04/29/compiegne-les-dep anneurs-offrent-un-concert-de-klaxons-pour-remercier-les-personnels-soignants/.

Ouest France (AFP). 2020. 'Toulouse. Les taxis klaxonnent une dernière fois en hommage aux soignants.' *Ouest France*, 12 May 2020. www.ouest-france.fr/region-occitanie/toulouse-31000/toulo use-les-taxis-klaxonnent-une-derniere-fois-en-hommage-aux-soignants-6833296.

Paris. 2019. 'Lutte contre le bruit.' 23 May 2019. www.paris.fr/pages/bruit-et-nuisances-sonores-162#le-plan-de-prevention-du-bruit-dans-l-environnement.

Renou, Aymeric. 2020. 'Coronavirus: à leurs fenêtres, de plus en plus de Français applaudissent les soignants.' *Le Parisien*, 19 March 2020. www.leparisien.fr/societe/coronavirus-a-leurs-fenetres-de-plus-en-plus-de-francais-applaudissent-les-soignants-19-03-2020-8283939.php.

Schafer, R. Murray. 1994. *The Soundscape: Our Sonic Environment and the Tuning of the World.* Rochester: Destiny Books.

Vavasseur, Pierre. 2020. 'Paris: une projection sauvage interrompue par la police.' *Le Parisien*, 26 October 2020. www.leparisien.fr/paris-75/paris-une-projection-sauvage-interrompue-par-la-pol ice-26-10-2020-8405103.php.

Victoroff, David. 1955. 'L'applaudissement: une conduite sociale.' *L'Année sociologique (1940/1948)* 8:131–171.

Voegelin, Salome. 2014. *Sonic Possible Worlds: Hearing the Continuum of Sound.* New York: Bloomsbury.

5 Pandemic Soundscaping

Rediscovering a New Aura in the Mediatised
Sonic Reality

Ludovico Peroni

Why So Surprised?

After several years living between Rome and Florence, I found myself living through the restrictions that led to the Italian pandemic lockdown – in its first, most drastic phase, between March and May 2020 – in the same small town where I grew up. Montappone is a small village of about 1,500 inhabitants in the Marche region of Italy, whose peculiarity lies in the presence of many small industries dedicated to the production of accessories (in particular, hats) in its territory, which are able to connote its culture, life, identity and, not secondarily, its soundscape. I have never personally paid attention to the latter, perhaps out of habit; the sounds from my home that I have listened to for years have never particularly impressed or excited me.

Something changed during the pandemic. One morning, when I woke up at five o'clock and looked out from my terrace, I realised that there had been a radical change in the components of the soundscape that I was implicitly familiar with: the noise of vehicles and businesses had given way to the sounds and noises of nature, fauna, and – perhaps – of the place itself. Trying not to let the moment pass me by, I recorded this soundscape in the best way I could, documenting the event with a video recording. Listening to the recording through headphones, I then discovered a definition of the sounds that I would never have expected; they seemed richer in details than the same sounds directly perceived (or rather, not mediated by sound recording technology).

On the same day, I shared the audiovisual product via my personal social network accounts, unexpectedly receiving very surprised comments from several friends, neighbours, and fellow citizens.[1]

Why so surprised?

Had none of my contacts yet paid attention to the new soundscape in their country?

Did the general amazement stem from the very nature of mediated sound, which can provide an even more immersive experience than non-mediated technology?

Or, perhaps, was it the peculiar situation of lockdown that had sensitised us to this kind of ecstatic experience?

Taken by the enthusiasm for this (new for me) field of research – explored, of course, through such questions – I started to travel, so to speak, through the soundscapes of the world by means of the only tools that the pandemic put at my disposal: a computer and a (good) headphone. Already after a first consultation of the web resources I understood that something was happening in the world in which I could recognise and inscribe my own experience.[2]

DOI: 10.4324/9781003200369-7

66 *Ludovico Peroni*

Towards Immersive Mediatisation

The COVID-19 pandemic has called us to live through a suspension that for some has represented an experiential unicum. Especially in the period of the strongest confinements, there is something that united the social experience of a very heterogeneous mass of individuals located in different parts of the world during the same historical moment:[3] the feeling of suspension, isolation, alienation from reality and daily life, and constant pandemic topos, widely found in speeches, surveys and research. In such a context of conditioned separation and social isolation, I would like to try to identify a particular and, perhaps privileged, use of experiential reality; this is not to diminish the tragic and pathological implications of the pandemic, which certainly still deserve attention from systematic studies,[4] but to try to investigate a new experience with which the subjects were confronted and which was able to occur thanks to the contribution of certain conditions that the pandemic itself favoured.

The technology that humankind has found at its disposal during the period of confinement has – in some cases surprisingly – worked not only to meet primary needs (such as work and social relations), but has also created new experiences that in a different cultural and historical context would certainly have taken on different meanings. In this propulsive thrust towards an increase in the proposal and use of online services, it is possible to assess the general tendency towards increasingly involving, immersive experiences that seek to compensate for the experience of being involved in the public health emergency by virtualising many aspects of everyday life. The changes that have taken place through the mediatisation of different spheres of experience can also be read by observing the movements of some large companies involved in digital network services: giants such as Google, which doubled the number of their Classroom users thanks to the pandemic (De Vynck and Bergen 2020); Zoom has put itself forward as the platform of choice for many spheres of life, beyond teleworking;[5] Microsoft has promoted its own work platform, Teams, through the progressive idea of remote working, projected even beyond the emergency situation of the coronavirus;[6] Webex, with its slogan 'supporting you during COVID-19' (Cisco Webex 2022), is sharing a communication strategy adopted by many other entities, aimed at empathically resonating with its customers.

In addition to communication and marketing strategies, many companies also implemented new features aimed at making their products more immersive right around the same lockdown period: WhatsApp (2020) started expanding the number of participants in video calls from four to eight in April 2020; Telegram (2021) introduced the possibility of group video calls in June 2021; and Tinder – a platform that makes the experience of a physical encounter between a couple of people the ultimate goal of its mission – announced in July 2020 that it would compensate for isolation by choosing to mediate the experience of direct acquaintance through the inclusion of video calls in its platform.[7] A final consideration is the launch of new digital products which, in a short period of time, gained prominence among users interested in a completely new category of use: one more focused on sound than on image. This is the case of Houseparty – which in the first days of lockdown allowed acquaintances to meet, anticipating business-type platforms – and Clubhouse, a social network based on sound and on the exchange of voice messages by users, which created a substantially new experience for the masses.

Online services have thus paid particular attention to recreating (sometimes by compensating, sometimes by emphasising) the emotions and atmospheres of reality not mediated by technology, immediately generating (more often than not) an important and shared response from users. Human habits – from the sphere of work to that of entertainment – have thus undergone a mediatisation unparalleled in history, not only because of the pandemic, but

Hic et Nunc of a Unique Experience

We can try to trace the outlines of some structural homologies between the development of technology in times of pandemic and some areas in the domain of aesthetic theory. Referring to the subject's *pathic sphere*, and framing it within a radical change in the environment, we can say that many have lived in an *atmosphere* – that is, a feeling diffused in space that causes tangible effects on the *body* of the person experiencing it (see Schmitz 2019; Böhme 2010; Griffero 2014a) – that has been totally reshaped.[8] As a consequence of this change, what informs general feeling has also undergone radical changes. In this scenario, an attempt was made to translate the physical experience of feeling, participating, and even working directly on the atmospheres into an *atmospheric mise-en-scène* (see Griffero 2014a, 83; Griffero 2014b, 174–175) that, together with the choices of digital service companies, directed everyone's actions.

The idea of presence and co-presence also certainly underwent a drastic reshaping. German philosopher Walter Benjamin's (2000) formula of the here and now provides us with an access to the reading of aesthetic reality on the level of the affections experienced by the subject, and also on the basis of what has been felt as an other, virtual reality, but is still made of an important material component such as domestic walls, family affections, sounds, lights and colours. In this experiential paradox, what remains indistinguishable is the idea of boundary (and confinement) between an inside and an outside, intimately connected with a before and an after, with an I and an Other. This turns out to be a 'disposition of mind' (Böhme 2010, 83), shared by many subjects in the world and delineated by affections such as melancholy, nostalgia, boredom, and fear that reverberate from body to body, from the surface of the senses to the depths of the *Leib*.

Affect is seen as a potential, a bodily capacity to affect and be affected. It is embodied in the automatic reaction manifested in the skin, on the surface of the body and in the heartbeat, but it is still something that goes beyond the body, a passage from one experiential state of the body to another (Hofman 2015, 36).

The modes of experience, lived through a process of mediatisation, are far from the *hic et nunc* that lead us to experience the aura, the atmosphere, of the normal (unmediated by electronic devices) life lived in Benjaminian presence; however, this does not mean that the ecstatic experience of the subject is to be read on a level of less intensity and that is 'more superficial' (Böhme 2010, 120). As with art, the experience of the pandemic has created 'privileged affective experiences' (Griffero 2014b, 166), lived in ways similar to *artistic atmospheres*, where what happens in them defines them more than they represent (ibid., 173). Art requires the observer to allow themself to be influenced by the atmospheric impressions that surround them, but without ever sinking into the impressions to the point of being completely subjugated (ibid., 166–167).

Continuing on the track of the aestheticisation of pathologically lived experience, we understand how the condition of immersion in the new reality – as well as the impossibility of detachment from it – has generated an ecstatic experience that coincides with the life of the percipient. For this reason, it can give access to a more cathartic experience than the contemporary artistic experience, which ultimately remains however in the domain of '(relatively) detached contemplation' (Griffero 2014b, 167) due to the extra-thing orientation of art itself (see Griffero 2014b, 166–167). In the pandemic, a procedure took place that was substantially different from what happens in art, but which drew elements from

68   *Ludovico Peroni*

art. The immersion of individuals in the all-encompassing pandemic situation – with technology as the privileged medium for informing experience and also reconfigured as a usable object (recordings, streaming, multimedia content) – has enabled the cathartic condition that pushes us to overcome the idea that the subject is in a state of separation from (unmediated) reality. In some cases, it also allows one to experience an ecstatic contact that is even enhanced by the very condition of separation.

Personally, writing even at a distance, I can clearly attribute strong aesthetic and ecstatic characteristics to the period of confinement: I lived immersed in a new atmosphere with a particular aura that characterised my existence and my relationships in that place and time. I attended, perhaps as never before, concerts, events, conferences, and round tables, and found groups of distant friends and old schoolmates. Through these mediated experiences I experienced very strong, even physically tangible emotions, which cannot be compared to the reality not mediated through technology using the same tools and yardstick. On a poietic level,[9] I had the possibility to deliberately choose what to make appear in my life, giving an *atmospheric-medial mis-en-scène* able to best represent what was being experienced. The outside world was present, perennially, even if experienced through screens. From this experience I developed the idea that the ecstatic result of coexistence with a certain *hic et nunc* had not dissolved, but had only been transformed and adapted to the spirit of the time of the COVID-19 pandemic. Thus many subjects were pushed to fully experience this new *hic et nunc* in the experience which, through the mediatised reality, not only did not passively manifest itself, but offered itself to be experienced, renewed with its own aesthetic categories, or at least to be discovered.

Pandemic Soundscaping

Just as we have witnessed over the years a habituation, through continuous exposure, to certain forms of medial action, the mediatisation of experience, as we have seen, was proposed as a tool to combat the disconnection of the subject with reality during the pandemic, due to the rapidity of the phenomenon and the situation of emotional urgency. This peculiar framework also includes the discourse on the activities of many composers, sound artists, and individuals who have been involved in the sharing and enjoyment of field recordings and soundscaping during the pandemic, as practices with their own aesthetics and operations. The desire to share a surprisingly euphoric feeling (in a dysphoric scenario) was, in the first instance, also what prompted me to record the sounds of my country. Unlike the webcam sharing of images of natural outdoor places, which finds interest on a more aesthesic level by including 'elements of nostalgia, connection, and a sense of freedom' (Jarratt 2021, 166), here I found interest primarily on the neutral level of the object itself. Certainly feelings of well-being and nostalgia are important components of the virtual journey through shared soundscapes from different parts of the world, but in my opinion these are not the predominant moods of the listener. The discourse can be pushed to a deeper level, where aesthetic issues can be investigated concerning peculiarities of the recorded and shared sound object, connected to the mediums that informed it. For these reasons I have chosen to speak of *pandemic soundscaping*.

Soundscapes and High-resolution Technology

During the lockdown, there was a culture shock due to the sonic shift that occurred in the surrounding environment.[10] This sensitised many individuals who, by recording, processing, and sharing ambient sounds, quickly transported the practice of soundscape composition from the domain of artistic products to the masses (virtually speaking). Many artists have exploited such concrete sounds – dense with symbols and meaning – by also creatively utilising them, instinctively overriding the documentary and informational purposes of ambient sound, often treating it as an aesthetically autonomous object.[11] This phenomenon reflects the prominence of certain acoustic characteristics that have connoted, with transient and evanescent properties, the recorded soundscapes, marking a clear difference between what happened before, during (see Bartalucci et al. 2021), and what will probably happen after.

Raymond Murray Schafer's pioneering studies provide us with elements that conduct us to a different interpretation of some of the qualities that have undoubtedly activated the subjects' attention to the soundscapes of the pandemic: the sudden lowering of the intensity of anthropogenic sounds in urban environments has in fact led to the emergence of a soundscape in which they have distinguished, some for the first time, a depth of acoustic field that was previously denied. Indeed, the pandemic has quite diffusely taken crowds out of squares, noise of work activities out of buildings, and transport from the streets, sometimes affecting perceptions related to the nature of the places themselves (see Jordan and Fiebig 2021). The process that historically transformed 'rural' landscapes into 'urban' landscapes (Schafer 1994, 43) has generally been reversed in the 2020s. The lo-fi soundscape has reverted to hi-fi (ibid., 43–44) and here in this expanded sound horizon, subjects – living in isolation and in communities suddenly reduced to the family unit or the singularity of the person – began to instinctively, and with an atavistic spirit, read some 'soundmark,' that is 'a community sound which is unique or possesses qualities which make it specially regarded or noticed by the people in that community' (ibid., 10).[12]

Curiously, the concepts of high and low resolution can be understood as devices to read the same relationship of the subject with technology in the digital contemporaneity and in relation to the qualities of perception: thus Massimo Mantellini (2018) reflects on how we can identify a low-resolution generation whose perception is informed by lo-fi mediums and objects of a different nature – from television, to the reproduction of music, to school, and social, political, and pedagogical relations. What was experienced during the pandemic can be placed in stark contrast to this assessment. The shift from lo-fi to hi-fi recorded in the pandemic soundscape is mirrored in the same direction by the process of refining the audio devices with which – increasingly naturally – we are equipped and which have also reverted to being hi-fi. The transmission and reception of recorded ambient sound has taken place through devices that tend to be of very high quality: devices such as smartphones that may also play a crucial role in the development of soundscape studies due to the fact that they can act on increasingly user-centred applications (see Radicchi 2018).

This is also the direction in which some of the technological choices made by manufacturers of digital consumer software and hardware seem to be heading in recent years. Audio devices have become complementary to the already developed visual enjoyment of smartphones and, at the same time, users are increasingly oriented towards a greater consumption of products strictly reserved to the auditory sphere. Reference can be made to Apple's experimentation with so-called 3D sound, which involves spatialising sound through its earphones, which are responsive to the position of the head in space;[13] Facebook's Soundbites project, which envisages a new type of post on the social network based on short voice recordings;[14] Apple's Podcast+ subscription service, which focuses on the use of audio podcasts (a format which has seen strong growth in recent years due to the large number of users); and Microsoft's Soundscape project, which focuses on immersion and the relationship between people in the sound of the environment.[15] Across the typological differences between the products, a basic focus on the development of the aural immersive dimension suggests an expansion of the audio dimension of the same

70 *Ludovico Peroni*

smartphones that in the last decade were predominantly oriented around their visual interface.

If we agree with Mantellini's thought – which describes a world in which the choices of information consumption, in the digital age, are progressively oriented more and more towards the low resolution (2018, 125–130) – we can assess how the fruition of a sound in high resolution may have really contributed to marking by difference new sensations of surprise developed on the ecstatic level.[16] Think of trends such as the ASMR (Autonomous Sensory Meridian Response); spread through all major social networks, this trend consists in the use and/or sharing of audiovisual recordings in which the sound recording is placed in the foreground. The ASMR videographic products are characterised by the presence of anthropic sounds (whispered words, chewing sounds, ordinary domestic actions) filmed in detail, designed mainly to stimulate a feeling of relaxation. The research and production of this type of video (made to be heard rather than seen) has been increasing over the last decade, especially during the pandemic.[17]

There seems to be a search for a kind of mediatised sound hyperrealism that effectively sums up the discourses in this paragraph, which tangentially addresses a fundamental aspect of the soundscape by looking at it from the outside. Thus, the production and enjoyment of pandemic soundscaping has developed in an environment that has emphasised such tendencies, pursuing high-resolution coordinates:

- because of the level of detail of the technology we have at our disposal, capable of recording the new hi-fi landscape;
- for the immersive possibilities presented by the user devices;
- for the promise of providing an experience that sometimes complements products that during the pandemic – also for reasons such as teleworking or distance learning – have saturated the visual dimension.

Aesthetics, Atmospheres, and the New Aura

As far as the artistic component is concerned, pandemic soundscaping should be seen in the context of the process that in recent years has seen a progressive increase in the use of ambient sound recordings for the constitution of heterogeneous musical works (see Ortega 2020, 104). Underlying this may certainly be a general focus on sound as a pathic meeting place, as well as an interest in the characteristics of 'concrete sound' to be reworked (ibid., 106). We can therefore see how, in some cases during the lockdown period, soundscaping became intimately connected to certain issues addressed in different words and with different intentions by Lena Ortega (2020):

- the decrease in creative interventions on recorded sound by artists;
- the willingness to connect emotionally with the listener or community of listeners;
- the inclusion of an autographic imprint given by the author's own experience of an interaction with the environment, in a product that rediscovers itself to be a composition.

We are within the contemporary aesthetic debate: Griffero (2014b) proposes the concept of 'atmospherology,' which places the activities of the subject at the centre of the process of aesthetic theorising, in its pathic dimension of *how one feels* in the co-presence of I and situations. It also investigates the competence of those professions that 'objectify sensitive knowledge through a real "aesthetic work"' (Griffero 2014b, 169), and are 'specialised precisely in the generation of atmospheres' (ibid.).

Pandemic Soundscaping 71

Recognising atmospheres as emotionally tonalised spaces (Böhme 2010, 84) or as the attributes of *half-things* (Schmitz 2019) capable of being witnessed through the listener's own bodily presence, one must also think that, in the face of an emotion felt by the subject, there must at least be some concrete responsibility on the part of the medium that embodied and/or evoked them: the *phonographic recording*. The very practice of sharing different soundscapes online makes us imagine that, in addition to evoking an atmosphere, authors are aware that their soundscapes are able to effectively engage the pathic sphere of different listeners and thus generate an atmosphere (sometimes even imagining it).[18] We should conclude that the *atmospheric*[19] – a characteristic that is distinct from the percipient self and tends to assume characteristics similar to those of things (see Böhme 2010, 99–115) – of ambient sounds, objectified through the medium of phonographic recording, is endowed with a renewed strength: a new aura capable of generating definite and definable ecstatic experiences.

This new aura has been described in the studies of Vincenzo Caporaletti who, dialoguing at a distance with Walter Benjamin's (2000) concept of aura, identifies what he calls *neo-auratic encoding* (Caporaletti 2005; 2014; 2015; 2019). This concept – also used by the musicologist to delineate the cultural and cognitive switch between societies with oral traditions and those informed by the medium of sound recording and reproduction, defined as 'audiotactile' (Caporaletti 2015, 241) – is particularly effective for reading the phenomenon of soundscaping in relation to the pandemic. The sound of the recorded environment succeeds in maintaining its atmosphere thanks precisely to the aesthetic characteristics of the medium of the neo-auratic encoding, while at the same time succeeding in triggering a deep pathic contact with the subject. The atmosphere of pandemic soundscaping – for these same ontological and phenomenological reasons – comes into contact with the listener's own bodily system, resonating with certain neo-auratically encoded characteristics.

In light of these theoretical coordinates, soundscaping opens up to a further important problem: the question of authorship projected onto the products of filming, assumed by the uncontrollable ambient sounds. Agreeing with Caporaletti (2015, 240) in this context, we can say that the same concept of individualised authorship of Western written art music is projected onto pandemic soundscaping through the phonofixation of sound experience. Today, some of the same cognitive implications that stand between the subject and the object are inherent in the possibilities conferred by the medium of phonographic recording that, as we have seen in the work of some sound artists, often hovers between documentary purpose and artistic practice (see Peirui 2021), projecting – immediately and naturally – their action in the domain of creative audiotactile processing (see Caporaletti 2015, 241): this is a process that is difficult to clearly frame even by resorting to the categories proposed by Thomas Turino for recording music, namely *high fidelity* and *studio audio art* (see Turino 2008, 66–92).

Soundscaping in times of pandemic has thus introduced the issue of the role of the subject who is physically charged with recording, processing, and sharing a soundscape. Such consequences of the textualisation of sound no longer necessarily pass for an exclusively intentional fact, but are objectively considered in the same mediological action of phonographic recording, filtered by the medium of neo-auratic encoding (Caporaletti 2015, 240). The eventuality of the field recordings that took place during the pandemic directed attention to the same subsyntactic emergencies central to audiotactile music (ibid., 239), of which the *atmosphere* and the *atmospheric* are necessary audiotactile components.[20] In the case of pandemic soundscaping, the producer almost always coincides with the same artist who works the concrete material, projecting his own figure within the work, going so far as to make his point of view (and listening) coincide with the action of documenting reality.

72 *Ludovico Peroni*

Through the medium of neo-auratic encoding, a subject – in a phenomenology of auto-graphic authorship (Caporaletti 2015, 242) – makes their pandemic history enter the soundscape. For this reason, when one listens to a pandemic soundscape, one feels the authorial impact of the person who recorded it, which is reflected poietically in the reading of their work as soundscape art.[21]

What is implemented through the practice of soundscaping in times of pandemic is not only a rational knowledge of the sound situation of the world in a peculiar historical-social phase, but also includes precise affective qualities diffused in a natural space, as a result of a human agency in countertendency to the linear development that history has experienced in recent years. An atmosphere is diffused and experienced that 're-tonalises the whole situation' (Griffero 2014a, 136) that the pandemic manages to produce, audiotactily, certain effects on even the bodies of those who listen to and receive the soundscape.[22]

For an Investigative Perspective

In this contribution, I have tried to present some features that I consider important of the phenomenon that I call pandemic soundscaping. I have tried to give shape to my personal experience through a combination of auto-ethnographic strategies, reflections on the technological state of the art, and some theoretical ideas coming from strictly philosophical-humanistic disciplines, with the hope – starting from the contact and hybridisation between different fields of knowledge – of offering a reflection on a phenomenon that is certainly complex, and that I do not intend to trivialise. One of the things that strikes me most is that some subjects, immersed in the sound environment of the pandemic, shared a certain *forma mentis* generated by a certain social and cultural set-up, influenced in turn – as well as by their own feelings – by the availability of technology. Such pandemic listeners met in a disposition of mind, filtered by a general form of mediological action conferred by the massive and unexpected fruition of digital technologies of connection and high-resolution audio.

We are now faced with an experience so common to many, and yet so delicate, to frame theoretically. We can, however, reduce the specifics by saying that during the pandemic many resonated with an atmospheric witnessed through hi-fi soundscapes, which in turn were newly codified thanks to high-resolution technologies, and experienced through a contrasting contact between the subject's mood and the surrounding atmosphere through the pandemic soundscape.[23] Such a perspective – which here is confined to a very narrow field of knowledge – can not only pave the way for a profound aesthetic reflection on the meaning of artistic works enjoyed through different technologies, but also rips apart the very ontological nature of the lived and mediatised life experience during the early days of the COVID-19 pandemic.

Notes

1 I published the product on the afternoon of 19 April 2020, first via Facebook, then Instagram. Thanks to social networks, I was able to interact publicly and privately with various users – friends and others – stimulated by the topic. The same video can now be accessed via YouTube (www.yout ube.com/watch?v=ncmNDf_r174).

2 The exchange of ideas via e-mail with Lena Ortega and Riccardo Chinni was fundamental to the realisation of this chapter, and I would like to thank them profoundly for the stimuli, materials, and food for thought they provided.

3 This periodisation must of course take into account the different legislative applications of the various local provisions. In this contribution, as already mentioned, I will mainly refer to the perspective I experienced in Italy. The different local experiences (think of the different legislative applications, of the preferred information and communication media, of the peculiar

Pandemic Soundscaping   73

artistic, social, and cultural phenomena) – although inserted in a globalising perspective and in a historical context that has similar features – must not be absolutised and generalised, but always considered in their social and cultural specificity. This is how I recommend reading this contribution, considering general and local characters as always in a dialectical and mobile relationship.

4 'The COVID-19 pandemic has taken a toll on people's mental health. Yet, the global extent of this impact remains largely unknown. By leveraging the best available data from surveys around the world with measurements of anxiety and depression both before and during the pandemic, and analysing these data using the Global Burden of Disease Study (GBD) model, the COVID-19 Mental Disorders Collaborators provide global insight into the burden of depression and anxiety disorders during the pandemic to date' (Taquet, Holmes, and Harrison 2021).

5 'This site is here to help you most effectively use Zoom as we all navigate the coronavirus pandemic … Educating Over Zoom … Effective Remote Working … Hosting Virtual Events … Telehealth' (Zoom 2021).

6 'The pandemic created a fully digital global workforce overnight. Ongoing research shows the distinct challenges and advantages of this momentous shift' (Microsoft 2022).

7 'With happy hours, concerts and coffee catch-ups becoming virtual by default over these past few months, our team had a lot of questions about how we could help our members get to know each other through in-app videos' (Tinder 2020).

8 In this essay the terms *pathic, atmosphere, body, atmospheric, atmospherology*, all have a specific aesthetic value in the *new phenomenology*, the *affective turn* or, more specifically, the *atmospheric turn*, in the different meanings they assume in the specific texts of the various authors cited.

9 For the notions of the aesthesic, poietic and neutral levels, see Molino 1990.

10 It is impossible to compile in this space a complete and exhaustive list of online resources arguing the importance of soundscaping during the pandemic period. The World Forum for Acoustic Ecology portal has one of the most comprehensive lists of resources on this topic to date (www.wfae.net/covid-19-soundscapes.html).

11 With her work *Pandemic Soundscapes*, the sound artist Mara Maracinescu aimed to inform, and make us reflect on certain sound peculiarities of certain parts of Europe. In this case, the fine line between the documentary value and the artistic value of the object is evident. If, on the one hand, Maracinescu's work brings out recordings that clearly document some peculiarities of the pandemic in relation to anthropic activities – which, also through the video, are intended to take the viewer on this virtual-sound tour of pandemic Europe – on the other hand, they are products that are conscious of being compositions, aesthetically autonomous, which in any case record or reflect on a change in the surrounding soundscape. 'The main purpose of the research and resulting composition was to trace the impact of the lockdown on European soundscapes: new rituals or types of gathering that appeared; changes in the ambient soundscape, due to lower traffic, or the amplified presence of the State's voice, for example through loudspeakers' (House of European History 2020).

12 The focus on these changes is evident and can be found in the fervent debate and new thinking related to soundscapes, such as soundscape ecology (see Pijanowski et al. 2021).

13 See https://support.apple.com/it-it/HT211775.

14 See https://about.fb.com/news/2021/04/bringing-social-audio-experiences-to-facebook/.

15 See www.microsoft.com/en-us/research/product/soundscape/.

16 https://locateyoursound.com/ is a testament to the high quality achieved and maintained in a collaborative collection of soundscapes.

17 The search graph provided by Google Trends documents how it reached one of the most significant peaks in user searches between 2020 and 2021 (https://trends.google.it/trends/explore?date=all&geo=IT&q=asmr).

18 In this respect, it is interesting to assess cases where soundscapes are entirely created by an AI, as is the case in www.imaginarysoundscape.net/, or created and customised for each user, as through the https://endel.io/ service.

19 Böhme distinguish *atmosphere* and the *atmospheric* by their stage of objectiveness (see Böhme 2017, 8–9).

74 *Ludovico Peroni*

20 Caporaletti himself (2020, 59–71) points out that atmospherology can be seen as a field of study that often converges with postulates that attribute audiotactile components to the psychosomatic system of subjects, and to certain characteristics of objects.

21 As an example of this, I would like to mention the Colombian project *VOZTERRA Sounds from Your Window*, which collects artistic and documentary contributions from the perspective of a private dimension. It strikes me, therefore, that one of the first discriminating factors giving aesthetic poignancy to soundscaping is the fact that an author's life, at their listening point, is projected within the soundscaping. For this reason, and because of the underlying authorial component, such recordings are rightfully classified as musical works (www.vozterra.com/capitulos/desde-tu-ventana/ventanas-del-mundo.php?lang=en).

22 The taxonomic relevance of soundscape composition as a genre of audiotactile music will not be addressed here. However, I would like to note how Caporaletti addresses the merits of the taxonomic pertinence of music that uses ambient sounds to construct aesthetically autonomous works: in an essay dedicated to the interpretation of musical Futurism, the musicologist reflects on the pertinence of the model of audiotactile formativity to frame such compositional experiences that also start from the study of noise (see Caporaletti 2013). In the extension of this musicological paradigm to the genre of soundscape composition, we can find significant recurrences between some themes we have mentioned in these paragraphs and a new field of investigation.

23 Böhme speaks more appropriately of *ingression* and *discrepancy* (2010, 82–83).

References

Bartalucci, Chiara, Raffaella Bellomini, Sergio Luzzi, Paola Pulella, and Giulia Torelli. 2021. 'A Survey on the Soundscape Perception Before and During the COVID-19 Pandemic in Italy.' *Noise Mapping* 8: 65–88. https://doi.org/10.1515/noise-2021-0005.

Benjamin, Walter. 2000. *L'opera d'arte nell'epoca della sua riproducibilità tecnica*. Translated by Enrico Filippini. Torino: Einaudi.

Böhme, Gernot. 2010. *Atmosfere, estasi, messe in scena. L'estetica come teoria generale della percezione*. Translated by Tonino Griffero. Milano: Christian Marinotti.

Böhme, Gernot. 2017. *Atmospheric Architectures: The Aesthetics of Felt Spaces*. Translated by A.-Chr. Engels-Schwarzpaul. London: Bloomsbury.

Caporaletti, Vincenzo. 2005. *I processi improvvisativi nella musica. Un approccio globale*. Lucca: LIM.

Caporaletti, Vincenzo. 2013. 'Dal Tattilismo all'Audiotattile: per un'interpretazione del Futurismo musicale.' In *I linguaggi del Futurismo*, edited by Diego Poli and Laura Melosi, 261–282. Macerata: EUM.

Caporaletti, Vincenzo. 2014. *Swing and Groove. Sui fondamenti estetici delle musiche audiotattili*. Lucca: LIM.

Caporaletti, Vincenzo. 2015. 'Neo-Auratic Encoding: Phenomenological Framework and Operational Patterns.' In *Musical Listening in the Age of Technological Reproduction*, edited by Gianmario Borio, 233–252. London: Routledge.

Caporaletti, Vincenzo. 2019. *Introduzione alla teoria delle musiche audiotattili. Un paradigma per il mondo contemporaneo*. Canterano: Aracne.

Caporaletti, Vincenzo. 2020. 'On Some Convergences between Atmospherology and the Theory of Audiotactile Music.' In *Resounding Spaces. Approaching Musical Atmospheres*, edited by Federica Scassillo, 59–71. San Giuliano Milanese: Mimesis International.

Cisco Webex. 2022. 'Supporting You during COVID-19.' COVID-19. www.webex.com/covid19.html.

De Vynck, Gerrit and Mark Bergen. 2020. 'Google Classroom Users Doubled as Quarantines Spread.' *Bloomberg*, 9 April 2020. www.bloomberg.com/news/articles/2020-04-09/google-widens-lead-in-education-market-as-students-rush-online.

Griffero, Tonino. 2014a. *Atmospheres: Aesthetic of Emotional Spaces*. Translated by Sarah De Sanctis. Farnham: Ashgate.

Griffero, Tonino. 2014b. 'Estetica patica. Appunti per un'atmosferologia neofenomenologica.' *Studi di estetica* 42 (1–2): 161–183.

Hofman, Ana. 2015. 'The Affective Turn in Ethnomusicology.' *Muzikologija* 18 (January): 35–55. https://doi.org/10.2298/MUZ1518035H.

House of European History. 2020. ' "Pandemic Soundscapes" – Sounds Across Europe During the Coronavirus Pandemic.' 28 December 2020. www.youtube.com/watch?v=0IY82EohslE.

Jarratt, David. 2021. 'An Exploration of Webcam-Travel: Connecting to Place and Nature through Webcams during the COVID-19 Lockdown of 2020.' *Tourism and Hospitality Research* 21 (2): 156–168. https://doi.org/10.1177/1467358420963370.

Jordan, Pamela and André Fiebig. 2021. 'COVID-19 Impacts on Historic Soundscape Perception and Site Usage.' *Acoustics* 3 (3): 594–610. https://doi.org/10.3390/acoustics3030038.

Mantellini, Massimo. 2018. *Bassa risoluzione*. Torino: Einaudi.

Microsoft. 2022. 'What We've Lost during the Pandemic … And What We've Gained.' Going Remote. www.microsoft.com/en-us/worklab/pandemic-lost-and-gained.

Molino, Jean. 1990. 'Musical Fact and the Semiology of Music.' *Music Analysis* 9 (2): 105–156. https://doi.org/10.2307/854225.

Ortega, Lena. 2020. 'Nature Sounds in Sound Art and Electroacoustic Music.' In *Resounding Spaces. Approaching Musical Atmospheres*, edited by Federica Scassillo, 105–121. San Giuliano Milanese: Mimesis International.

Peirui, Yang. 2021. 'Sound Art and Pandemic: A Documentary Soundscape.' Paper presented at the Music and Art in Pandemic International Conference, University of Arts, Târgu Mureş, March 2021.

Pijanowski, Bryan C., Almo Farina, Stuart H. Gage, Sarah L. Dumyahn, and Bernie L. Krause. 2011. 'What Is Soundscape Ecology? An Introduction and Overview of an Emerging New Science.' *Landscape Ecology* 26: 1213–1232. https://doi.org/10.1007/s10980-011-9600-8.

Radicchi, Antonella. 2018. 'The Use of Mobile Applications in Soundscape Research: Open Questions in Standardisation.' Paper presented at Euronoise 2018: 11th European Congress and Exhibition on Noise Control Engineering, Heraklion, May 2018. www.euronoise2018.eu/docs/papers/409_Euronoise2018.pdf.

Schafer, R. Murray. 1994. *The Soundscape: Our Sonic Environment and the Tuning of the World*. Rochester: Destiny Books.

Schmitz, Hermann. 2019. *New Phenomenology. A Brief Introduction*. Translated by Rudolf Owen Müllan, with support from Martin Bastert. San Giuliano Milanese: Mimesis International.

Taquet, Maxime, Emily A. Holmes, and Paul J. Harrison. 2021. 'Comment: Depression and Anxiety Disorders during the COVID-19 Pandemic: Knowns and Unknowns.' *The Lancet* 398 (10312): 1995–1666. https://doi.org/10.1016/S0140-6736(21)02221-2.

Telegram. 2021. 'Group Video Calls.' *Telegram News* (blog). 25 June 2021. https://telegram.org/blog/group-video-calls?ln=r.

Tinder. 2020. 'Tinder Begins Testing Face to Face Video.' *Tinder Newsroom*. 8 July 2020. https://uk.tinderpressroom.com/2020-07-08-Tinder-Begins-Testing-Face-to-Face-Video.

Turino, Thomas. 2008. *Music as Social Life: The Politics of Participation*. Chicago: University of Chicago Press.

WhatsApp. 2020. 'Group Video and Voice Calls Now Support 8 Participants.' *WhatsApp* (blog). 28 April 2020. https://blog.whatsapp.com/group-video-and-voice-calls-now-support-8-participants.

Zoom. 2021. 'COVID-19 Support.' Zoom. https://explore.zoom.us/en/covid19.

6   Not People but a Sound

Virtual Audio and the Appropriation of Fandom Practices in Pandemic Football

Giulia Sarno

> Cheers of elation and encouragement, as well as booing and whistling and fan chants are all inserted by means of a mixing desk. Alternatively, they are simply transmitted to the stadium from the public squares hosting [public viewing] events. No imaginative choreography, no original fan songs, no banners are present in this arrangement; they are shown on the video walls every now and then as relics.
>
> (Schulke 2010, 71)

In 2010, a book edited by Sybille Frank and Silke Steets took stock of the recent transformations of the football stadium

> not only as a built, but also as a social space, connected to specific social norms and practices, where not only characteristics of national and social culture but also global economic developments, as well as media and design trends congregate and are expressed.
>
> (2010, 1)

In one of the chapters, Hans-Jürgen Schulke painted a 'Scenario of the Future, 2020' (2010, 70) in which the stadium as we know it would essentially disappear: taking some trends evident in European football over the last 30 years to the extreme, Schulke imagined a future generation of sports facilities characterised, above all, by pervasive technological equipment (screens, sensors, headsets, buttons changing the seats colour) and a drastic reduction of audiences, with cheering and chanting replaced by mediatised forms. His predictions were certainly dystopian enough, highlighting certain tendencies of so-called modern football with a rhetorical device more than outlining a concrete reality. Nonetheless, Schulke's scenario has partly materialised: as a matter of fact, the quote I offered seems to depict the peculiar forms that football fandom took in 2020.

This is more than an unsettling coincidence. Indeed, the restrictions associated with the COVID-19 pandemic prompted European football leagues and broadcasters to experiment with new ways of broadcasting matches, based – as we shall see – on novel sound design solutions, which are in line with broader processes labelled as modern football. Observing these experiments allows, first, to highlight the importance of sound as an atmospheric vehicle and veridiction tool in the rewriting of sports on television. It also adds to the understanding of modern football by considering the clash of different ideas of support and participation within its framework: these emerge clearly from the strategies adopted by the major European leagues in order to resume the football championships 'safely,' as well as from the reactions that these have provoked. Finally, this study suggests that mediatisation, triggered by the pandemic, can act in problematic and ethically dubious ways on participatory performing practices (Turino 2008), such as those of organised fandom.[1]

DOI: 10.4324/9781003200369-8

I will consider the peculiarities of televised football in the time of the pandemic, that is, its rewriting in the media regime of *audio-vision*, focusing on the audio pole of the 'audio-visual chain' (Chion 1994, 47). In particular, I will show how Virtual Audio[2] solutions based on top-down, imposed mediatisation imply a divide between an idea of fandom as social agency and a vision that sees it as a mere ingredient (albeit a major one) of a media product. These reflections are conducted against the backdrop of an ethnography of the sonic and musical practices of organised football fandom conducted in Florence, Italy (see Sarno 2022). In my work, these practices are conceptualised as an expressive platform for the identity performance of specific *ultras* groups, which – although involved in powerful transcultural dynamics – generate distinct local traditions. It is from this perspective that I observed the phenomena that emerged when football seasons resumed in Europe and I was prompted to adapt my ethnography to changed conditions (for example, watching television versus going to the stadium, analysing social media and press content versus participant observation), and extend the dialogue to new actors.[3]

Behind Closed Doors: Restarting Safely and the Atmosphere Problem

> As they say, gentlemen, 'The show must go on.' If you resume the season, you'll be doing it alone. And along with your profits you will have our contempt, for this crazy choice.
>
> (Curva Fiesole press release, Florence, 14 May 2020)[4]

Since the 1990s, European football has undergone a number of dramatic transformations which have radically increased its financial value. While in 1985 the *Sunday Times* could still describe English football as 'a slum sport played in slum stadiums increasingly watched by slum people' (quoted in Webber 2017, 9), the game is now an impressive industry that contributes considerably to the GDP of several countries on the continent: the English Premier League (EPL) plays a leading role, as 'one of Britain's wealthiest industries and the only one with a truly global reach' (Webber 2017, 9). Television coverage has played a crucial function in the process of football's transformation: the establishment of live broadcasting in the 1990s had far-reaching effects on the whole system.[5] Television rights to the top divisions of the major European leagues (the EPL, the German Bundesliga, the Spanish LaLiga, the French Ligue 1 and the Italian Lega Serie A) have now reached exorbitant levels. The consolidation of football as a global business thus relies substantially on the sale of television rights, which today 'constitutes in many countries a primary and growing component of football revenues' (Preta 2016, 103).

Given the industrial dimension football has acquired, it is not surprising that the suspension of national championships in the face of the health crisis caused great concern, and that the European leagues (except Ligue 1) took immediate action to resume them safely. The commonly adopted solution, which is playing matches 'behind closed doors' (without stadium audiences) and broadcasting them live on television, is only conceivable against the backdrop of the 'intermeshing of sport and media' (Rowe 2004, 13) that has become established over the last 30 years: for most European fans, football *coincides with* its televised version (Crawford 2004, 7–9). Perhaps it would have been unthinkable without the most recent developments in digital technologies, allowing for the manipulation of audio and video signals with great agility.

Between 10 and 13 March 2020, football seasons stopped for two to three months, until mid-May (Bundesliga) or mid-June 2020 (LaLiga, EPL, and Serie A). The decision to resume them raised discussions echoing the resumption of all professional and commercial activities suspended during the first lockdowns, with governments looking for the balance

78 *Giulia Sarno*

between containing the contagion and reactivating local and global economies. More specifically, as shown by the quoted press release of the Fiorentina supporters, this choice aroused bitter criticism from some groups of fans. This criticism reflects a friction between football as a platform for sociability, embedded in the physical reality of the stadium, and football as show business, offered as mediatised entertainment to television viewers, that is at the heart of the game's contemporary development of the Against Modern Football movement (see Webber 2017). For the *ultras*, and also for fans who do not identify as such but who attend stadiums regularly and passionately, the pandemic materialised a dystopian scenario in which football completely lost its participatory aspects and ended up a definitive media spectacle, to which they contribute (despite themselves) as purveyors of *atmosphere*. After all, this is what emerges emblematically in a statement from EPL Chief Executive Richard Scudamore: 'We can't be clearer. Unless the show is a good show, with the best talent and played in decent stadia with full crowds, then it isn't a show you can sell' (quoted in Lawn 2020, 187–188). This vision is embedded in the choices that leagues and broadcasters made regarding televised football in times of COVID-19: the pandemic confirmed that the presence of full crowds in stadiums matter not so much for ticket sales revenue, but for the atmosphere supporters provide to the television show (Zinganel 2010, 78). Faced with the absence of fans, there was an urgent need to find solutions to overcome the emptiness undermining the broadcast's poignancy.

The term atmosphere recurs in discourse associated with televised football, not only in times of pandemic, and with particular reference to the aural experience;[6] for these reasons, I prefer it, as a hermeneutical tool, to other options such as soundscape, which appears to circulate far less.[7] Atmosphere, as it is used in this study, is an intensely debated concept. Recent studies have highlighted its spatial qualities, how it identifies a 'sensorial and affective quality widespread in space,'[8] or better, 'in any spatial (non-geometric) dimension' (Griffero 2020, 15). In a televised game, two spaces are simultaneously involved, with each permeated by an atmosphere and its real-time mediatised form, namely the stadium and any environment in which the television viewer happens to be watching the game – as 'media use is always "located" somewhere' (Crawford 2004, 137). This *atmospheric transfer* from the stadium to the metonymic living room seems to be accomplished mainly by the vehicle of sound, which with its incorporeal quality 'holds some kind of affective power to penetrate situations, collectives and selves, and manifests *as* environmental atmosphere among them' (Riedel 2020, 3). And it is especially *the sound of fandom*, mediatised and reconstructed in its immersive potential,[9] that allows stadium atmosphere to penetrate the spaces of television consumption. This emerges clearly in pandemic football broadcasting: although new visual solutions were also experimented with,[10] the game of virtual atmosphere was played in the acoustic sphere.

Rewriting Football on Television in Times of Pandemic

As literature on the 'marriage of interests' (Iozzia and Minerva 1992) between football and television has long pointed out, televised football has an ambiguous and complex relationship with its counterpart in physical reality: what we see on the screen is the result of

> a work of mediation that can be placed halfway between referentiality and simulacralisation. . . . The evidence of the manifest image has a testimonial function, but the more sophisticated technical apparatuses are used, the more distant is the reality to which it refers. Truth, therefore, gives way to strategies of veridiction.
>
> (Abbiezzi 2007, 54–55)

Virtual Audio and Fandom Practices in Pandemic Football 79

On television, the football event is not so much *hosted*, but rather *rewritten* 'through the potential of [the] linguistic autonomy' (ibid., 99) of the medium, according to its technical and ethical rules and in line with the objectives the broadcast proposes: if an 'informative pretext ... underlies every media representation of sport,' it is the 'entertainment function' (ibid., 43) that seems to have increased and refined over time. In this sense, 'the emotional dimension ... takes on a fundamental weight with the attempt to transmit the intensity of participation and the emotion of cheering' (ibid., 119).

This transmission implies a rewriting in which the supporters' autonomous expressiveness can and must be regulated in accordance with the ethical regime of television: for example, '[in the UK] "offensive" language or chants emanating from football crowds tends to be silenced on television by production teams' (Giulianotti 2011, 3301). The space of personal expression and, especially for some types of fans, collective agency through participatory sonic and musical practices in the physical reality of the stadium becomes, through mediatisation by televisual rewriting, an audiovisual text owned by the broadcasters. And broadcasters, '[delimiting] the universe of viewing that "the audience" inhabits' (Born 2000, 416), organise and amend it not on the basis of a presumed testimonial fidelity, but of linguistic effectiveness, commercial attractiveness, and ethical-political correctness. Thus, performative practices and the resulting media texts relate to different and in some ways irreconcilable agencies, one incontrovertibly dominant over the other. Games behind closed doors have made the rewriting operation opaque, revealing what is normally quite transparent: the direction, which is the exercise of power underpinning the discursive assemblage that textualises live events.

Before the pandemic, audio rewriting of games involved only sound mixing. With the matches behind closed doors, new issues and strategies have arisen that complicate how televised football is situated within the audiovisual regime, 'a specific perceptual mode of reception' based on a 'contract' between the two senses of sight and hearing, whereby 'one perception influences the other and transforms it' (Chion 1994, xxv–xxvi).

These new strategies can be gathered into two groups: those in the first group lean more towards the testimonial-informative pole of the television broadcast, while those in the second group stress emotional involvement and seem more directly intended to maximise its entertainment potential, resulting – as I argue – in considerable problems. This distinction reflects the two 'anthropological points of focus' that Schulke identifies as constitutive of the spectator experience of sport:

> On the one hand there is the human quest to experience the unknown, here and now – humans are characterised by a life-long, life-saving and unquenchable curiosity. They wish to experience what is happening with their own eyes and ears. ... On the other hand there is the endless yearning for an all-embracing commonality, for concordance with as many other people as possible, which crowd psychologists refer to as infectious emotion.
>
> (2010, 56)

Each of these groups manages differently the elements that make up the game's onscreen audio, of which Nicolai Graakjær distinguishes two types: first, *football television sounds* that 'appear to originate from the television channel' (e.g. the commentators' voices), and second, *televised football sounds* that 'seem to originate from the execution and organisation of the soccer match at the venue' (2020, 148). The latter can be further divided into three categories: 'performer sounds (e.g. the referee's whistle and players shouting and striking the ball), organiser sounds (e.g. the stadium announcer), and spectator sounds' (ibid.).[11]

The testimonial-informative strategies in the first group work exclusively with the signals collected from the physical event, therefore on the leftovers of what normally makes up televised football sounds. In the new acoustic conditions of the stadium, directors optimised the use of microphones to play with sound balance and compose the soundscape of the broadcast. This is what Dazn did with the Italian Serie A matches it was assigned to broadcast: the game's soundscape, which was much more hi-fi than usual due to the absence of the great keynote of fandom,[12] was mainly occupied by performer sounds coming from the pitch and, in a peculiar way, from the benches. Dazn creatively compensated for the drastic decrease in sound density by giving unprecedented prominence and legibility to sounds that are normally masked in the overall media acoustic environment. Coaches, staff, and players on the bench, as well as active players and referees, took on a new sonic protagonism; the stadium's soundscape, emptied of its features as a vehicle of collective emotion (the mass of supporters singing and shouting), manifested itself to the television viewer as a space of relationship and communication between identifiable subjects, in particular between players and coaches. Dazn seems to have exploited the informative potential inherent in the audio signals coming from the stadium, tickling the viewers' unquenchable curiosity about aspects of the game's dynamics that are typically not very evident.

Strategies of emotional engagement in the second group were aimed at filling the great sonic void caused by the lack of a stadium audience by *adding something*, by which I mean inserting some virtual elements into the audiovisual text. Specifically, spectator sounds were reconstructed in order to keep the viewers' emotional involvement intact and provide them with a satisfying and entertaining product. The four leagues adopted somewhat different approaches: on one hand, the EPL and LaLiga, with their respective broadcasters and producers Sky and Mediapro,[13] relied on the expertise of EA Sports, a division of the US video game company Electronic Arts, which developed an ad hoc system from the sound library of their *FIFA* series – an immense collection of high quality recordings made in stadiums around the world. On the other hand, Sky Deutschland and Sky Italia managed the process in-house, resorting to a calculated reuse of sections from audio tracks of older matches in their archives. But beyond the differences in approaches[14] and outcomes,[15] the sense of these operations is similar – to recreate spectator sounds through systems that combine three elements: recordings of fans' chants captured before the stadiums closed; a background buzz that conveys the sense of the presence of a large crowd; and a series of reactions, namely pre-constituted effects that reflect the typical sound actions of fans (cheers, applause, boos, and so on) in relation to the events on the pitch.

These three elements can be roughly associated with the three functions that Graakjær identifies for spectator sounds: chants mainly 'embody an indication of atmosphere' (2020, 154); the crowd buzz mainly contributes to 'establish an impression of continuity and liveness' (ibid., 155); the reactions mainly 'contribute to informing viewers how the match is going' (ibid.).

In physical reality, these three elements (chants, buzz, reactions) are inseparable and organically connected to the unfolding of the game; in the normal media reality, this connection remains even if the three elements are channelled into audio signals and rewritten by means of mixing. In the virtual-media reality they are completely separated, both from each other and from the event: in order to effectively carry out the functions identified by Graakjær, they must therefore be *composed* with sound design and re-associated (Chion 1994) with the images. This requires, at the very least, an understanding of the basic dynamics of fandom,[16] but also a knowledge of and respect for the specificities of local traditions. For example, in Italy, *ultras* offer a solid musical performance, created according to a precise set of criteria (Sarno 2022), with chants hardly ever repeating over the course of 90 minutes; on the contrary, many Serie A matches that I *listened to* on Sky

Italia focused on a few repeated chants, revealing the sound track as based on a loop:[17] this undermined the sense of liveness and continuity of the broadcast. Again, a precise correspondence between action on the pitch and audio reaction is essential for spectator sounds to function as indicators of the game's progress: in the virtual system this can be problematic both because it depends on the shrewdness of the person pushing the reaction buttons, and because of possible latency. As a result, virtual audio can have alienating effects on the viewer. Furthermore, it implies a series of *choices* (e.g., which chants to include in matches and which not) at different times and at different levels of the production chain, which constitute a new scenario in the audiovisual regime of live broadcasting: sound technicians become responsible for interpreting the match and rendering the supporters' reactions according to a certain rational logic, so that they are congruent.[18] Through this, they acquire unprecedented agency, becoming able to influence the meaning of the audiovisual text.

The easiest function of spectator sounds to fulfil with virtual audio is the first of the three identified by Graakjær (2020): conveying the atmosphere of the football event. This can be considered their crucial function, given the relevance of atmosphere in the discourse around football. From this perspective, we can frame the experiments with sound design devised during the pandemic as 'atmospheric practices' aimed at virtually reconstructing those 'atmospheric relations' (Riedel 2020, 5)[19] that would normally occur between the stadium and the living room to give meaning to television consumption, the absence of which affectively deprives viewing. Matches behind closed doors and virtual audio strategies show how, by transmitting the atmosphere of the stadium, spectator sounds function as a veridiction platform of the mediatised sporting event: this atmospheric transfer acts as an important affective medium for the television audience, carrying the sense of involvement that characterises the stadium's 'all-embracing commonality' (Schulke 2010, 56). Connecting to the feeling of those participating in the football event, whether they are actually present in the stadium or virtually evoked by their mediatised sound, television viewers will to a certain extent, or partially, or intermittently share their affective values, validating their perception that what is shown on television 'represents something important and attractive' (Graakjær 2020, 154). Spectator sounds contribute not only to making televised football engaging, but also to making it *real* – without the audience and its aural marks, it is difficult to escape the feeling of just watching 22 guys kicking a ball around.[20]

Restrictions imposed by the pandemic have dramatically altered the stadium's acoustic environment, which is crucial to televised football. These conditions have shifted the balance between referentiality and simulacralisation that is typical of televised sports towards the simulacral pole: the lack of spectator sounds has prompted broadcasters to act on televised football more as a fictional audiovisual product to be constructed than as an event to be documented. Thus, in order to salvage the show, the agency of fans is replaced by that of the sound technicians, who reuse its simulacrum (the media texts derived from the living practices of supporters) to virtually reconstruct the effects and affects of the atmospheric transfer.

Not People but a Sound: Ethical Issues of Virtual Audio

The virtual audio strategies I have described shift spectator sounds into a middle ground between televised football sounds and football television sounds. In fact, the voices and noises we hear now in virtual audio are recordings being transmitted live by the channel's sound technician as football television sounds; but at some point in the past they were emitted by fans, and broadcast as televised football sounds: they were recorded, and are now being reproduced in a new context without any explicit consent on the fans' part.

82 *Giulia Sarno*

All of the actors involved in these experiments insisted on the content-specificity of the solutions they developed: using material from the different stadiums and the sounds of real supporters of the different teams, instead of generic audio content, was deemed necessary (and sufficient) to guarantee a certain degree of authenticity to virtual crowds. This raises a crucial question: to whom do those voices *belong*? To the fan community, or to the broadcaster or company that owns the recordings and can, apparently, reuse them to make the television product more attractive, ultimately for its own profit? So far, I have considered the sounds of pandemic football in terms of their effectiveness or ineffectiveness, from a neutral point of view. From here, I underline the ethically dubious aspects of these operations,[21] as they represent the virtual translation of very real expressive practices: these are engulfed without appeal in a remediation process that betrays their participatory aspect, turning them into commodities and distorting their representation.[22] This is particularly problematic considering that in many contexts, including Italy, specific groups of supporters (the *ultras*) are almost exclusively responsible for the musical performance of the stadium. For them, football chanting represents the ground for a performance of collective identity and, potentially, a platform for social and political contestation. If the sound of the fans' agency (the chants) can be freely taken away and remobilised with no ethical qualms, the stadium's 'closed doors' become the symbol of an exclusion from the decision-making processes that govern football, and virtual crowds reveal themselves as the outcome of an appropriation – of social agency, but also more concretely, of the expressiveness and creativity of more or less circumscribed groups of people.

The pandemic did not invent modern football, but it acted as an extraordinary spotlight on some of its perhaps less visible features. In fact, the appropriation and commodification of football chants that was already taking place in the context of video games and countless audiovisual products is now being amplified by its transfer to live television broadcasting, with the ghostly re-association of the sounds of fandom with the images of the stands emptied of fans' bodies. The exploitation of what can be considered the supporters' cultural traditions – a participatory performance practice that does not crystallise into 'original' and 'authorial' products recognised by intellectual property legislation, nor represent a designated intangible heritage that warrants protection – recalls issues associated with the circulation of local musics in the context of World Music, brought into focus during the 1990s; to paraphrase Steven Feld (2000, 165), from the standpoint of the sampler, football fans are *not people but a sound*.[23] This is echoed in the statements of the general secretary of FASFE (Federación de Accionistas y Socios del Fútbol Español) Emilio Abejon:

> The use of virtual images and sounds in the television broadcasts of what they are trying to sell as the return of football feels to us like a lack of respect to the fans they want to replace. We've spent years protesting against the fact that La Liga wants to make us the props for their broadcasts but with this move they have gone one step further. Now all they need to do is replace the players and they'll have reinvented video games.
>
> (quoted in Martin 2020)

FASFE's reaction can be likened to 'the anger of artists and communities around the world about the use, and misuse, of their cultural traditions in commodified products' (Seeger 2004, 158), mobilising similar 'issues relating to access, ownership, intellectual property, control of knowledge, control of commodification, fair compensation, and damaging appropriation' (Diamond et al. 2017, 19) that arise in the context of Indigenous traditional knowledge. Moreover, the fact that fan practices are the outcome of *contrafactum* procedures that are often applied to so-called original and authorial products (typically popular songs) adds a

Virtual Audio and Fandom Practices in Pandemic Football   83

layer of complexity that cannot be untangled here, but which results in the paradox that, in order to reuse chant recordings, broadcasters have had to pay copyright to the original song owners, while being able to ignore supporters altogether.[24] The creative action of the fans, although considered essential for the stadium atmosphere, gets no recognition; they are the only ones excluded from the distribution of the revenue generated by the football industry.

Concluding Remarks

This case study shows how, in the context of the pandemic, some sound practices, which can engage the digital media landscape in a variety of forms (social networks are full of amateur channels dedicated to football chanting), were *subjected to* forms of mediatisation that have crystallised and distorted them, over which performers exercised no control, and which, moreover, reflect an economic and cultural system that they even explicitly oppose. On the other hand, football behind closed doors marked the development of other solutions that instead insisted on the sense of *sonic participation* that fandom has: apps were created that would allow supporters to send their sound reactions in real time directly to television stations, which could then integrate them into their broadcasting stream. While I cannot dwell on the description of these products, it is important to note that they do not seem to have been successful, at least in Europe:[25] this suggests that the physical dimension of collective participation is indispensable for these kinds of musical practices to unfold, and irreducible to media surrogates.[26] In this sense, then, these experiments intersect with broader issues concerning participation in times of pandemic, given the extraordinary growth of the remote and medial dimension over the physical and in-presence one that characterised lockdown and beyond, at least for those social groups with access to technology and the web. This warns us about the risk of detecting only those sounds of the pandemic that have successfully adjusted to the medial dimension and forgetting what has lagged behind.

Notes

1 On Italian organised fandom, a model widespread in many countries, see Marchi 2015 and Doidge, Kossakowski, and Mintert 2020.
2 Virtual Audio is the name of the new feature developed by Sky for matches behind closed doors.
3 Beside several members of *ultras* groups (especially Andrea Chelazzi, Curva Fiesole's *lanciacori*), I dialogued with Paul Boechler, sound designer at EA Sports (Skype interview, 27 August 2020), and Tiziano Mantovani of Sky Italia (Skype interview, 21 July 2020). I thank all of them for providing precious information and for deepening my understanding of this subject.
4 The press release can be found on Curva Fiesole's Facebook page (www.facebook.com/fuori dalcoro1926). Curva Fiesole is both a section of Florence's stadium and the collective name encompassing the organised supporters of ACF Fiorentina.
5 In England, Sky won the exclusive right to live broadcast the EPL in 1992; in Italy, it was the newly created Telepiù which offered the 1993–1994 season by subscription. On televised football, see the recent literature cited in Graakjær 2020.
6 Recent examples include EA Sports naming its virtual audio system for matches behind closed doors 'Atmospheric Audio,' while Sky Deutschland's video presentation states: 'We create football atmosphere ... until the fans are back' (McCue et al. 2020).
7 On the other hand, I will occasionally use 'acoustic/sonic environment' and 'soundscape' referring respectively to 'sound at the receiver from all sound sources as modified by the environment' and 'acoustic environment as perceived or experienced and/or understood by a person or people, in context' (ISO 2014).
8 This definition is found in the description of Mimesis International editorial series *Atmospheric Spaces*, directed by Tonino Griffero, one of the scholars who made major contributions to the concept's systematisation.

84   *Giulia Sarno*

9   Increasingly explored thanks to recent technologies such as Dolby Atmos (atmospheres again …), a surround sound technology that is becoming the new standard for broadcasting sports events.

10 From simple changes in camera angles that show less of the empty stands, to more radical approaches, such as LaLiga's 3D computer graphics stands, which provoked mixed reactions from commentators and fans (see www.vizrt.com/vizrtv/on-demand/la-ligas-virtual-fan-experience).

11 I adopt this terminology for the sake of simplicity, although I believe it to be problematic: conceptualising the *ultras* as spectators is inaccurate, since they play a *performative* role that I argue is largely independent of what happens on the pitch (Sarno 2022).

12 The terms 'hi-fi' and 'keynote' are derived from Raymond M. Schafer's (1977) well known proposals.

13 It is not easy to retrace the network of accountability behind these initiatives, which may involve football leagues, audiovisual production companies, and television broadcasters to varying degrees on a case-by-case basis.

14 These were clarified in November 2020 during a panel of the Next Generation Audio online summit, which featured key figures from the German and British divisions of Sky (see McCue et al. 2020).

15 These can be very problematic, as I have observed and as is evident from social media commentary and the press (e.g., Dipollina 2020). However, it is possible to find sources advocating for the effectiveness of virtual audio (e.g., Boyd 2020). Virtual systems depending on novel technicians' performances may contribute to a certain fluctuation in outcomes and reactions.

16 For Martin Bergmann (Senior Manager of EVS at Sky Deutschland and responsible for virtual audio for the broadcaster) it is essential that sound technicians in charge of the system are proper football enthusiasts (McCue et al. 2020).

17 Sky Deutschland also seems to have had this kind of problem, caused by the presence of chants in the loop used as a 'sound carpet' (McCue et al. 2020). In this sense, EA Sports' system appears to be based on design features that minimise the possibility of a loop sensation occurring, by incorporating a rather high number of chants that are randomly grafted into the audio fabric. While this may perhaps be sufficient to faithfully represent the spontaneous chanting of English supporters, which is interspersed with long interruptions, it does not seem adequate to account for the performance of the Italian *ultras*, which unfolds consistently over the course of 90 minutes.

18 Interesting in this regard is the very brief testimony of a sound technician from Mediapro, the audiovisual production company responsible for LaLiga matches: 'Work is stressful during the game because you have to pay attention to what's happening, you have to follow the game. If there is an offensive play the public reacts. You have to interpret an action and it's difficult, fast and tense' (Daniel Ruibal in the television documentary *LaLiga: Behind the Cameras*, broadcast on 17 July 2020 on Movistar LaLiga).

19 Riedel defines atmospheric relations as 'the kind of modalities, structures, relations and mediations that are vital to atmospherically charged situations,' and atmospheric practices as 'the musical and auditory operations of constructing, manipulating or curbing atmospheric relations' (Riedel 2020, 5).

20 Of course, this is also due to habit: television audiences have become accustomed to a particular sound dimension (including a certain overall microphone and mixing set-up), so its absence has a disruptive effect on the perception of the broadcast event. However, my own experience during the closed-door period showed me how quickly habits can change: after the first weeks of bewilderment, the silence of the television broadcast no longer seemed so strange and the matches started to *feel* important again, even without virtual audio.

21 Other commentators have underlined yet another ethically dubious aspect of virtual audio in pandemic football, stating that, although the work of sound engineers is 'commendable' and the result 'impressive,' fake crowd noise is 'disingenuous. There's no need to paper over what's happening. Clubs are playing in empty stadiums because of a pandemic that has claimed hundreds of thousands of lives and impacted millions of others. … So why are we pretending like it's business as usual on the broadcast when these times are anything but that?' (Creditor 2020).

22 This is the case of a match that I listened to in the early stages of the restart: the looped sound carpet consisted mainly of a chant from the home fans (which I will not mention) where the insult

Virtual Audio and Fandom Practices in Pandemic Football   85

terrone (commonly addressed to fans from southern Italy) was repeated. While during a regular match that chant would have been sung once or twice, for a maximum of about five minutes, and would therefore have had a relative value in the overall economy of the event, hearing the insult repeat incessantly for 90 minutes provides a distorted and extreme representation of the supporters in question, as well as being a source of discomfort for television viewers. On the other hand, the strategy commonly employed by EA Sports, and thus in the virtual audio of EPL and LaLiga, produces equally distorting effects, since each recorded chant is evaluated and possibly censored if it includes profanity or content deemed offensive – which, at least in Italian practices, considerably reduces the number of passable chants and radically alters the supporters' expressiveness.

23 This is the original quote: 'From the initial standpoint of the sampler, Afunakwa is not a person but a sound' (Feld 2000, 165).

24 See the testimony of Kevin McCue, Director of Technical Operations at Sky Sports, in McCue et al. 2020. This is not the case with normal live matches, where broadcasters have no control over what supporters will sing – they just transmit what happens in the reality of the stadium; whereas virtual audio implies a conscious construction of an audiovisual product with its own soundtrack.

25 Several apps and systems allowing spectators at home to send audio (sound effects or real-time recordings) to broadcasters do not seem to have been taken forward: e.g., the announced Italian app *Il Dodicesimo*, which was supposed to 'collect, mix, and transmit audio from the smartphones of fans watching the live event' (ANSA 2020). Although unable to verify the results, I will also mention *Remote Cheerer*, an app designed by the Japanese company SoundUD in collaboration with Yamaha, which reportedly allows fans to send sound effects to the PA systems of sports facilities, thus adding to the physical acoustic environment of the event (then televised). These instruments face latency problems that are difficult to solve.

26 Nevertheless, several clubs have also acted independently of the leagues to fill the physical spaces of their stadiums with signs and simulacra of the supporters' presence such as scarves and posters, but also screens with live video broadcasts of fans at home. These actions were often associated with fundraising for the health crisis: some examples from the EPL can be found in its 'BCD [Behind Closed Doors] Guide' (www.premierleague.com/news/1682755). These initiatives are clearly meant to give fans a sense of participation and to stimulate, especially through the televising of these signs, a reflexive form of gratification. Interesting reflections could emerge from the analysis of these examples on the emotional and social needs aroused by pandemic restrictions.

References

ANSA. 2020. 'Arriva "Il Dodicesimo," app che 'porta' il tifo allo stadio.' 1 June 2020. www.ansa.it/sito/notizie/cultura/tv/2020/06/01/arriva-il-dodicesimo-app-che-porta-il-tifo-allo-stadio_94fbbddb-b87a-4cad-8b80-ba01196559e7.html.

Abbiezzi, Paola. 2007. *La televisione dello sport: teorie, storie, generi*. Cantalupa: Effatà.

Born, Georgina. 2000. 'Inside Television: Television Studies and the Sociology of Culture.' *Screen*, 41 (4) (Winter): 404–424. https://doi.org/10.1093/screen/41.4.404.

Boyd, Sam. 2020. 'Spanish Soccer Returns with Computer-Generated Crowds, and It Actually Works.' *The Verge*, 12 June 2020. www.theverge.com/2020/6/12/21288963/la-liga-fake-crowd-noise-betis-sevilla.

Chion, Michel. 1994. *Audio-Vision: Sound on Screen*. Edited and translated by Claudia Gorbman. New York: Columbia University Press.

Crawford, Garry. 2004. *Consuming Sport: Fans, Sports and Culture*. London: Routledge.

Creditor, Avi. 2020. 'Fake Crowd Noise on Soccer Broadcasts Provides Comfort, but It's Disingenuous.' *Sports Illustrated*, 27 May 2020. www.si.com/soccer/2020/05/27/fake-crowd-noise-soccer-tv-broadcasts-bundesliga.

Diamond, Beverly, Aaron Corn, Frode Fjellheim, Cheryl L'Hirondelle, Moana Maniapoto, Allan Marett, Taqralik Partridge, John Carlos Perea, Ulla Pirttijärvi, and Per Niila Stålka.

2018. 'Performing Protocol: Indigenous Traditional Knowledge as/and Intellectual Property.' In *Ethnomusicology: A Contemporary Reader*. Volume 2, edited by Jennifer C. Post, 17–34. New York: Routledge.

Dipollina, Antonio. 2020. 'Su Sky l'audio virtuale, ma c'è il rischio "fuorigioco".' *La Repubblica*, 23 May 2020. www.repubblica.it/sport/calcio/esteri/2020/05/23/news/su_sky_l_audio_virtuale_ma_c_e_il_rischio_fuorigioco_-257478363/?__vfz=medium%3Dsharebar.

Doidge, Mark, Radosław Kossakowski, and Svenja Mintert. 2020. *Ultras: The Passion and Performance of Contemporary Football Fandom*. Manchester: Manchester University Press.

Feld, Steven. 2000. 'A Sweet Lullaby for World Music.' *Public Culture* 12 (1): 145–171. https://doi.org/10.1215/08992363-12-1-145.

Frank, Sybille and Silke Steets, eds. 2010. *Stadium Worlds: Football, Space and the Built Environment*. London: Routledge.

Giulianotti, Richard. 2011. 'Sport Mega Events, Urban Football Carnivals and Securitised Commodification: The Case of the English Premier League.' *Urban Studies* 48 (15): 3293–3310. https://doi.org/10.1177/0042098011422395.

Graakjær, Nicolai Jørgensgaard. 2020. 'Sounds of Soccer On-Screen: A Critical Re-Evaluation of the Role of Spectator Sounds.' *Journal of Popular Television* 8 (2): 143–158. https://doi.org/10.1386/jptv_00015_1.

Griffero, Tonino. 2020. 'Introductory Remarks: Where Do We Stand on Atmospheres?' In *Resounding Spaces. Approaching Musical Atmospheres*, edited by Federica Scassillo, 11–24. Milano: Mimesis International.

Iozzia, Giovanni and Luciano Minerva. 1992. *Un matrimonio d'interesse: sport e televisione*. Roma: Rai Libri.

ISO (International Organization for Standardization). 2014. *Acoustics – Soundscape – Part 1: Definition and Conceptual Framework*. ISO 12913-1:2014. Paris: ISO.

Lawn, Andrew. 2020. *We Lose Every Week: The History of Football Chanting*. Huddersfield: Ockley Books.

Marchi, Valerio. 2015. *Ultrà: le sottoculture giovanili negli stadi d'Europa*. Roma: Hellnation Libri.

Martin, Richard. 2020. 'Virtual Crowds and Stadium Noise Greet Return of La Liga.' *Reuters*, 10 June 2020. www.reuters.com/article/soccer-spain-virtualcrowds-idINKBN23H2RE.

McCue, Kevin, George Wiles, Alessandro Reitano, and Martin Bergmann. 2020. 'Taking the Lead: Remote Production, Augmented Stadium Sound and Broadcast Audio in the "New Normal".' Panel presented at the SVG Europe Next Generation Audio Summit 2020, 5 November 2020. www.youtube.com/watch?v=oYc3v8NoeGI.

Preta, Augusto. 2016. 'Diritti TV, l'anomalia italiana.' *limes* 5 (May): 103–116.

Riedel, Friedlind. 2020. 'Atmospheric Relations: Theorising Music and Sound as Atmosphere.' In *Music as Atmosphere: Collective Feelings and Affective Sounds*, edited by Friedlind Riedel and Juha Torvinen, 1–42. London: Routledge.

Rowe, David. 2004. *Sport, Culture and the Media: The Unruly Trinity*. Second edition. Maidenhead: Open University Press.

Sarno, Giulia. 2022. '"Noi si fa i cori": note dalla curva Fiesole sulle pratiche musicali del tifo organizzato.' *Acusfere* 1 : 61–91.

Schafer, R. Murray. 1977. *The Tuning of the World*. New York: Knopf.

Schulke, Hans-Jürgen. 2010. 'Challenging the Stadium: Watching Sport Events in Public.' In *Stadium Worlds: Football, Space and the Built Environment*, edited by Sybille Frank and Silke Steets, 56–73. London: Routledge.

Seeger, Anthony. 2004. 'Traditional Music Ownership in a Commodified World.' In *Music and Copyright*, edited by Simon Frith and Lee Marshall, 157–170. Second edition. New York: Routledge.

Turino, Thomas. 2008. *Music as Social Life: The Politics of Participation*. Chicago: University of Chicago Press.

Webber, David M. 2017. '*The Great Transformation* of the English Game: Karl Polanyi and the Double Movement "Against Modern Football".' In *Football and Supporter Activism in Europe: Whose Game Is It?*, edited by Borja García and Jinming Zheng, 9–26. Cham: Palgrave Macmillan.

Zinganel, Michael. 2010. 'The Stadium as Cash Machine.' In *Stadium Worlds: Football, Space and the Built Environment*, edited by Sybille Frank and Silke Steets, 77–97. London: Routledge.

7 A Digital Archive of Participatory Location Rhythm Performances

Listening as a Way of Attending to the Pandemic

Marcel Zaes Sagesser

The global COVID-19 pandemic lastingly changed how we relate to the auditory in our daily lives. The various lockdowns in many regions around the world required many of us to shelter in place with family, loved ones, flatmates, or alone. Therefore, it incentivised us to develop social skills that exist outside of face-to-face social life. 'Sociality,' the plastic and fragile product of relations among humans, capable of 'tak[ing] many forms' (Long and Moore 2012, 2), changed – perhaps permanently. Since lockdowns, communication among humans has increasingly shifted to digital tools. In addition, daily entertainment and media are consumed from home. And for those of us privileged enough to hold on to our jobs, schools, and university programs, the entire work or study day shifted online. Whether we were talking to friends and family members, meeting with coworkers, working on a group project for school, streaming a film, listening to the radio, reading the news, or attending a virtual exhibition, many of us found ourselves listening to some sort of media that included streamed audio – sound that is picked up on one end with a microphone, and played back on the other via a speaker. I argue that the increased amount of mediated content in our everyday experiences, and the changes involved in our auditory realities, lastingly affect the ways in which we listen. These novel listening modalities will stay with us for a long time because we will henceforth keep using most of the digital communication and streaming tools.

In this chapter, I will first discuss the changes in our listening modalities that the pandemic has given rise to, before presenting the arts piece *#otherbeats* – a website that archives rhythm recordings – which I made in direct response to the altered auditory realities that the lockdowns have fostered. The discussion will analyse how *#otherbeats* constitutes a device with which to attend to the pandemic; but it will also include how it emphasises an altered listening practice that holds value even outside of, and after, the global pandemic.

Listening in Quarantine: Hybrid Auditory Realities

Hybrid auditory realities is what I call the complex blend of streamed and actual audio that many of us experienced increasingly during lockdown and ever since. I call them hybrid because they blend sonic events from the virtual and physical worlds with one another in the same place (my room, my ear) and at the same time (now). In other words, the sonic composite made up of streamed content – made audible in a set of speakers, headphones, or earbuds – and the actual acoustic surroundings that 'leak' (Stanyek and Piekut 2010, 20) over to my mediated auditory space: traffic from the street, my neighbours watching TV, my flatmate washing the dishes, and the water noise from my radiators. Such a hybrid auditory reality is a complex acoustic space, because it is made up of several physical and virtual architectures that intersect at my very preceptor of the auditory: my ear. This reality

DOI: 10.4324/9781003200369-9

88 *Marcel Zaes Sagesser*

is further complicated given that my counterpart, with whom I am on a video call, also has their outdoors leak in their microphone and over to me – through all the time-based and spectrum-based algorithms that the software developer may have introduced between the two of us. What arrives at my ear, and what I am listening to as a result, is a multidimensional collage in which different architectures, spaces, places, times, algorithms, and artefacts blend with one another. It is effectively impossible to distinguish the heard sources, tools and algorithms from one another, or even tell the heard spaces from one another – in part because their auditory artefacts sound similar, and in part because everyday listening largely functions unconsciously and is subordinated to vision, since in 'Western culture, thought, and philosophy … seeing equals knowing and knowing equals consciousness and rationality' (Pais 2017, 237). Because the heard sonic spaces are not accompanied by a visual counterpart, it might be difficult to perceive them as distinct spaces. But it is also impossible to distinguish between the different levels of mediated and immediate sonic matter because all sound – regardless of its mediation or immediacy – reaches my ear as nothing more than vibratory matter. Even if we are covered under headphones, virtual and actual sounds alike reach us as physical sound material on the eardrum: following Nina Sun Eidsheim's 'reconception of sound as event through the practice of vibration' (2015, 3), they are both vibratory matter.

The lockdowns presented us with increased hybrid auditory realities, which in turn made us listen differently – consciously or not. Some of the effects were perhaps more evident or visible than others. Access to the outdoors, in many places, was temporally restricted so that windows and balconies constituted the last sonic openings through which to perceive the outdoors – almost in the sense of what the Renaissance painting theorist Alberti described as the 'open window' – the 'quadrangle' that acts as a visual frame through which 'historia' is observed, and according to which the painting is arranged on the canvas (2011, 33). Opening the window framed the acoustic outdoors for many of us during lockdown, recalling the 'leakage effect' described by Jason Stanyek and Benjamin Piekut in the case of the recording studio, for which 'an activity in one area expands unexpectedly into another area' (2010, 20). A sonic leakage from the outdoors into a domestic space is always there, but it is the window (or balcony door) as a device that brings to my attention the leakage's agency over my perception. It might have surprised the average citizen in quarantine how crucial a sonic change is produced by the simple fact of opening or closing a domestic window.

More so, the acoustic outdoors we consumed were changed in their texture from what we might have heard before the pandemic, just because fewer pedestrians and less traffic were on the roads in many places, and offices, institutions, and sports and cultural venues remained closed. This effect perhaps was less durable than the mediated realities that the pandemic brought about, and that might last for a long time, but nonetheless it is an important effect because it made many humans – particularly those living in densely populated urban areas – aware of modifications happening to their sonic worlds. In urban and suburban spaces, less human activity might have been compensated for by birds and wild animals taking up more space, altering the sonic texture not only by a louder presence, but also by enlarging their frequency range within the texture: their sonic 'niches' (Krause 1993, 6) might have grown. The assemblage of sounds that leaked through our windows was thus a changed one, and defamiliarised many of us with what we knew as our sonic surroundings before the pandemic. The window might be an important device in the shaping of our auditory reality because it constitutes the point of contact between metaphorical and actual architecture. As a transparent opening it allows my view through, letting sight guide my listening. As a movable device it shows the leakage. And as a part of the physical architecture surrounding

A Digital Archive of Participatory Location Rhythm Performances   89

me, it informs my listening as it tells me something about the materials, walls, and textures surrounding me (Blesser and Salter 2007). The window is a mediator in my auditory reality.

The point is that the window as the device serves as a good simile for sonic leakage when our auditory realities extend into the virtual. In digital communication tools, multiple sonic windows into different times and spaces are available to me simultaneously or in fast succession, such as on a video conferencing call with multiple people. Every participant's surroundings leak in, and in contrast to my actual window, I cannot close out my coworkers' sonic leakage. Moreover, the participant's surrounding architecture leaks in just as much – for example, if someone joins the call from a bathroom or a hallway, a heavily reverberated version of their voice reaches me. The leakage effect thus occurs on many levels in the virtual as well as in the physical world. What arrives as physical matter at my eardrum is what I call a hybrid auditory reality.

My argument is that the lockdowns have hugely increased the presence and importance of these hybrid auditory realities, and as a result, the lockdowns have led to novel modalities of listening. A listening modality that grew out of the lockdowns might pay less attention to sonic markers of physical motion in space, and instead put emphasis on the sonic traces of leakage as an index for the architectural, mediated, temporal, and social spaces that constitute the hybrid auditory reality in my ear.

Listening is inseparably linked to time, and to our perception thereof. We have heard people claim that their perception of time passing while in lockdown has changed, and research has indeed shown that in quarantine 'people encounter lack of motivation and are plunged into an impression that the sense of time is blurred' (Grondin, Mendoza-Duran, and Rioux 2020, 4). The connections between the sonic, listening, and time are manifold. Sound exclusively unfolds in time – as vibratory energy in physical matter – and so does the activity of listening. It is thus not farfetched to conclude that this blurred sense of time in quarantine may be, at least partially, linked to the altered hybrid auditory realities to which we are subjected. James Gibson argues that time is not perceivable as such, but only via the perception and awareness of single, discrete events (1975, 297–299). Because some events come to us via sound, and the awareness thereof via listening to sound – such as sound notifications on our electronic devices – these sonic markers constitute events that then act as perceptual stimuli that shape our sense of time passing in the sense of Gibson's information-based perception theory. The connection between the sonic and the sense of time increasingly relies on digital communication tools, during and beyond the pandemic.

Moreover, digital communication tools, particularly video conferencing systems, add their distinct temporal distortion to the streamed audio: voices arrive sometimes fragmented, truncated, sped up, or slowed down. The use of these tool makes their users aware of these – often artificial sounding – artefacts; users read and understand them as what they are: markers of imperfect connections and algorithms. These distortions on the time level become particularly audible when they appear with material for which these tools are not intended; for example, if my conversation partner at the other end of the connection activates a coffee machine close to the microphone while their dog barks in the background. Tools that are programmed and tailored to clean human voices will produce all sorts of surprising temporal compositions. As a part of my perceived hybrid auditory reality, I read these distortions as a marker for the temporal mismatch between what I am hearing and what my conversation partner might be hearing, and as marker for the mediation process proper. Listening thus makes me aware of the changed temporal conditions, and it does so in the realm of time in which my listening takes place.

Yet sound, even though it takes place in time, also carries information about the space from which it emits. According to Barry Blesser and Linda-Ruth Salter (2007), hearing and listening are processes in which the physical qualities of the architecture surrounding us are put to use in order to gain additional information about these spaces. They argue that the echo of a hand clap is perceived simultaneously in two ways, first as just an 'additional sound (sonic perception),' and second 'as a wall (passive acoustic object),' an example in which the reflecting (silent) wall becomes a sonic cue even though it is not itself emitting the original sound (Blesser and Salter 2007, 2). It is here where spatial and temporal signals blend: we may perceive a sound with a temporal dimension, such as an echo, as a spatial cue giving us insight into the echo-causing space. When applied to the hybrid auditory realities that the pandemic has increasingly given rise to, the temporal and spatial dimensions of sound are further confused. Cues for temporal distortion and cues for sonic architectures reach us both from our physical surroundings and from our mediated audio streams.

As a modality to attend this overly complex sonic assemblage, I understand listening to be a tool that – if we practice enough – can help us in decoding some of the assemblage's information, particularly its temporal and spatial distortions, in order to gain knowledge about that which is mediated and that which leaks in. Many of us have knowingly or unknowingly practiced the decoding of such auditory cues in quarantine (and outside of it). Going back to the window metaphor, I believe that the pandemic-related listening modalities have caused many to become increasingly aware of the complex and manifold (virtual) windows within our auditory worlds and media landscapes – or to put it differently, how our hybrid auditory realities resemble fractals made of multiple windows nested within one another.

An Archive of Listening: #*otherbeats*

As a scholar and an artist, at the onset of the pandemic I developed a creative intervention: a project published under my artist moniker *Marcel Zaes*, entitled #*otherbeats*, with which I offer a device that puts emphasis on that thing that I see as the most characteristic of the lasting changes to our everyday: listening. With #*otherbeats*, I created a tool that gives its users the possibility to listen to temporal distortion, spatial distortion, to the leakage effect, and to the artefacts of mediation through an array of virtual sonic windows. My main goal in making this intervention was to explore the increased hybrid auditory realities that the pandemic has brought about (and which are not exclusively bound to the pandemic). Because the pandemic has lastingly changed the everyday and the work or school routines for many of us, I argue that we can understand #*otherbeats* as a tool that remains relevant outside of and after the pandemic. Hence, the thoughts and knowledge about listening that we might gain from analysing #*otherbeats* hold value beyond the global COVID-19 pandemic.

For the project, I prompted volunteer participants around the globe to send me audio recordings (or if they preferred, video) that they made from their sheltering-in-place locations or, ideally, from adjacent outdoor spaces – if the current policies allowed access to them (see M. Sagesser 2021; M.Z. Sagesser 2021; Zaes 2021). I emailed friends and acquaintances encouraging them to participate, and I spread the prompts on social media. I also encouraged participants who ended up recording indoors to keep a window open, so as to allow for a more pronounced outdoor leakage effect. In other words, I wanted them to capture the sonic realities that best described their sheltering locations, while also capturing the artefacts of their mediation and leakage effects. What I circulated among potential participants were four small rhythm exercises. In their nature, they drew directly from my interest in time and rhythm, and their effects on mediation. I asked participants to imagine and clap a 'dance tempo,' to improvise a 'beat with any means,' to find or invent what they thought is an 'alternative metronome,' and lastly, to sonically portray their 'space,' that is, their location. These recordings were made with low-key technologies such as mobile

A Digital Archive of Participatory Location Rhythm Performances 91

phones, and the collected media artefacts – roughly 150 files – I collected and organised on a website that I published in September 2020 at https://otherbeats.net (Zaes 2020). This website stores and makes the archive of collected recordings available to the public.

The website archives the collected files, yet it is not a conventional archive because it does not satisfy those criteria that would otherwise make for a valid archive: it removes all metadata from the recordings, it omits author attribution by anonymising them (while listing all participants in the credits section, detached from their respective media), and it leaves the listener groping in the dark about which audio event had been captured in response to which rhythm prompt. Instead of an archive that would make connections between metadata and media objects, *#otherbeats* produces more of what Salomé Voegelin might call a 'sonic volume' (2018, 47) or an 'indivisible volume' (ibid., 60) – a fluid expanse of (sonic) matter in which the listener is immersed, and in which the distinct sonic windows into the participants' individual recordings gradually merge with one another. Left alone, given little information beyond the sonic content, the user might start to engage this aural architecture solely with their ears – by listening. It is a volume full of sound that still holds and stores the participants' sonic traces, even if it is otherwise devoid of archival characteristics. Considering Wendy Chun's definition of the archive as the 'relation between memory and repetition' (2011, 98) and as 'of indexing and organizing' (ibid., 100), I ask in *#otherbeats* where the memory is if all overt indexing is omitted. Is it perhaps, in the vein of Alexander Weheliye's argument about the African diasporic disc jockey practice of 'the mix' (2005, 73), which 'highlights the amalgamation of its components ... as much as it accentuates the individual parts from which it springs' (ibid.), that 'the archival and the sonic' (ibid., 88) exist in close relationship (see also M.Z. Sagesser 2021)?

#otherbeats constructs a sonic window through which users can listen to a slice of other people's sonic realities in lockdown: they can focus on individual recordings or on the composite mix alternatingly, or even simultaneously – the latter being an affordance of our listening, as Ingrid Monson (2008) argues. Analysing how we perceive a solo in a jazz tune,

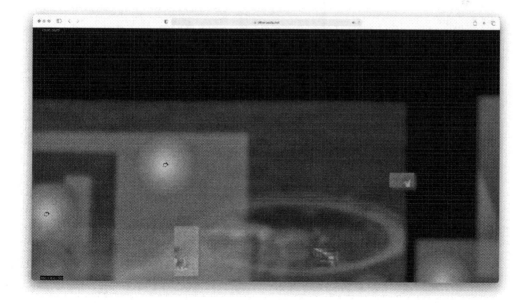

Figure 7.1 Screenshot of *#otherbeats* website with several colour dots on a dark background, with mouse cursor.

92 *Marcel Zaes Sagesser*

she claims that 'the process of listening is deepened by noticing the ear's capacity to shift attention while simultaneously hearing the ensemble as a whole' (Monson 2008, 41). The negotiation between individual collected ephemera and emerging composite sound mix is what primarily characterises *#otherbeats*. This negotiation is animated via an idiosyncratic audio playback system that allows playing back many files at once while focusing on one. This playback system is graphically represented and taught to the user via a custom-made user interface (see Figure 7.1). In the following section, I will discuss both the playback system and the user interface in depth.

Playing as Listening: The Playback System and User Interface of *#otherbeats*

On the *#otherbeats* website, the collected sound files are randomly spread across a large visual space – a two-dimensional plane coded in HTML. Algorithms coded in JavaScript are used to randomly distribute the collected sound recordings anew each time the website is loaded. That means that each time a user visits the site or clicks the refresh button, they will face a different arrangement of sounds, making it ultimately impossible to learn the specific geography of the website. Each sound file is visually represented by a colour dot. The user is given agency to interact with these colour dots and their respective sound files: by scrolling left and right and up and down, they navigate through a visual field of dots and sounds, thereby choosing which files they are listening to. While scrolling and clicking, they create their own mix (see M. Sagesser 2021). To put it differently, when they move around on this two-dimensional picture plane, they choose which sonic 'windows' to listen to – they navigate around and between multiple windows and thereby control how much one sound dot leaks into the other. But for the user, the leakage effect is multidimensional: while they choose which sounding colour dots they allow into their mix, they also face the ambient outdoor leakage of the original participant's recording situation. Similar to the hybrid auditory realities that I described at the beginning of this chapter, the *#otherbeats* users find themselves presented with leakage on many immediate and mediated levels.

The playback system with which this sonic reality is constructed is coded in the Web Audio API and embedded within the website's HTML framework. Each reload produces a set of x and y coordinates for every sound and colour dot. The website continuously captures the user's scrolling position and measures, for each sound dot, the respective distance between the dot and the user's scrolling position (the virtual location they are listening to). This constant calculation takes place using JavaScript. Sounds that are closer to the user's listening position play at a louder volume and at a clearer timbre, while sounds further away become more muffled, more reverberated, and softer. The user's speed of scrolling thus controls the pace of the mix that they instantaneously produce (see M. Sagesser 2021). It is up to them whether they scroll fast and explore the dynamically changing composite texture, or whether they scroll slowly, stopping at a given colour dot in order to attend to the respective individual sound file. When they do so, they hear the sound dot in the middle of their screen clearly and loudly, while other sound dots in the proximity around it softly and blurrily leak in. The playback system not only mimics the physics of how vibratory matter behaves in the actual world, particularly the effects of proximity and distance, but it also mimics multidimensional hybrid auditory realities as many of us know them from our everyday lives in quarantine and beyond. It applies proximity and distance effects to recorded audio files that themselves contain signal and noise, rhythmic content and outdoor leakage, acoustic foreground and background. In addition, some graphic objects on the website are clickable, upon which they will move the user to a randomly produced new listening position within the two-dimensional space, resulting in a sudden change in the sonic mix.

A Digital Archive of Participatory Location Rhythm Performances 93

Given that the sonic processes triggered by the user's scrolling and clicking are modelled after real-world physics, the user might intuitively understand the playback system and the involved user interface – an important detail, given that the website otherwise offers little to no instructions. Although there is a credits section with a short description of what the website does, users find themselves thrown into this vast, immersive site, and have to experiment and find out what to do with it. Once they have learned that scrolling and clicking – the idiomatic controls to steer a computer and navigate the internet since its mainstream advent – will alter the auditory reality that they are listening to, they may start composing their own sound piece, or their own mix (see M. Sagesser 2021).

There is a music compositional interest to the sonic world that emerges from interacting with the website. The many sound dots that I am scrolling by as a user, they are not static objects. Because each sound dot is nothing more than a participant's rhythm or ambient recording, it has a temporal dimension: there is an inner time to each audio file. The files mostly contain some version of rhythm: slightly unsteady tapped tempos, recorded dripping water from a water faucet, or the leaves on a tree irregularly shaking in the wind that a participant recorded for the project as their version of an alternative metronome. A still of the latter is printed in Figure 7.2, while Figure 7.3 graphically illustrates the rhythms emitted from these tree leaves. The point is, all these rhythms run asynchronously with one another, and the duration of the audio files is irregular too. No attempt at normalising or synchronising these files was made. Because there will always be some files leaking in from a distance, the user will always hear multiple files playing at any given moment. And because they all contain different versions of rhythm, the composite whole is necessarily a rhythmic cacophony. With the exemplary rhythm of a tree leaf shaking in the wind, depicted in Figures 7.2 and 7.3, one can easily imagine how a composite of several such rhythms unavoidably leads to a densely populated sonic texture full of fine-grained, varied rhythms. The individual rhythms may be more or less regular, with some of them resembling pulses, hence the composite texture will always remind its listener of multiple pulses running at once. According to Kathryn Kalinak, pulses act as markers for

Figure 7.2 Still from video recording showing a tree leaf in nature (courtesy of Annie Rüfenacht).

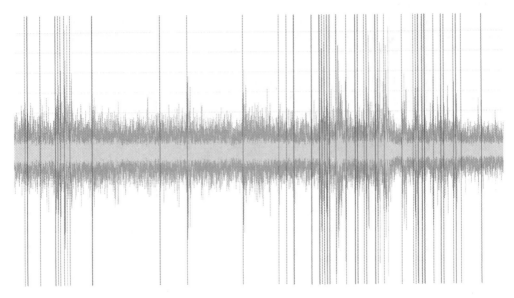

Figure 7.3 Graphic representation of the rhythm produced by the tree leaf irregularly shaking in the wind.

the 'passage of time' (2010, 13); they are often deployed as such in film soundtracks and the like. #*otherbeats* confuses our notion of time passing with its manifold pulses running simultaneously yet asynchronously, since it evades being perceivable as a single main pulse with subdivisions and variations thereof.

To further complicate the situation, the participants' files are between one and four minutes in length (they were prompted to record for roughly three minutes). This duration is too long for humans to perceive as repetition, and looping these irregular and long files produces a composite texture that never exactly repeats. Yet, it always sounds alike. The relation between memory and repetition that Chun (2011) described as characteristic for archives is activated: users might wonder which rhythms or sounds they have already encountered, which ones they are remembering, and which ones they are hearing for the first time.

The music compositional part of #*otherbeats* is built around this tension between memory and repetition on the one hand, and the tension between the individual and the whole on the other, as we have already seen. The rhythmic nature of the files amplifies both these tensions because many of these files' inner rhythms are meaningful when listened to in isolation, but they gradually lose their articulation, which is to say their meaning, as they become part of a sonic ensemble and blend with others. This tension is similar to that which occurs when I am subject to digital communication tools and find myself in a video conferencing call trying to identify individual layers and markers of mediation and leakage out of a sonic whole.

My interest in creating cacophonic rhythmic worlds draws on my observations of how humans are increasingly subjected to hybrid auditory realities as a direct effect of the pandemic lockdowns, and it also speaks to the fact that many have experienced a blurred sense of time in the pandemic: their everyday rhythms were as cacophonous as #*otherbeats* is. Instead of a primary rhythmic pulse, there were many or none. Grondin, Mendoza-Duran, and Rioux argue that humans are 'untrained to deal with so few temporal landmarks,' as we had during lockdown, and therefore they create a sense of 'confusion' about the perception of time (2020, 5). The music compositional strategy of #*otherbeats* takes on this argument by presenting the user with a huge number of small-grained, lively rhythms that are blended, blurred, and confused with one another, while it lacks those 'few temporal landmarks' (ibid.). It is these larger sonic events that would act as clear sonic events in our sensing of time passing, and which Grondin, Mendoza-Duran, and Rioux describe as a prerequisite to perceive time in more traditional ways – I am using traditional here in reference to that which used to be before the global pandemic.

#*otherbeats*' idiosyncratic playback system and user interface thus draws inspiration from, and might even reproduce, a sensation of how we lastingly listen differently since quarantine, thereby reiterating that listening and the perception of time passing are connected. The piece produces an opportunity to play with sonic artefacts that are mediated, leak, and which are temporally and spatially distorted. Thus, it produces an opportunity to think about the (rhythmic) intersections of memory and repetition, individual and whole. One could thus argue that the website #*otherbeats* archives the sounds of the pandemic, making them available long after many public restrictions have been lifted. But I instead want to put forward a notion that is slightly yet crucially different: #*otherbeats* does not archive the pandemic, but it archives a specific modality of listening, that is listening more consciously to the markers of mediation, leakage, and to the markers of temporal and spatial distortion. It presents to its users the same unwanted 'lags, leaks, and perforations' that Jason Stanyek and Benjamin Piekut have identified for the recording studio (2010, 22). If there is an archival element to #*otherbeats*, then it is one that stems from how its users reinterpret and remix these 'lags, leaks, and perforations' (ibid.) that were stored in it in the first place by the volunteering recording participants. And because the use of digital communication tools and the importance of hybrid auditory realities most likely will remain high also after the pandemic, the listening modality that is stored into #*otherbeats* holds value beyond the pandemic, as a part of our increasingly technology-driven everyday. It is my hope that its users will practice and learn such listening skills, carry them to their everyday, and that they will be able to attend to auditory hybridity.

References

Alberti, Leon Battista. 2011. *On Painting: A New Translation and Critical Edition*. Translated and edited by Rocco Sinisgalli. Cambridge: Cambridge University Press.

Blesser, Barry and Linda-Ruth Salter. 2007. *Spaces Speak, Are You Listening? Experiencing Aural Architecture*. Cambridge: MIT Press.

Chun, Wendy Hui Kyong. 2011. *Programmed Visions: Software and Memory*. Cambridge: MIT Press.

Eidsheim, Nina Sun. 2015. *Sensing Sound: Singing and Listening as Vibrational Practice*. Durham: Duke University Press.

Gibson, James J. 1975. 'Events Are Perceivable but Time Is Not.' In *The Study of Time II*, edited by J.T. Fraser and N. Lawrence, 295–301. Berlin: Springer. https://doi.org/10.1007/978-3-642-50121-0_22.

Grondin, Simon, Esteban Mendoza-Duran, and Pier-Alexandre Rioux. 2020. 'Pandemic, Quarantine, and Psychological Time.' *Frontiers in Psychology* 11. https://doi.org/10.3389/fpsyg.2020.581036.

Kalinak, Kathryn. 2010. *Film Music: A Very Short Introduction*. Oxford: Oxford University Press. https://doi.org/10.1093/actrade/9780195370874.001.0001.

Krause, Bernard L. 1993. 'The Niche Hypothesis: A Virtual Symphony of Animal Sounds, the Origins of Musical Expression and the Health of Habitats.' *The Soundscape Newsletter* 6: 6–10.

Long, Nicholas J. and Henrietta L. Moore, eds. 2012. *Sociality: New Directions*. New York: Berghahn Books.

Monson, Ingrid. 2008. 'Hearing, Seeing, and Perceptual Agency.' *Critical Inquiry* 34 (S2) (Winter): S36–S58. https://doi.org/10.2307/20184424.

Pais, Ana. 2017. 'Almost Imperceptible Rhythms and Stuff Like That: The Power of Affect in Live Performance.' In *Theorizing Sound Writing*, edited by Deborah Kapchan, 233–250. Middletown: Wesleyan University Press.

Sagesser, Marcel. 2021. 'Within the Grid: Inquiries in the Socio-Rhythmic Ambiguities of Mechanical Time.' PhD diss., Brown University.

Sagesser, Marcel Zaes. 2021. '#otherbeats: A Hypermediated Playback System in Web Audio.' *Proceedings of the International Computer Music Conference 2021*, 185–191. Santiago: Pontificia Universidad Católica de Chile.

Stanyek, Jason and Benjamin Piekut. 2010. 'Deadness: Technologies of the Intermundane.' *TDR/The Drama Review* 54 (1): 14–38. https://doi.org/10.1162/dram.2010.54.1.14.

Voegelin, Salomé. 2018. *The Political Possibility of Sound: Fragments of Listening*. London: Bloomsbury Academic.

Weheliye, Alexander G. 2005. *Phonographies: Grooves in Sonic Afro-Modernity*. Durham: Duke University Press.

Zaes, Marcel. 2020. *#otherbeats*. https://otherbeats.net/.

Zaes, Marcel. 2021. '#otherbeats: Performing a Participatory Archive of Social Rhythm.' Paper presented at the New Interfaces for Musical Expression conference, online and New York University, Shanghai, June 2021. http://nime2021.org/program/#/music/178.

Part II

Experiences

Musicking in the Face of the Pandemic

8 *Huapanguitos pa seguir aguantando en cuarentena*

Mexican SonTube Channels as Emergent Digital Spaces of Music and Community during COVID-19

Daniel Margolies and J. A. Strub

During the COVID-19 pandemic, YouTube channels featuring music livestreams and streamed video queues dedicated to *son*, a complex family of music and dance traditions from Mexico (Sturman 2015; González Paraíso 2014), became critical spaces of social gathering, community fellowship, and musical self-expression. The communities that emerged within the chatrooms of SonTube, the term used here to describe this network of content creators, performers, and participant-consumers, constituted new and distinctive digital spaces during the pandemic where needs for social intimacy and cultural connection were met in the absence of physical gatherings. SonTube formed a virtual musical ecosystem encompassing sites of artistic production and dissemination, spaces of social and emotional sustenance, and for some, engines of income during a time of recession.

YouTube is one of the principal music streaming services across the globe, and it also operates as a repository of culturally significant music. As music has become increasingly consumed online, platforms – the virtual spaces of its dispersal – have become a topic of wide scholarly interest. Researchers and listeners alike are only beginning to fully comprehend how digital platforms 'have substantially restructured the ways in which music is distributed, discovered and consumed' (Airoldi, Beraldo, and Gandini 2016, 2; Tepper and Hargittai 2009). David Hesmondhalgh, Ellis Jones, and Andreas Rauh have dissected the ways the new streaming system has had an impact on music making and consumption, a process they call the 'platformisation of cultural production' (2019, 2). YouTube shares similar organising corporate sensibilities and other convergences with platforms like Soundcloud or Spotify, but these authors argue correctly that 'its extraordinary multiplicity' places it into a separate category (ibid., 10).

One key connection between all platformised music consumption, however, is the reliance on algorithms to shape listening and to trigger and maintain viewer interest and time on the site. This serves YouTube's commercial interests while also shaping consumption habits in fundamental ways (Rieder and Coromina 2018; Bishop 2019; Kaiser and Rauchfleisch 2020). Consequently, 'YouTube videos can be seen as nodes in a network ... [in which] the weight of the edges is determined by the users' aggregated consumption practices on the platform' (Airoldi et al. 2016, 4).

Other scholars have observed that YouTube's algorithmic approach transforms the entire notion of the platform into an ongoing content and user-generated *flow*. Sheenagh Pietrobruno and David Berry have emphasised the concept of 'real-time streams as dispersed narratives' (Pietrobruno 2018, 527). The streams themselves, in this analysis, are combined with user data and 'evaluated, combined, and rechanneled back to the user as well as other users in the form of patterns of data' (Berry 2011, 144). Importantly for scholars interested

DOI: 10.4324/9781003200369-11

in vernacular music communities, Pietrobruno argues provocatively that this rechannelled mass stream, built off the distinct makeup of individual user-generated streams, could in fact enable expanded expression of cultural heritage and 'the potential to challenge the homogenizing of culture' (2018, 528).

The SonTube channels considered in this chapter occupy a distinctive cultural eco-system within YouTube's wider platform. While new viewers often may have found their way to SonTube channels via YouTube's algorithms, the regularity, consistency, and length of streamed queues allowed pandemic-era participants to ease into these spaces for long periods of time night after night. SonTube channels became, effectively, virtual locales founded on music consumption where outside time became fluid, individuals interacted with one another in shared spaces, and the chatbox community became the focus of social interaction to replace that lost in mandatory pandemic-era isolation.

Considering YouTube 'as an unfiltered, bottom-up cultural archive,' has become fairly well established by scholars (Burgess and Green 2018, 137). YouTube 'creates spaces for engagement and community-formation' with 'spillover into other sites of everyday culture, meaning, identity, and practice' (ibid., 80). This 'participatory turn' has consequently transformed 'the relationship between the individual and a global, culturally diverse idea of community' (ibid., 125). YouTube is of immense significance to participatory and global Do-It-Yourself (DIY) culture, and to the documentation of intangible cultural heritage. The platform's growth is congruent with global moves toward both deliberate and implicit community archival efforts (Baker and Collins 2017). Yet the impact and significance of this archival project remains a matter of both interest and dispute to scholars in media studies, archival studies, ethnomusicology, critical heritage studies, and other fields. Some have argued that as an archive, YouTube has facilitated the transnational spread of vernacular cultural forms (Pietrobruno 2013). Alternatively, John H. McDowell wrote that if indeed the site is 'an archive of expressive culture, it is … an unusual archive that makes special demands on those who would use it' because of its user-generated nature and its lack of 'signposts' (2015, 263). Many of these DIY archival efforts started as a simple expression of fan culture without structure, but have transformed into culturally significant, and even profitable, enterprises as the organisers developed their craft and expanded their viewer base. Content creation on YouTube such as can be seen on SonTube can thus be characterised as both project and process. Perhaps the virtual community spaces themselves act as the key signposts to the broader archival project of preserving and exhibiting cultural heritage.

Researchers have highlighted the advantages of this decentered and DIY character of YouTube archiving, which has allowed the development of informal cultural arbiters to build idiosyncratic spaces and attract transnational communities of shared affinity. Sheenagh Pietrobruno argues that 'YouTube's archives of intangible heritage seem to be forging a new form of structure that absorbs both dominant and marginal perspectives and is produced by the efforts of the human and machine' (Pietrobruno 2013, 1263). Scholarship on virtual communities has frequently stressed connections to the textures and structures of existing communities, arguing that 'what the Internet does not do is create a community if there are no pre-existing common interests' (Waldron 2013, 91). In Latin America in particular, various studies have highlighted the creative use and reappropriation of new social technologies and social media platforms like YouTube among artists and musicians (Martens, Tapuy, and Venegas 2020; Valcarce and Mallero 2020; Mitchell 2011).

SonTube's creation of virtual spaces for community gathering, self-support, and fellowship constituted a distinctive coping mechanism throughout the pandemic for quarantined consumers of Mexican music. Taken as a whole, these channels and their streams served as gathering places built around cultural connections for diverse and transnational groups of individuals seeking new forms of social interaction and sources of individual sustenance and wellbeing during the pandemic (Margolies and Strub 2021b). These virtual spaces operate unrestrained by state interdiction, time, or physical place, notwithstanding the potentially undermining policies of automated copyright enforcement and the right of full removal retained by YouTube itself (Pietrobruno 2013). Nonetheless, these channels successfully developed an expansive atmosphere of co-creative placemaking in virtual space.

SonTube can be usefully viewed through the framework developed by Doris Elena Pinos Calderón and Cristina Venegas to describe artistic and cultural projects for 'local and co-creative work' (2020, 55–56). Instead of focusing on the redevelopment or reclaiming of actual physical spaces in cities, these channels used the platform of YouTube at the critical time-space moment of the coronavirus pandemic to build unique virtual spaces of cultural and communitarian-minded connection. Pinos Calderón and Venegas argue

> that effective forms of technological appropriation require co-creative approaches that can contribute to new socio-technical and communicational processes that make local initiatives visible, strengthen cultural identity, and re-articulate urban space as a place of social, community innovation, and political organization.
>
> (2020, 55–56)

The pandemic-era DIY SonTube channels capitalised on new technologies and nascent infrastructures to transcend physical space and time, reimagine, and foster anew virtual spaces of community when social gathering was impossible during the time of pandemic isolation, crisis, and fear.

The Streamed Queues of 'SonTube' during COVID-19

YouTube channels dedicated to *son* jointly coalesce into an overlapping and interrelated musicultural ecosystem that this project terms 'SonTube.' While SonTube includes numerous channels managed by individual musicians, performing groups, and official cultural organisations, this chapter focuses on three of the most significant channels run by DIY archivists and documentarians that came to constitute core sites of virtual gathering during COVID-19.

GavBroadcast, the oldest and most popular of the three channels, was founded by communications engineer Gabino Vera in 2007 as a personal repository of travel videos.[1] Vera, who is originally from Santa Maria Ixcatepec in the Huasteca region of Veracruz, began to use his channel to showcase musicians from his hometown. As of August 2021, GavBroadcast is an impressive repository of over 4,000 videos of contemporary *son huasteco* performance, with nearly a third of a million subscribers. Another channel largely dedicated to *son huasteco* is QuerrequeFilms, run by husband-and-wife team Hector Manuel Delgado Flores and Maria del Carmen Camarena Torres. Delgado Flores and Camarena Torres met as undergraduates in the philosophy department at the University of Michoacán of San Nicolás de Hidalgo, and began producing documentary content for QuerrequeFilms in 2009.[2] The channel currently showcases *son huasteco* and *son arribeño* performance videos ranging from formal presentations to casual get-togethers of musicians, as well as documentary and ethnographic footage from the Huasteca and Sierra Gorda regions. A classmate of Delgado Flores named Gilberto Salvador Perez Baeza worked alongside the couple in the early years of QuerrequeFilms, but ultimately set out to found his own channel dedicated to the music of his home state. In 2014, he launched Son Michoacán, which differed in the regional focus of its musical content but modelled its format and style on those pioneered by GavBroadcast and QuerrequeFilms.[3]

102   *Daniel Margolies and J.A. Strub*

The production approaches of all three SonTube channels are interlinked but distinct. In each, content creators travel to sites of musical production, record performances, edit audiovisual material, and release videos in a variety of static and streamed formats. All three channels are monetised, meaning that creators are paid by the platform based on views. The dollar value of a view varies depending on a variety of factors, including, perhaps most significantly, the location of the viewer. These content producers typically do not charge groups for recordings, nor do they pay groups for views. Content producers justify this by noting the costs implicit in travelling, purchasing and maintaining equipment, and dedicating work time to post-production and promotion (Gilberto Salvador Perez Baeza, pers. comm, 25 August 2021; Gabino Vera, pers. comm., 16 May 2020). The content creators behind each of these three channels are quick to distance themselves from generalised characterisations that draw extensive comparisons between their wider bodies of work as documentarians. Nonetheless, queued streams, parallels in format and content, and overlapping audiences during the pandemic produced undeniable linkages between these channels.

By the outset of the COVID-19 pandemic shutdowns in March 2020, all three channels were being operated as full-time projects by their founders, who previously had maintained their channels as hobbies or secondary jobs while working in other settings. Taken as a whole, SonTube channels constituted a linked system because of their related content and overlapping support base. The pandemic shutdowns produced increased screen use and enforced stretches of time at home, while also provoking the social anxieties implicit in living through a life-threatening pandemic while being physically isolated from the wider world. Quarantine thus generated both the prerequisite conditions for increased SonTube consumption and the elevated needs for cultural connection and community intimacy that SonTube's content creators would ultimately seek to meet.

Perhaps the most successful YouTube livestream format for these channels throughout the initial months of the COVID-19 pandemic was that of the streamed queue. This format involved the establishment of a shuffled lineup of pre-recorded videos sourced from the channels' own collections of footage. All three channels eventually adopted the streamed queue, the automated nature of which allows for content creators to invest relatively little attention while retaining constant engagement with their audience and reusing legacy footage in new ways. While content creators note that streamed queues do not generate much income compared to livestreamed performances or popular static videos (Gilberto Salvador Perez Baeza, pers. comm, 25 August 2021; Gabino Vera, pers. comm., 16 May 2020), the income they do generate is passive. For viewers, the appeal of this format was the chat feature running alongside the streamed queues, which created a novel virtual space for social engagement and cultural participation during a time of mandatory self-isolation. Indeed, while the streamed queue format predated the pandemic, the conditions of quarantine contributed to its growth as an essential component of the SonTube ecosystem. All three channels began to run streamed queues nightly, with growing audiences and lively chats of regular visitors.

As living under pandemic conditions became the new normal throughout 2020, the titles of streamed queues began to reflect the shared experience of participants. Streamed queues from QuerrequeFilms began to take on names such as *huapanguitos para una noche de lunes aburrido y en cuarentena* (huapanguitos for a boring Monday night during quarantine, 18 May 2020),[4] and *Huapanguitos pa seguir echando flojera en esta cuarentena* (Huapanguitos to stay lazy to in this quarantine, 6 August 2020).[5] Streamed queues on the Son Michoacán channel made regular note of the pandemic, featuring phrases such as *en año covid-19* and *en tiempos de covid* in their titles, even while presenting music not topically connected to coronavirus. In February 2021, the channel featured a stream entitled *Toritos de Petate Morelia Carnaval 2021 vs Covid nos vemos en el 2022 ánimo Morelia* (the Toritos de Petate Carnival of Morelia v. Covid, see you in Morelia in 2022, cheer up!),[6] a stream of festival performances which included the regular appearance of cartoon coronaviruses superimposed on the corners of the screen.

The use of streamed queues as opposed to uploaded videos serves multiple utilitarian and communitarian functions. Videos broadcast on YouTube as livestreams, whether performances or queues, can be pre-advertised as premieres. They are also prioritised in personalised feeds by YouTube's structural algorithms, which in turn make streams more visible to viewers and ensures that they are highlighted in push notifications for channel subscribers. This potentially could drive larger audiences than even the 'semantic core' cluster networks of YouTube algorithms (Airoldi, Beraldo, and Gandini 2016, 6). Streams also maintained a sense of momentum for viewers. This was facilitated by a built-in timestamp of when broadcasting began, and a ticker showing the changing number of present viewers.

The monotony of self-isolation and the erosion of social routines during the early stages of the pandemic changed, and possibly warped, the way that individuals experienced time. Whether a stream was 74 minutes or five hours, viewers consistently exhibited a willingness to actively engage, transforming what would have otherwise been time wasted into time well spent. Scholars in a variety of fields have recently begun to explore this concept of the fluid temporal dimensions of life during the pandemic. 'Pandemic time' altered the experience of time's passage. As Patricia Arés-Muzio wrote,

> The coronavirus, with its peculiar capacity to disrupt all points of reference, makes time appear to go by swiftly and slowly at once. We live through a distorted sort of time that seems slow if taken day-by-day…paused, stalled. But if we look back, then we see time has vanished as if in an instant.
>
> (2021, 80)

The experience of being involved in a four- or eight-hour livestream on a phone or computer screen, where stationary or while doing some other task, indeed produced this feeling of simultaneous but suspended experience. 'With space disappearing as a coordination device for many people' as a result of the pandemic-era shutdowns and social isolation, 'time is often the most salient coordination mechanism that structures days and weeks' (Kunisch, Blagoev, and Bartunek 2021, 1411). Researchers have therefore suggested focusing on the ways that 'the situated enactment of time enables scholars to recognize the dynamics that underpin the emergence, change, continuity, and persistence of socio-temporal orders' (ibid., 1413).

The streamed queues of SonTube provided a specifically sonic and social variant of this situated enactment of time, as elisions of temporality in virtual spaces. Music continues to play without interruption, seemingly without end or beginning. Recording contexts ranged among performances indoors, outdoors in plazas and raucous backyard parties, in small jams, and on huge stages. Shuffled videos recorded during both day and night, in varied settings, and featuring musicians at distinct stages of their performing careers further disrupt the viewer's sense of temporal order. Queued videos from crowded festivals inspire nostalgia for pre-COVID gatherings. Some clips feature musicians who are no longer living, promoting remembrance and providing a reminder of the ever-presence of death and the looming possibility of loss.

In GavBroadcast's streamed queues, a static frame for a few seconds featuring the channel's name and its social media handles is inserted between videos, whereas Son Michoacán's streamed queues feature no transitions between videos, producing a wholly uninterrupted flow of audiovisual content. While both GavBroadcast and Son Michoacán feature streamed queues of pre-recorded performance videos, QuerrequeFilms has pioneered an approach to streaming in which queued musical recordings are overlaid with ethnographic footage featuring syncretic religious rituals, the elaborate preparation of festival foods such as *zacahuil* (an enormous tamal) and barbacoa, scenes of *tianguis* (outdoor farmer's markets), drone-captured aerial shots of towns, and clips of driving through the Huasteca region recorded on a dashboard-mounted camera. Viewers tuned into such streamed queues thus watch video clips and listen to musical recordings that were each produced during a distinct time and in a unique location.

At the same time, such viewers observed or participated in a real-time chat where themes often ranged far beyond the scope of the presented audiovisual content. As listeners arrived to the streams, numbers of views increased, and the chat grew and evolved, incorporating new voices and forming an essential and desirable aspect of the livestream or streamed queue viewing experience. Indeed, one important difference between streamed live performances and streamed queues was the nature of the concurrent chats. In streamed performances, the chat was frequently used to make requests of the band, to greet the performers, and for the band to thank the audiences and various individuals involved in production (Margolies and Strub 2021a). Livestreamed performances were bounded events, wherein participants and viewers arrived at an appointed time to engage in a specific staged presentation. In contrast, streamed queues were imbued with the essence of an unbounded space; the automation of the shuffled videos for hours on end generated a significantly less proscribed environment than that of a performance. The chats associated with SonTube streamed queues were casual and topically diverse. While comments regarding the performances remained common, other threads of discussion also emerged that highlight the importance of the space as a site not only of musical consumption but of social encounter. It was common for participants to introduce and excuse themselves when entering and exiting the chat, extend greetings to familiar names, and ask after regular participants who had not been seen in some time.

During a streamed queue from QuerrequeFilms on 14 August 2020 entitled *Huapanguitos en un domingo de cuarentena* (Huapanguitos on a quarantine Sunday, 14 August 2020),[7] Chat Participant 1 asked after a loved one in the chat. Chat Participant 1 directed the question at Chat Participant 2, an active chat participant and moderator, and referred to the sought individual not by name, but as *mi chulo* (my cutie), signalling the typically high level of intimacy and familiarity among the chat participants. Chat Participant 2 responded by noting that she had not seen him today, but that 'maybe he'll connect a little later.' Chat Participant 1 later mentions with concern to Chat Participant 2 that she hasn't seen her *chulo* since Thursday. As this conversation takes place, the QuerrequeFilms moderator greeted individual participants who recently entered the chat by name. While the audiovisual recordings in the QuerrequeFilms stream queue were all recorded in the field by Flores Delgado, Camarera Torres is most often the individual behind QuerrequeFilms' accounts, including in YouTube chats (Maria del Carmen Camarera Torres, pers. comm., 29 August 2021). Moderator 1, another active participant who acted as a moderator in GavBroadcast chats but not in those of QuerrequeFilms, greeted Chat Participant 1 and cheekily asked her to serve him some pulque, an alcoholic beverage made from fermented agave.[8]

Frivolous as they may seem, these interactions signalled a deep sense of embodiment on the part of participants articulated within the virtual spaces of the streamed queue and their chats. Kilten, Groten, and Slater propose a framework of 'sense of embodiment' that rests upon three features: 'sense of self-location,' 'sense of agency,' and 'sense of body-ownership' (2012, 375). In the streamed queue chat setting, participants exhibited both corporeal and virtual self-location by simultaneously announcing their physical location (i.e., 'greetings from Pachuca Hidalgo,' or the screen handle 'Edgar el de Torreón'), and reified their participation in the digital space, for example by engaging in play surrounding the sharing of virtual pulque. Agency and body-ownership were also expressed by the activity of moderators, whose screen names were highlighted in blue in the chats and accompanied by a wrench icon to distinguish them from the other participants. People in the chats utilised screen names that later discussions confirmed were either their real names, or anonymous, playful, or cryptic handles. Some people changed their chat names over the course of the pandemic; it is possible, but not confirmed, that people even appeared in chats under more than one name. Just as with spaces filled with physical bodies, presences and absences were actively felt and reinforced by participants and constituted a fundamental component of the chat's character. The disappearance of a single regular attendee could be cause for concern and gossip, and the return of a former participant who no longer frequented the chatroom served as pretext for celebration and community intrigue. The names of QuerrequeFilms' streamed queues began to reflect the central importance afforded to the chat feature, with titles such as *Huapanguitos para un martes chatero* (Huapanguitos for a chatty Tuesday, 26 January 2021),[9] and *Huapanguitos para desvelarse con los amigos* (Huapanguitos for staying up late with friends, 26 April 2021).[10]

The chats of the Son Michoacán streamed queues during the COVID-19 pandemic were unique in that they lacked moderators or rules, and in turn evolved to demonstrate a unique style of exchange and community standards. In several instances, Son Michoacán's administrator noted that messages were never deleted from the chat.[11] Much of the chat's content was not textual, but rather consisted of long strings of emojis. However, the lack of vibrant textual messaging should not be confused for a lack of semiotic content. Emoji use is a complex form of communication invoking multiple valences of gender, age, race, and region where the default has often been images rendered with white skin (Pohl, Domin, and Rohs 2017, 6). The scholarly literature on the uses and emotional resonance of emojis indicates levels of commitment as well reflections of emotional states (Butterworth et al. 2019; Herring and Dainas 2020; Prada 2018; Aldunate and González-Ibáñez 2017; Derks, Bos, and Grumbkow 2007). Some scholars have emphasised situating emojis in 'race critical code studies' in order to illuminate the coding of racial ideas into supposedly neutral technologies (Ruha 2019). Instead, it is clear that 'the values of technological objectivity and neutrality that are so deeply prized by computer scientists and other technical workers are also deeply instrumental in the perpetuation of discriminatory technological systems' in the development of emojis (Miltner 2021, 529). The racialised character of emojis, dissected at length by academics, was not ever commented upon in the chats observed for this project during the pandemic-era streams, leaving significant room for future study on the relationship between emoji use and ethnic self-representation in the context of participatory media platforms and their use in the global south.

The use of emojis was fairly well governed in GavBroadcast chats, with users encouraged to use only single emojis and cautioned when use becomes excessive by the established, idiosyncratic standards of the channel. According to the channel's founder and administrator, this limitation on emoji use produced a neater looking and more conversational chat alongside the videos, and, importantly, resulted in fewer streams being automatically marked as active sites of spam. Additionally, while Son Michoacán advertises that chat messages are never deleted, GavBroadcast and QuerrequeFilms support robust groups of volunteer moderators who sometimes delete messages that are evaluated as disrespectful, hateful, or potential spam (Gabino Vera, pers. comm., 19 June 2020).

There are no such limits placed on the Son Michoacán chats, which sometimes appear to consist almost entirely of emojis. Administrator Perez Baeza notes that his lax approach toward the chat stemmed from his personal distaste for the chat feature, noting that 'many people who follow the channel love the chat, keep coming for the chat, but I do not always feel comfortable speaking as myself in there, around people I don't know' (pers.comm.). Some of the most frequently used emojis are 'woman dancing' (💃), 'guitar' (🎸), 'violin' (🎻), 'two hearts' (💕), 'musical notes' (🎶), 'cowboy hat face' (🤠), and 'clinking beer mugs' (🍻). The 'guitar' emoji, which is modelled on the Fender Stratocaster electric guitar, presumably stood in for a variety of Mexican stringed instruments commonly seen on SonTube streams such as *jaranas*, *guitarras del golpe*, and *vihuelas*. The 'woman dancing' emoji, perhaps the most common emoji in the chats, appears in a red dress, high stepping to the right, her flowing dress lifting to reveal some of her leg. Often the individual responses in these chats were a string of six or more emojis, as a signal of pure enthusiasm. Self-identified male viewers tended to post male emojis, and self-identified women used emojis coded as female, from the dancer to kissing lips, hearts, and other symbols. Also quite common are textual 'shouts' indicating happiness, laughter, and expressions of joy, such as 'ayayyyyy *mi canción ajuajuaaaaaaaa.*'[12]

Taken as a whole, these chats united a community of listeners and enthusiasts seeking space to connect with the music and also, importantly, with each other during the pandemic time. Participants in streamed queue chats flirted, joked, and encouraged each other. After a three-and-a-half hour streamed queue from Son Michoacán on 30 November 2020, Chat Participant 3 said goodbye to her friend Chat Participant 4. It was after midnight, so Chat Participant 3 signed off with 'Good morning, my friend, Roberta, rest, blessings, sweet dreams, have a nice awakening, thank you also for dancing with me.'[13]

In a 30 August 2020 Son Michoacán stream entitled *Música de Tzitzio 30 de agosto año COVID 19* (Music of Tzitio 30 August in the COVID-19 Year), Chat Participant 5, who was a regular in the channel's chats throughout the pandemic, welcomed everyone to end their Sunday with good music (*Buenas noches. Vamos a terminar bien el domingo con buena música* 😊😊). An active user in the chat using the anonymous handle 'im_sure_its_true' identified himself as living in Austin, Texas and a former resident of Huetamo, Michoacán. 'And I miss that beautiful land,' he wrote. 'One day I'll marry myself to that music hehehe.' He asked to hear a song called 'La Tortolita,' but was told by chat regular Chat Participant 5 that all the videos were pre-recorded because of the circumstances of quarantine. Im_sure_its_true continued to dominate the chat with questions and comments. He asked for forgiveness for his many comments, citing his excitement as a new subscriber and participant. Son Michoacán was not concerned. 'It's fine, that's why we're here,' he wrote.[14] During a stream on 13 September 2020, the chat continued between the same people in this familiar and welcoming form. 'In spite of hearing La Tortolita a thousand times, I still love it,' (*Aunque oiga mil veces La Tortolita me sigue encantando* 😊) enthused Chat Participant 5. Arriving late and not wanting to miss the interaction, im_sure_its_true asked 'where is everyone?'[15]

The understanding that the chats would feature friends and regulars was a common aspect of the experience. 'Everyone' referred to by im_sure_its_true were the people who showed up night after night to share space together to listen and have a manner of conversation and fellowship. This welcoming approach to participants, new and returning, was even extended to the authors of this chapter, who logged hundreds of hours in the queue streams on all of the SonTube channels throughout the pandemic, listening and engaging with the chats and their participants over the course of 18 months. The constant stream of emojis, greetings, and light conversation indicated that the people were indeed listening and enjoying, providing a form of social affirmation and audiovisual stimulation during a time of mass isolation. Son Michoacán's New Year's Eve stream on 31 December 2020 featured the usual array of music making, as well as a video of the preparation of *carne de res*. The chat was almost entirely a string of emojis, signalling a generalised sense of excitement. Chat Participant 6 posted a singular stream of eighteen emojis featuring boots (for dancing), dancing men, music notes, and violins. Chat Participant 7 fired off dozens of separate messages into the chat, each with four to six emojis of dancing women, music notes, and instruments.[16] Filled with emojis and written cries of joy, this colourful chat matched the exuberance of this moment: the end of the first year of the pandemic, although not the end of the COVID era, nor the musicultural responses to it.

Notes

1 www.youtube.com/user/GaVbroadcast.
2 www.youtube.com/user/moldovianful.
3 www.youtube.com/channel/UC_58Ux10plXm8RAPa_GIzJA.
4 www.youtube.com/watch?v=XAMfxvXd7OY.
5 www.youtube.com/watch?v=ZHZqEVlcZqY.
6 www.youtube.com/watch?v=SQuI2x-TnpI.
7 www.youtube.com/watch?v=v-73LJqZBqU.
8 www.youtube.com/watch?v=v-73LJqZBqU.
9 www.youtube.com/watch?v=engL9S-ZBH4.
10 www.youtube.com/watch?v=8WE4J1J9UhY.
11 www.youtube.com/watch?v=Zw2srEnCofo.
12 www.youtube.com/watch?v=t1OR69Xn9Qg.
13 www.youtube.com/watch?v=NhtXS67eQTs.
14 www.youtube.com/watch?v=9Cd0pLTQY5o.
15 www.youtube.com/watch?v=Zw2srEnCofo.
16 www.youtube.com/watch?v=_MHfFfcmrEU.

References

Airoldi, Massimo, Davide Beraldo, and Alessandro Gandini. 2016. 'Follow the Algorithm: An Exploratory Investigation of Music on YouTube.' *Poetics* 57: 1–13. https://doi.org/10.1016/j.poetic.2016.05.001.
Aldunate, Nerea and Roberto González-Ibáñez. 2017. 'An Integrated Review of Emoticons in Computer-Mediated Communication.' *Frontiers in Psychology* 7. https://doi.org/10.3389/fpsyg.2016.02061.
Arés-Muzio, Patricia. 2021. 'The Mysteries of "Pandemic Time."' *MEDICC Review* 23 (2): 80. https://doi.org/10.37757/MR2021.V23.N2.6.
Baker, Sarah and Jez Collins. 2017. 'Popular Music Heritage, Community Archives, and the Challenge of Sustainability.' *International Journal of Cultural Studies* 20 (5): 476–491. https://doi.org/10.1177/1367877916637150.
Berry, David M. 2011. *The Philosophy of Software: Code and Mediation in the Digital Age.* London: Palgrave Macmillan.
Bishop, Sophie. 2019. 'Managing Visibility on YouTube through Algorithmic Gossip.' *New Media & Society* 21 (11–12): 2589–2606. https://doi.org/10.1177/1461444819854731.
Burgess, Jean and Joshua Green. 2018. *YouTube: Online Video and Participatory Culture*, second edition. Cambridge: Polity.
Butterworth, Sarah E., Traci A. Giuliano, Justin White, Lizette Cantu, and Kyle C. Fraser. 2019. 'Sender Gender Influences Emoji Interpretation in Text Messages.' *Frontiers in Psychology* 10. https://doi.org/10.3389/fpsyg.2019.00784.
Derks, Daantje, Arjan E.R. Bos, and Jasper von Grumbkow. 2007. 'Emoticons and Social Interaction on the Internet: The Importance of Social Context.' *Computers in Human Behavior* 23 (1): 842–849. https://doi.org/10.1016/j.chb.2004.11.013.
González Paraíso, Raquel. 2014. 'Re-Contextualizing Traditions: The Performance of Identity in Festivals of Huasteco, Jarocho, and Terracalenteño Sones in Mexico.' Ph.D. diss., University of Wisconsin-Madison: Wisconsin. ProQuest (ATT 3626003).

Herring, Susan and Ashley Dainas. 2020. 'Gender and Age Influences on Interpretation of Emoji Functions.' *ACM Transactions on Social Computing* 3 (2) (June): 1–26. https://doi.org/10.1145/3375629.

Hesmondhalgh David, Ellis Jones, and Andreas Rauh. 2019. 'SoundCloud and Bandcamp as Alternative Music Platforms.' *Social Media + Society* 5 (4) (October–December). https://doi.org/10.1177/2056305119883429.

Kaiser, Jonas and Adrian Rauchfleisch. 2020. 'Birds of a Feather Get Recommended Together: Algorithmic Homophily in YouTube's Channel Recommendations in the United States and Germany.' *Social Media + Society* 6 (4) (October). https://doi.org/10.1177/2056305120969914.

Kilten, Konstantina, Raphaela Groten, and Mel Slater. 2012. 'The Sense of Embodiment in Virtual Reality.' *PRESENCE: Teleoperators & Virtual Environments* 21 (4) (August): 373–387. https://doi.org/10.1162/PRES_a_00124.

Kunisch, Sven, Blagoy Blagoev, and Jean M. Bartunek. 2021. 'Complex Times, Complex Time: The Pandemic, Time-Based Theorizing and Temporal Research in Management and Organization Studies.' *Journal of Management Studies* 58 (5) (July): 1411–1415. https://doi.org/10.1111/joms.12703.

Margolies, Daniel and J.A. Strub. 2021a. 'Music Community, Improvisation, and Social Technologies in COVID-Era Música Huasteca.' *Frontiers in Psychology* 12. https://doi.org/10.3389/fpsyg.2021.648010.

Margolies, Daniel and J.A. Strub. 2021b. '#QuédateEnCasa y Huapango! Diasporic Community and Musical Wellbeing in Streamed Live Performances of Son Huasteco Music.' *Journal of Music, Health, and Wellbeing* (Autumn). http://dx.doi.org/10.26153/tsw/35052.

Martens, Cheryl, Etsa Franklin Savio Sharupi Tapuy, and Cristina Venegas. 2020. 'Transforming Digital Media and Technology in Latin America.' In *Digital Activism, Community Media, and Sustainable Communication in Latin America*, edited by Cheryl Martens, Cristina Venegas, and Etsa Franklin Salvio Sharupi Tapuy, 1–24. Cham: Palgrave Macmillan.

McDowell, John Holmes. 2015. '"Surfing the Tube" for Latin American Song: The Blessings (and Curses) of YouTube.' *Journal of American Folklore* 128 (509): 260–272.

Miltner, Kate M. 2021. '"One Part Politics, One Part Technology, One Part History": Racial Representation in the Unicode 7.0 Emoji Set.' *New Media and Society* 23 (3): 515–534. https://doi.org/10.1177/1461444819899623.

Mitchell, Gail. 2011. 'Gerardo Ortiz Was Unknown in Regional Mexican Music until He Was Discovered on YouTube Several Years Ago.' *Billboard* 123 (38): 58.

Pietrobruno, Sheenagh. 2013. 'YouTube and the Social Archiving of Intangible Heritage.' *New Media and Society* 15 (8): 1259–1276. https://doi.org/10.1177/1461444812469598.

Pietrobruno, Sheenagh. 2018. 'YouTube Flow and the Transmission of Heritage: The Interplay of Users, Content, and Algorithms.' *Convergence* 24 (6): 523–537. https://doi.org/10.1177/1354856516680339.

Pinos Calderón, Doris Elena, and Cristina Venegas. 2020. 'Sounds of the Neighborhood: Innovation, Hybrid Urban Space, and Sound Trajectories.' In *Digital Activism, Community Media, and Sustainable Communication in Latin America*, edited by Cheryl Martens, Cristina Venegas, and Etsa Franklin Salvio Sharupi Tapuy, 53–79. Cham: Palgrave Macmillan.

Pohl, Henning, Christian Domin, and Michael Rohs. 2017. 'Beyond Just Text: Semantic Emoji Similarity Modeling to Support Expressive Communication.' *ACM Transactions on Computer-Human Interaction* 24 (1): 1–42. https://doi.org/10.1145/3039685.

Prada, Marília, David L. Rodrigues, Margarida V. Garrido, Diniz Lopes, Bernardo Cavalheiro, and Rui Gaspar. 2018. 'Motives, Frequency and Attitudes toward Emoji and Emoticon Use.' *Telematics and Informatics* 35 (7): 1925–1934. https://doi.org/10.1016/j.tele.2018.06.005.

Rieder, Bernhard, Ariadna Matamoros-Fernández, and Òscar Coromina. 2018. 'From Ranking Algorithms to "Ranking Cultures": Investigating the Modulation of Visibility in YouTube Search Results.' *Convergence* 24 (1): 50–68. https://doi.org/10.1177/1354856517736982.

Ruha, Benjamin. 2019. *Race After Technology: Abolitionist Tools for the New Jim Code*. Hoboken: John Wiley.

Sturman, Janet. 2015. *The Course of Mexican Music*, second edition. New York: Routledge.

Tepper, Steven J. and Eszter Hargittai. 2009. 'Pathways to Music Exploration in a Digital Age.' *Poetics* 37 (3): 227–249. https://doi.org/10.1016/j.poetic.2009.03.003.

Valcarce, David Parra and Charo Onieva Mallero. 2020. 'El uso del podcast para la difusión del patrimonio cultural en el entorno hispanoparlante: análisis de las plataformas iVoox y SoundCloud.' *Naveg@mérica* 24: 1–31. https://doi.org/10.6018/nav.416541.

Waldron, Janice. 2013. 'YouTube, Fanvids, Forums, Vlogs and Blogs: Informal Music Learning in a Convergent On- and Offline Music Community.' *International Journal of Music Education* 31 (1) (February): 91–105. https://doi.org/10.1177/0255761411434861.

9 'WHY DO THEY DANCE IN THE MIDDLE OF THE PANDEMIC?'

Post-Pandemic Cumbia, Mediated Live Music, and Digital Heritage from Mexico City

Michaël Spanu

Like most of the world's megapolises, Mexico City used to have a vibrant music industry before the coronavirus crisis. It had numerous live music scenes, from reggaeton to metal, indie rock to hip hop, neo-folk to *sonideros*, to name a few. In many parts of the Global North, these music scenes have shaped an image of so-called creative cities, serving urban (re)development, commercial and touristic ends (Picaud 2021; Shaw 2010). However, in Mexico City, although cultural experts have mapped and commented on some of the vibrancy of local music, academic research points to the difficulties that many venues and practitioners face because of the lack of official recognition (e.g., Román García 2016). Such a divide played a crucial role during the pandemic, as most countries from the Global North deployed considerable efforts to save and support the creative sectors that were particularly hard hit by the pandemic, whereas Mexican artists and cultural practitioners received almost no help: the lack of industry advocates and the polarised relationships with the political field did not allow for any structural support.

The pandemic revealed and emphasised the pre-existing polarisation of the Mexican music sector. Indeed, Mexican music has good media coverage, even on the major American media such as NPR and KEXP, but clearly lacks public and structural support (Kanai and Ortega-Alcázar 2009) and, in the case of live music, adapted legal frameworks. Informality, fragmentation, and, in some cases, drug trafficking-related insecurity remain essential factors within Mexico City's local music scenes, while most formal live music activities are concentrated in mainstream entertainment companies, especially OCESA (the third largest live music promoter worldwide), leaving little room for the independent sector (Mercado-Celis 2017; Spanu 2022a). On a more global scale, the pandemic stressed and questioned the mediated nature of live music, prompting the development of new modalities such as live streaming concerts, intimate live video performances from home, and Zoom parties. Platforms such as YouTube, Twitch, and Facebook worked as digital venues. However, the possibilities they offered quickly fell short, as no sustainable business model emerged from the pandemic experiment, nor were people engaged in live music virtual content satisfied with it as a replacement or proxy for in-person events. In a way, such a failure questions the roles and values of digital platforms related to music scenes and thus requires more detailed investigation. This chapter explores these questions through the example of *cumbia sonidera* in Mexico, a practice that challenges familiar notions of live music while simultaneously showcasing a particular notion of heritage on digital platforms.

An Introduction to Pre-Pandemic Mexican *Sonideros*

Like Jamaican or Colombian sound systems, Mexican *sonideros* refer to a culture rooted in the live performance of recorded music, especially cumbia. The *sonidero* is the person at the

DOI: 10.4324/9781003200369-12

Post-Pandemic Cumbia, Mediated Live Music, and Digital Heritage   111

centre of this culture, playing the role of both DJ and MC. *Sonideros* are characterised by the quality and size of their sound and lighting system – as portrayed Mirjam Wirz (2013) in the photobook *Sonidero City* – their capacity to curate music, to create a festive atmosphere, and to vocalise messages that people from the audience dedicate to their loved ones (Ragland 2012; Blanco Arboleda 2012). Some *sonideros* can be considered as crate digging pioneers, as they used to travel to Caribbean countries in search of rare vinyl, while others compose their music nowadays. Their intervention on the sound of the records is also notable: they usually play them at slower speeds, creating a unique and deep soundscape known as *cumbia rebajada*. Traditionally, *sonideros* are men, even though more and more women are participating nowadays. A recent form of gentrification of its practitioners and audiences is also at play, but massive in-person *sonidero* events still mostly involve nomad gatherings in poor neighbourhoods, generally in informal spaces such as parking lots or large streets or, more rarely, in brick-and-mortar venues.

The nomad component of *sonidero* corresponds to its informal economy and the necessity to find cheap places to keep the entrance accessible. Playing songs at these events is not related to a formal system of copyright, allowing for more structural flexibility and certain forms of creativity, as also seen in other contexts (Power and Hallencreutz 2002). However, the increasing transfer of *sonidero* culture to digital platforms challenges this aspect, as we shall see. Local scholars have defined *sonidero* as a socio-musical culture given the centrality of music collecting, listening, and dancing, on the one hand; and because of the increasing reference to *sonidero* as an autonomous field and music genre, on the other hand (Ramirez Paredes 2021). The recent movie *I'm no longer here* (*Ya no estoy aquí*) (Canclini, Urteaga, and Cruces 2012) portrays how a fragment of the underprivileged youth from Monterrey (in the north of Mexico) lives almost entirely through *sonidero* music, especially *cumbia rebajada*, very similarly to other youth cultures such as goths, emo and punks in Mexico and the rest of the world. In this sense, *sonidero* music constitutes a powerful identification artefact, a practice between citizenship and consumerism and, as suggested in the movie, an escape from local criminal activities.

Sonidero events engage many social, aesthetic, and choreographic relationships: musical nostalgia, re-recording and other creative sound exploration, couple dancing, sending messages to the Mexican diaspora, and the construction of Latin American identity. In concrete terms, it differs from more mainstream forms of popular music, as the live performance model is not rooted in the physical encounter between the original performing artist and the audience but rather in gathering people around the *sonidero*. Dancers at these events create new steps and styles, for instance small jumps, breaking the rules of more traditional styles related to *salsa* and *cumbia*. In this sense, *sonidero* culture can be compared to many electronic music practices, from house to dancehall, where the audience's participation in the form of innovative performance is crucial (Rietveld 2013).

Sonidero culture, as we define it, emerged in the 1960s in Mexico City's poor neighbourhoods before expanding to other cities such as Puebla, Monterrey, and Tijuana, and also the United States, through the working-class diaspora. Before the pandemic, one of its pioneers, *Sonidero* La Changa, had been scheduled at the massively popular music festival Primavera Sound in Spain, showing the increasing recognition of and interest in *sonidero*. Such a process goes hand in hand with the *cumbia* revival in Mexico, with artists like Los Angeles Azules and Celso Piña, and more indirectly with the worldwide ascent of Latino music over the last decades (Bénistant 2018). *Cumbia sonidera* not only represents a contact zone between Latin American cultures, as most of the *sonidero* music comes from other countries (especially Colombia, Peru, and Cuba), but also between social classes. Based on his investigation in Monterrey, Jesús A. Ramos-Kittrell frames *sonidero* culture as follows:

112 *Michaël Spanu*

> On the one hand, cumbia audiences and musicians ... sought to find ... symbols to make sense out of ... their invisibility to the state, and their marginalization as undesired subjects that did not conform to the ideal image of the middle-class, bourgeois citizen ... The working class made this music a site for self-recognition. On the other hand, the activity of middle-class youths who did not find in the city's corporate materialistic culture expressive spaces that reflected their new socioeconomic cultural experience after NAFTA triggered concerns for cultural origins. ... Far from democratic, dialogue between these two frames of cultural action is necessarily uneven, for such asymmetry makes ideas about roots and urban 'base culture' viable.
>
> (2019, 187)

These observations easily work for other Mexican cities and soundly synthesise an essential part of the complex *cumbia sonidera* cultural assemblage. However, they do not address the impact of new mediations enabled by digital platforms. In this chapter, I am interested in these mediations and particularly their role during the pandemic. I should also emphasise that despite this contact zone between social classes, the historical and current links between *sonidero* events and underprivileged populations often lead to classist and racist reactions from the cultural elite, or even persecution by the Mexican authorities. Its status remains ambivalent today, as no specific cultural institution has structurally recognised and supported this cultural practice. This lack of institutionalisation also explains the absence of financial support to *sonideros* during the pandemic.

Because of the lack of institutionalisation, many pre-pandemic commentators saw *sonidero* culture mainly as a form of resistance in a context of social segregation and lack of access to public space and culture (Lippman 2018; Blanco Arboleda 2012). Scholars classified it as fundamentally translocal, along with other Mexican musical traditions (Rinaudo 2019). Josh Kun identifies it as an aesthetic of 'going beyond' (2015, 539), not only national frontiers but also oneself. He notes, for instance, that *sonidero* aesthetics are

> a way of using the arts – specifically, here recorded sound and audio technology – to engage with spatial politics in the age of asymmetrical economic globalization, naturalized systems of deportation and border militarization, and intensive global migration and displacement.
>
> (ibid.)

Indeed, people commonly send messages to their relatives in the United States through the *sonidero*'s voice during events in Mexico, while US-based Mexicans sometimes financially contribute to the organisation of events in Mexico (Ragland 2012). In a similar vein, Alexandra Lippman (2018) sees it as a practice that conveys emotion, relations, and memory and produces co-presence across increasingly militarised borders. More generally, if some of the conventions and meanings of *sonidero* culture are already well described, their digital mediations remain to be assessed, not only because the streaming platforms constitute a crucial infrastructure for music scenes today (Magaudda 2020), but also because the pandemic made this phenomenon even more apparent.

The Pandemic and the Social Stigma

As a culture primarily rooted in live and in-person gatherings, *sonidero* was hit particularly hard by the pandemic. Most events were cancelled. Music events were not considered essential, and no specific protocol was designed to imagine how people could dance with limited risks, even in outdoor spaces where *sonidero* events usually happen. As a global

Post-Pandemic Cumbia, Mediated Live Music, and Digital Heritage   113

phenomenon, the pandemic reinforced, and in some cases created, a rigid hierarchy of cultural practices. For instance, massive gatherings of people were allowed at pop music concerts in some countries, whereas dancing in a nightclub was still forbidden. Mainstream live music is generally more recognised as a cultural artefact than DJing and other music based gatherings that rely on dancing practices (Picaud 2021). In Mexico City, through the REABRE program designed by the municipality, music venues were allowed to reopen as restaurants and people had to eat to hear live music. As a cheap and attractive food, the burger temporarily replaced the concert ticket, especially for pop and rock venues. However, such measures were very much unsuited to *sonidero* culture.

I should mention that almost no support was given to the population to face the pandemic in Mexico, which explains why the sanitary measures were less restrictive than in most countries. There was tolerance towards effectively illegal events, especially in private places, which was unfair to both formal businesses, and *sonideros* that needed access to public spaces. Mexican authorities have repeatedly taken advantage of crises such as earthquakes to reinforce their vision of legitimate cultural practices, especially in public spaces, and the cultural agenda of the Mexican elite has long been conservative, focused on heteronormative, familial, bourgeois, and nationalist conceptions (Macías 2021). In a way, the pandemic has reinforced its capacity to limit the proliferation of *sonidero* culture. However, because *sonideros* never received financial help, holding illegal events became an issue of survival issue at a certain point, which triggered controversy around the role that *sonideros* and poor neighbourhoods were playing in the collective battle against the circulation of coronavirus. Even within the *sonidero* community, the topic was addressed through YouTube videos such as Richard TV's 'WHY DO THEY HAVE A DANCE IN THE MIDDLE OF THE PANDEMIC? THIS IS WHAT HE SAID' (RICHARD TV 2020c).[1]

Sonidero culture gained a particular momentum during the pandemic when some of them performed on their rooftops, attracting a great deal of media attention.[2] Both general and *sonidero*-oriented media coverage of the performance featured neighbours dancing and enjoying the music, showing a form of resilience to the dramatic situation that the country was experiencing as one of the worst in the world for deaths, after the United States and Brazil. The shape of this musical resilience not only conveyed a more positive image of *sonidero* culture than the regular reports of violence, but it also publicly reaffirmed the immense popularity of the genre in Mexico City. From the perspective of a media narrative, *sonidero* worked as one of the capital's sonic identities that needed to be claimed in the face of the crisis. Paradoxically, *sonidero* events kept spreading beyond Mexico City during the pandemic, especially in rural regions where the sanitary measures were less strict. Other cities were struggling to attract famous *sonideros* and to promote their own events before the pandemic, a situation that changed dramatically and triggered a sort of battle between regions, especially regarding who had the best dancers and atmosphere.

The only support to *sonidero* culture during the pandemic came from commercial brands. La Changa and other *sonideros* were scheduled in heavily branded livestreams and in-person events related to the Mexican beer empire Victoria.[3] These events were designed to have a strong impact on social media, especially Instagram. In this sense, the pandemic had multiple effects and marked a new wave of commercial interests in *sonideros* through a marketed regime of visibility that mainly benefited the most prominent ones. Besides, as in many other music cultures, some *sonideros* started to broadcast performances from home on digital platforms such as Facebook and YouTube. One interesting aspect of the *sonidero* streaming movement is that most of the important digital channels are not from *sonideros*, but from digital broadcasters or YouTubers specialised in *sonidero* music, such as Richard TV (848,000 followers, active since 2008; see Figure 9.2) and Sonidero Latino TV (439,000 followers, active since 2016). There are countless channels like these on YouTube, some

114 *Michaël Spanu*

Figure 9.1 Richard TV's YouTube video 'FROM THE ROOFTOP HE ORGANIZES HIS DANCES AND THIS IS HOW HIS NEIGHBORS RESPONDED' (RICHARD TV 2020b).

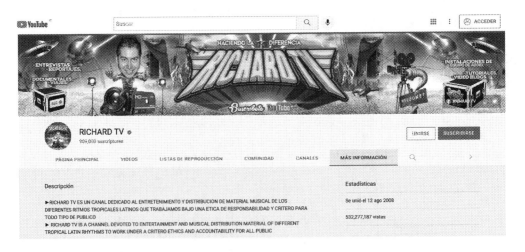

Figure 9.2 Richard TV's official YouTube page.

more professional than others. The most important ones are based in Mexico City, where the *sonidero* scene is massive. New amateur channels started during the pandemic, whereas others, for instance Sonideros Locales SD in San Diego, relaunched their activity to support their local *sonidero* scene (Freddy Ortiz, pers. comm., 8 October 2021).[4]

Post-Pandemic Cumbia, Mediated Live Music, and Digital Heritage   115

Clicking on a *sonidero* video on YouTube means entering into a whole new universe entirely dedicated to *sonidero* culture, with specific audiovisual conventions, including the characteristic use of capital letters in the videos' titles. The audience for these videos was already significant before the pandemic. Many of them reached several million views,[5] but they experienced a tremendous boom during the pandemic (Ricardo Mendez, pers. comm., 6 October 2021).[6] In other words, the pandemic reinforced a process that was already in place, challenging the very definition of *sonidero* culture as a primarily in-person practice. In this chapter, I want to take the pandemic as a starting point to understand better how the *sonidero* culture's digital dimension works, how it relates to the *sonidero* culture as a whole, and what kind of values it fosters.

Live Music, *Sonideros*, and the Media Economy

Live music events are often portrayed as a cultural activity that contrasts with the excessive mediatisation of modern urban society and so-called functional daytime activities such as work (Bottà and Stahl 2019). Most music events, especially *sonidero* events, represent a separate sphere from work, at least for the audience. Despite their live, in-person, and collective dimensions, they are not an unmediated cultural activity, as they rely on record players, DJ devices, and huge speakers. Though more informally than the mainstream live music sector (Holt, 2010, 2020), *sonidero* culture is located at the crossroads of the music industry (records), the media industry (social media and streaming platforms), and the entertainment industry (cultural and local promoters). However, *sonidero* culture did more than emerge from the global circulation of records and media broadcasting, it favoured the global expansion of *cumbia* through more mainstream forms of live music such as *cumbia* concerts in brick-and-mortar venues.

Fans started to record *sonidero* events systematically when a new generation of cell phones equipped with audiovisual recording was commercialised at the end of the 2000s. The platform MySpace represented the main channel of diffusion, and was later replaced by Facebook and YouTube. Ricardo from Richard TV was one of these fans who quickly became interested in producing higher quality videos. Both passion and economic necessity drove his interest. On the one hand, his uncle was a *sonidero* himself. Ricardo had an emotional bond with sonidero music from early in his life and was eager to share his passion on a global scale via the internet. On the other hand, Ricardo was unemployed and managed to get paid a small fee to make videos for the older generation of *sonideros*. Informal at the beginning, the relationship between *sonideros* and video broadcasters became increasingly organic, and even crucial to the new *sonidero*'s online presence, very similar to the *videoconciertos* from Colombian sound systems (*pícos*) (Paulhiac 2014). These videos increased the *sonideros*' visibility and were used to promote events, securing the audience's presence. In some cases, video producers diversified their activity, covering other types of events such as weddings and birthdays (especially *quinceañeras*).

How can we understand these videos? The concept of liveness embodies the idea of a media continuum between in-person activities and cultural artefacts such as recorded music (Auslander 2008). Although it tends to diminish the relevance of live music institutions and instead focus on media institutions (Holt 2020), liveness helps to understand the great variety of music performances and their entanglement with media practices, especially in the digital age. Even before the pandemic, the reception and consumption of live music events were highly mediated by smartphones and social media, paradoxically reinforcing the value of being physically present (Bennett 2012). This phenomenon is evident in the case of *sonideros*: there have never been so many events now that they are being broadcast online and promoted on social media. In a way, *sonidero* videos work as a promotional device

116 *Michaël Spanu*

for in-person events, as in most music cultures. However, despite the massive presence of live performance videos (including live streaming) produced and broadcast online by both amateurs and professionals,[7] live music videos hardly constitute a proper field, especially in economic terms. There have been historical difficulties in monetising visual content related to popular music (e.g. music videos, see Buxton 2018), and the *sonideros* are no exception.[8] This difficulty became all the more clear during the pandemic, as the live streaming boom did not enable any substantial income for artists. If they are not just commercial assets, then we need to look deeper into the other values related to live music videos on digital platforms (Spanu 2022b; Spanu and Rudent, forthcoming).

In aesthetic terms, not only do most *sonidero* video producers film the event in real-time and from different angles, but they also try to capture and recreate the specific liveness of the event, especially from the perspective of the audience: people dancing and other people watching the dancers, while the *sonidero* plays music and sends messages. For instance, Richard TV and Sonidero Latino TV's videos generally focus on the dancers, placing them in the centre of the frame while the rest of the audience remains on the margin. The *sonidero* is also a central figure of the videos, the only identifiable one (visually and in the title or description of the video), but more for his omnipresence in sending messages than as a visual performer. He is generally at the centre of the frame, surrounded by people showing their messages.

Figure 9.3 Sonidero Latino TV's YouTube video 'THIS IS HOW WE DANCE CUMBIA IN MEXICO CITY'S NEIGHBORHOODS – ((SAMPUESANA SIBONERA)) SIBONEY – LA MERCED.' Siboney is the name of the *sonidero*, 'Sampuesana Sibonera' is the name of the song, and La Merced the name of the neighbourhood (Sonidero Latino TV 2019a).

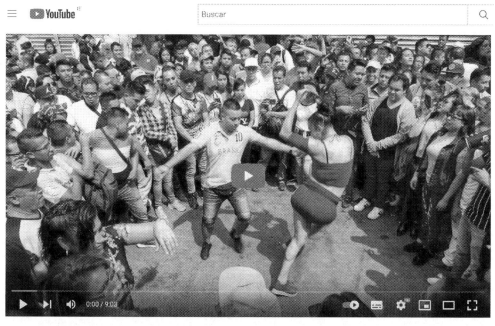

Figure 9.4 Richard TV's YouTube video '(((THE LAST SAMPUESANA))) FROM SONIDO CONDOR IN LA MERCE[D] 62th ANNIVERSARY EQUIPO RETRO' (RICHARD TV 2019).

In a way, these videos reproduce and expand the audiovisual nature of the in-person events, where people go to see other people dancing, be seen dancing, send messages through the *sonidero*, and listen to other people sending messages. They do not create a specific setting for the musical performance, as in other popular video music channels such as Boiler Room, where the audience performs in front of the camera and triggers a specific gaze (Heuguet 2016). Different observers told me that people have always gathered around dancers, creating a sort of micro-spectacle within the event, long before the systematic presence of smartphone cameras. The same happens with the *sonidero*'s central figure: his name is everywhere in capital letters, from the in-person event to the video. In other words, the media economy that penetrates *sonidero* culture reinforces the 'spectacularisation' within the audience rather than creating it (Djebbari 2018; Aterianus-Owanga 2020).

In this sense, this audiovisual format functions mainly as a sort of testimony, a document similar to certain concert films and screens that convey a form of mediated live authenticity (Spanu 2022b; Sarno 2018; Bratus 2016), rather than a simple advertising tool or a music streaming service. It gives a strong sense of the audience's experience, but also rearticulates the controversy around *sonideros*' relationship to music, as their messages can be interpreted as noise. In the comment section, some people complain about the *sonidero*'s voice and messages, whereas others celebrate the whole atmosphere of the event or even the *sonidero*'s speech talents. Celebrations of the atmosphere in the video often comes from Mexicans

118 *Michaël Spanu*

Figure 9.5 Richard TV's YouTube video 'THE SONG THAT TOOK OVER THE NIGHT PERUVIAN CUMBIA ((SIN TU AMOR)) SONIDO CARIBE 66 SAN JUAN DE ARAGON.' San Juan de Aragon is the name of a neighbourhood in the north-eastern part of Mexico City (RICHARD TV 2018).

living in the United States or Canada, who miss their country's culture and conviviality. Other comments come from people in other Latin American countries celebrating Mexican culture.

Because of this controversy, *sonidero* YouTube channels sometimes upload so-called clean (*limpio*) versions of songs, which are versions recorded without the presence of the *sonidero*'s voice and with a visual emphasis on dancers. Richard TV's most viewed video (with almost 27 million views) is one of these clean versions: 'CUMBIA EL FINAL DE NUESTRA HISTORIA TEMA LIMPIO – SONIDO CONDOR' (RICHARD TV 2020a). These clean versions function as regular music videos and can be as popular as the more immersive and documentary-like videos. They can also be compared to so-called aftermovies[9] referring to nostalgia and advertising, that became very popular in the mainstream festival economy (Holt 2018). Indeed, these *sonidero* videos usually feature a global impression of the event, triggering nostalgic reactions or interest in the next events in the comments. Other comments refer to the pleasure of listening to the song repeatedly, which emphasises the musical value of the video and, less directly, interest in musical creativity and curation.

Until the implementation of automatic content identification tools, digital platforms such as YouTube reproduced informal ways of performing and sharing music publicly. In this sense, music cultures that thrived on the margin of formal copyright management found a sort of natural environment to grow in these platforms. Scholars such as Paulhiac feared that internet regulations seeking to preserve international copyright business models might eventually affect cultural practices such as *sonideros*. However, in the present case, the opposite happened: in fact, the platform economy increased the visibility and respectability of *sonideros*, primarily through the audiovisual emphasis on dancing and conviviality.

Moreover, videos and social media have been used as promotion tools that made the information more readily available, which increased the number of events and people assisting with them. Automatic content identification tools, such as YouTube's ContentID, have mainly affected the broadcasters' economy, as most of the digital monetisation went to copyright owners. So far, they have not fundamentally changed the economy of in-person events. ContentID relies on a conceptualisation of music as fixed data (Heuguet 2019). Therefore it does not work for all *sonidero* videos, especially those where *sonideros* speak over the music they play, or when the video's sound quality is too low. In these cases, the broadcasters cannot monetise their videos, but they still can share their passion for *sonidero* music and events.

While artists and professionals from a stabilised copyright context consider the platform economy a threat, it also constitutes a real opportunity to reach greater levels of visibility in more informal economies. Interestingly, producing content depends more on the in-person economy: *sonideros* generally hire audiovisual producers to gain visibility. Digital revenues from video streaming account for a tiny part of a *sonidero* broadcaster's revenue (Ricardo Mendez, pers. comm., 6 October 2021). Other broadcasters that I interviewed do not even get paid; they do it mainly out of a passion for *sonidero* culture and to support local *sonideros*, as for instance in the case of Sonideros Locales SD from San Diego, United States. Their videos do not reach the quality and professionalism of Richard TV, which shows how a great deal of the *sonidero* video production remains amateur and not economically driven. Filming and sharing *sonidero* videos is a performance embedded in the *sonidero* culture that echoes the attempt to build or consolidate an affective community through music in a context of social stigmatisation, diasporisation, and cultural creativity.

Sonidero Digital Channels as Challenged Archivists and Niche Heritage

Despite their precarity, *sonidero* audiovisual producers still enjoy a certain degree of autonomy. They can decide to film an event on their own and in their own way. Not only does their audiovisual style refer generally to a sort of testimony, as I argued above, but *sonidero* video producers often think of themselves as archivists. For instance, Ricardo told me: 'To me what I do is to make history. The videos are meant for posterity, for the future, for people to remember it, to know about it. Like a library' (Ricardo Mendez, pers. comm., 6 October 2021). The Richard TV channel is full of behind-the-scenes videos related to the organisation of *sonidero* events and sound systems. The context of the pandemic and the momentary disappearance of *sonidero* events urges us scholars to consider this video production more seriously as an archive and, ultimately, as heritage.

On the one hand, academic interest in popular music heritage has dramatically increased over the last decade (Roberts and Cohen 2014), contributing to the recognition of a diversity of heritage participants, practices, and objects related to popular music memory that were previously ignored or marginalised within cultural institutions and industries. Richard TV falls into this category, as it represents an insider gaze on *sonidero* culture, simultaneously fixing and creating a new version of it through the mediation of digital platforms. On the other hand, scholars also consider YouTube a site of translocal community building (Olvera et al. 2015), and new heritage practices related to dominant and marginalised cultures, official and non-official sources, professionals, and amateurs (Pietrobruno 2013, 2018). It is a performative archival tool, as any video can be stored for an undetermined duration, which contrasts with platforms such as Twitch that are more focused on live streaming and interactivity. For instance, Richard TV's oldest videos are still accessible, and anyone can appreciate the evolution of his visual style as much as the growing impact of *sonidero* culture in terms of views and comments. His channel accounts for 4,498 videos as of November 2021, which constitute a unique contribution to the history of *sonidero* culture.

However, the immense popularity of Richard TV's content (474,779,109 total views) contrasts with the fragility of its preservation, as the channel relies entirely on YouTube's private infrastructure. YouTube's role as a digital repository is very ambivalent. The platform uses *sonidero* content (and virtually all content) as a means to redirect attention to advertising content. In other words, YouTube simultaneously enhances and appropriates cultures such as *sonideros* (Nowak 2018). After being bought by Google, YouTube went from being a user-generated content platform to a more professional and commercial one that increasingly incorporates traditional media patterns (Kim 2012), and pushes content creators to monetise all their content. This change triggered a race to attract views that was very much embedded in the attention economy, and that challenged the heritage value of these videos. In the case of *sonidero*, not only are the videos surrounded by ads, but most channels (re)orientate their content in what they perceive as an attractive way. The most noticeable aspects are the exoticisation of poor neighbourhoods and the instrumentalisation of women's bodies in a sexually assertive way, which reinforces the bourgeois, heteronormative, male gaze on this culture. The first aspect results from the particular way in which the neighbourhood is systematically mentioned in the title, for example "THOSE FROM THE TEPITO NEIGHBORHOOD KNOW HOW TO DANCE" (RICHARD TV 2021) and visually staged in the video. The insistence on the name of the neighbourhoods functions, for the audience, as a way to recognise themselves and, for the viewer, as a form of social voyeurism. This is particularly visible in the comments, where some users point out the bad reputation of a particular neighbourhood; one example is a video mentioning La Merced, where one user jokes in the comments that his wallet disappeared when he pressed play. The second aspect relies on showing a sexually assertive female dancer on the fixed image of the miniature video (see Figure 9.6) and using a video title that stages her as the main protagonist, e.g. "THIS IS WHAT HAPPENS WHEN SHE DANCES CUMBIA" (Sonidero Latino TV 2019b), whereas men are generally depicted in a more casual way. Sometimes, the female dancer turns out to be a transgender person, tricking the viewer out of their expectation. Indeed, in poor neighbourhoods of Mexico City, it is common for transgender people to participate in *sonidero* events.

Ricardo from Richard TV raised his voice against sexual instrumentalisation initially, but claims that his videos started to lose many viewers compared to other channels, so he decided to use the same 'recipe' (pers. comm., 6 October 2021). However, in general, the content of his videos does not specifically focus on women's bodies, nor does it amplify

Figure 9.6 Richard TV's list of videos that emphasise women's assertive bodies.

the already heteronormative regime of the in-person *sonidero* culture. Rather, it is the platform itself that triggers gendered representations of the videos' potential attractiveness (Bishop 2019), questioning the public and heritage value of these videos on YouTube.

Considering the heritage value of these videos means describing how they translate in-person events into audiovisual content beyond commodification. In the absence of institutional support, *sonidero* YouTube channels rely on the attention economy and the in-person *sonidero* economy, but it does not mean that their value is limited to promotion and attention. Richard TV's signature is unique and creative in portraying different sides of *sonidero* culture, from the immersive view of the dancing practices to the interviews with *sonideros* about technical, legal, and even cultural aspects of *sonidero* events. Such creativity properly reflects the *sonideros'* cultural creativity, their controversial ability to talk over the music, send messages, and comment on the event with a particular touch and sense of the rhythm. The controversial creativity also works for dancers who always invent new steps. The comments section of YouTube showcases these controversies and participates in creating a specific heritage value, as it keeps a record of the different social representations and values operating within the *sonidero* culture.

But cultural heritage is also defined by its relation to the concept of curation, which designates the selection of content. In a certain way, YouTube belongs to the same genealogy as other archive apparatuses that store a vast majority of content and only display a minority of it through the action of a curatorial authority (Prelinger 2009). If most traditional heritage institutions rely on the concept of significance to curate content and perform heritage (Nowak and Baker 2018), YouTube's curation of content is shared between users and algorithms. As a high metric channel, Richard TV's videos quickly appear in YouTube's recommendations when a user starts watching *sonidero* videos, which creates a snowball effect. The experience of *sonidero* videos is mediated by the number of views, especially when the number is high, which contrasts with the idea of a marginalised culture. In other words, YouTube's curation appears to reinforce channels dedicated to *sonidero* music and their insider gaze on the one hand, but polarises its relationship with other forms of institutional heritage practices on the other hand, which raises deep questions about *sonidero* channels' potential to advocate and articulate public values in a critical context such as the pandemic.

Despite a certain number of constraints, YouTube *sonidero* channels embody a specific and vernacular form of heritage, as they simultaneously document, promote, and recreate this culture. From their perspective, there is no culture to save or fix, and they are more likely to reflect a collective cultural group in both mundane and spectacular ways, showing its complexities and even its internal debates. The lack of institutional support and the biases introduced by the digital platforms give a particular shape to these practices and contents that contrasts with common definitions of heritage (Knifton 2012). Nevertheless, they also represent a unique opportunity for *sonidero* practitioners to create a new version of their digital selves, and partially control the gaze upon their culture in a moment of global and upper-class interest in cumbia music. Indirectly, they also foster an alternative image of Mexico City and Mexico in general, far from the mainstream cultural centres and providing accounts of the city and country's cultural vibrancy in marginalised neighbourhoods, despite the new threats that emerged with the pandemic.

Notes

1 All videos' original titles are in Spanish. Translations by the author.
2 For instance, *Sonidero Conquistador Latino* in Mexico City's suburban neighborhood of Naucalpan.

122 *Michaël Spanu*

3 For example, the Posadón event that took place in Ex-Fabrica de harina in Mexico City on 16 December 2021, aimed at promoting the sweet beer brand Vickys.
4 Freddy Ortiz is the owner of the Sonideros Locales SD channel.
5 Richard TV and Sonidero Latino TV's most viewed videos reached 26 and 20 million views, respectively, as of October 2021.
6 Ricardo Mendez is the owner of Richard TV's channel.
7 Fabian Holt identifies this phenomenon as a video turn in popular music that resulted from the internet boom in the 2000s, and significantly affected music cultures (Holt 2011).
8 The only exceptions are the institutional contexts where live music videos are considered as cultural heritage, such as France (Guibert, Spanu and Rudent 2021).
9 An aftermovie is a video broadcasted after an event that aims to tell the story of the event in a synthetic and dynamic way. This video generally represents the environment of the event, but also its values.

References

Aterianus-Owanga, Alice. 2020. 'Une culture de "bougeuses de fesses": Spectacle du pouvoir et incorporation genrée dans les groupes d'animation culturelle du Gabon.' In *Sexe et genre des mondes culturels*, edited by Sylvie Octobre and Frédérique Patureau, 195–207. Lyon: ENS Éditions.
Auslander, Philip. 2008. *Liveness: Performance in a Mediatized Culture*. London: Routledge.
Bénistant, Alix. 2018. 'La transnationalisation de l'industrie musicale de Miami.' *Communication* 35 (1). https://doi.org/10.4000/communication.7607.
Bennett, Lucy. 2012. 'Patterns of Listening Through Social Media: Online Fan Engagement with the Live Music Experience.' *Social Semiotics* 22 (5): 545–557. https://doi.org/10.1080/10350 330.2012.731897.
Bishop, Sophie. 2019. 'Managing Visibility on YouTube through Algorithmic Gossip.' *New Media and Society* 21 (11–12): 2589–2606. https://doi.org/10.1177/1461444819854731.
Blanco Arboleda, Darío. 2012. 'Los Bailes Sonideros: Identidad y resistencia de los grupos populares mexicanos ante los embates de la modernidad.' In *Sonidos en las aceras, vengase la gozadera*, edited by Mariana Delgado and Marco Ramírez Cornejo, 53–82. Mexico City: Tumbona Ediciones.
Bottà, Giacomo and Geoff Stahl, eds. 2019. *Nocturnes: Popular Music and the Night*. Berlin: Springer.
Bratus, Alessandro. 2016. 'In-Between Performance and Mediatization: Authentication and (Re)-Live(d) Concert Experience.' *Rock Music Studies* 3 (1): 41–61. https://doi.org/10.1080/19401 159.2015.1129112.
Buxton, David. 2018. 'La vidéo musicale comme "marchandise échouée."' *Volume!* 14 (1): 193–200. https://doi.org/10.4000/volume.5591.
Canclini, Nestor G., Maritza Urteaga, and Francisco Cruces, eds. 2012. *Jóvenes, culturas urbanas y redes digitales*. Barcelona: Ariel.
Djebbari, Élina. 2018. 'Vidéochoréomorphose: danses et vidéo-clips au Mali.' *Volume!* 14 (2): 137–160. https://doi.org/10.4000/volume.5577.
Guibert, Gérôme, Michaël Spanu, and Catherine Rudent. 2021. 'Beyond Live Shows: Regulation and Innovation in the French Live Music Video Economy.' In *Researching Live Music: Gigs, Tours, Concerts and Festivals*, edited by Chris Anderton and Sergio Pisfil, 238–249. London: Focal Press.
Heuguet, Guillaume. 2019. 'Vers une micropolitique des formats. Content ID et l'administration du sonore.' *Revue d'anthropologie des connaissances* 13 (3): 817–848.
Heuguet, Guillaume. 2016. 'When Club Culture Goes Online: The Case of Boiler Room.' *Dancecult: Journal of Electronic Dance Music Culture* 8 (1): 73–87. https://doi.org/10.12801/1947-5403.2016.08.01.04.
Holt, Fabian. 2010. 'The Economy of Live Music in the Digital Age.' *European Journal of Cultural Studies* 13 (2): 243–261. https://doi.org/10.1177/1367549409352277.
Holt, Fabian. 2011. 'Is Music Becoming More Visual? Online Video Content in the Music Industry.' *Visual Studies* 26 (1): 50–61. https://doi.org/10.1080/1472586X.2011.548489.
Holt, Fabian. 2018. 'Les vidéos de festivals de musique: une approche "cérémonielle" de la musique en contexte médiatique.' *Volume!* 14 (2): 211–225. https://doi.org/10.4000/volume.5599.

Post-Pandemic Cumbia, Mediated Live Music, and Digital Heritage   123

Holt, Fabian. 2020. *Everyone Loves Live Music*. Chicago: University of Chicago Press.

Kanai, Miguel and Ilana Ortega-Alcázar. 2009. 'The Prospects for Progressive Culture-Led Urban Regeneration in Latin America: Cases from Mexico City and Buenos Aires.' *International Journal of Urban and Regional Research* 33 (2): 483–501. https://doi.org/10.1111/j.1468-2427.2009.00865.x.

Kim, Jin. 2012. 'The Institutionalization of YouTube: From User-Generated Content to Professionally Generated Content.' *Media, Culture and Society* 34 (1): 53–67. https://doi.org/10.1177/016344371 1427199.

Knifton, Robert. 2012. 'La musique, la mémoire et l'objet absent dans les archives numériques.' *Questions de communication* 22 (2): 45–56. https://doi.org/10.4000/questionsdecommunicat ion.6822.

Kun, Josh. 2015. 'Allá in the Mix: Mexican Sonideros and the Musical Politics of Migrancy.' *Public Culture* 27 (3): 533–555. https://doi.org/10.1215/08992363-2896219.

Lippman, Alexandra. 2018. 'Listening across Borders: Migration, Dedications, and Voice in Cumbia Sonidera.' *Tapuya: Latin American Science, Technology and Society* 1 (1): 201–215. https://doi.org/ 10.1080/25729861.2018.1497273.

Macías, Yolanda. 2021. 'La noche oficial de la Ciudad de México. ¿Para qué y para quién?' In *Noche urbana y economía nocturna en América del Norte*, edited by Alejandro Mercado Celis and Edna Hernandez Gonzalez, 153–176. Mexico City: CISAN-UNAM.

Magaudda, Paolo. 2020. 'Music Scenes as Infrastructures: From Live Venues to Algorithmic Data.' In *Popular Music, Technology, and the Changing Media Ecosystem*, edited by Emilia Barna and Tamas Tofalvy, 23–41. New York: Palgrave Macmillan.

Mercado-Celis, Alejandro. 2017. 'Districts and Networks in the Digital Generation Music Scene in Mexico City.' *Area Development and Policy* 2 (1): 55–70. https://doi.org/10.1080/23792 949.2016.1248455.

Nowak, Florence. 2018. 'Les clips de musique régionale indienne, *dreamcatchers* d'une économie de l'attention.' *Volume!* 14 (2): 161–173. https://doi.org/10.4000/volume.5578.

Nowak, Raphaël and Sarah Baker. 2018. 'Popular Music Halls of Fame as Institutions of Cultural Heritage.' In *The Routledge Companion to Popular Music History and Heritage*, edited by Sarah Baker, Catherine Strong, Lauren Istvandity, and Zelmarie Cantillon, 283–293. London: Routledge.

Olvera, José Juan, José Carlos Zarazúa, Velasco Rodríguez Heydi, and Ruth Castro Yenifer. 2015. 'Música, migración y redes sociales digitales en tres comunidades mexicanas.' *Trace* 67: 62–91.

Paulhiac, Juan. 2014. 'La scène musicale de la *champeta* face à Internet.' *Volume!* 10 (2): 131–149. https://doi.org/10.4000/volume.4073.

Picaud, Myrtille. 2021. *Mettre la ville en musique (Paris-Berlin)*. Vincennes: Presses Universitaires de Vincennes.

Pietrobruno, Sheenagh. 2013. 'YouTube and the Social Archiving of Intangible Heritage.' *New Media and Society* 15 (8): 1259–1276. https://doi.org/10.1177/1461444812469598.

Pietrobruno, Sheenagh. 2018. 'YouTube Flow and the Transmission of Heritage: The Interplay of Users, Content, and Algorithms.' *Convergence* 24 (6): 523–537. https://doi.org/10.1177/135485651 6680339.

Power, Dominic and Daniel Hallencreutz. 2002. 'Profiting from Creativity? The Music Industry in Stockholm, Sweden and Kingston, Jamaica.' *Environment and Planning A: Economy and Space* 34 (10): 1833–1854. https://doi.org/10.1068/a3529.

Prelinger, Rick. 2009. 'The Appearance of Archives.' In *The YouTube Reader*, edited by Pelle Snickars and Patrick Vonderau, 268–274. New York: Wallflower Press.

Ragland, Cathy. 2012. 'Comunicando la imaginación colectiva: el mundo socio-espacial del sonidero mexicano.' In *Sonidos en las aceras, vengase la gozadera*, edited by Mariana Delgado and Marco Ramírez Cornejo, 83–90. Mexico City: Tumbona Ediciones.

Ramirez, Paredes and Juan Rogelio. 2021. 'Noche de luz, sonido y aroma en la Ciudad de México.' In *Noche urbana y economía nocturna en América del Norte*, edited by Alejandro Mercado Celis and Edna Hernandez Gonzalez, 153–176. Mexico City: CISAN-UNAM.

Ramos-Kittrel, Jesús A. 2019. 'Sounding Cumbia: Past and Present in a Globalized Mexican Periphery.' In *Decentering the Nation. Music, Mexicanidad, and Globalization*, edited by Jesús A. Ramos-Kittrell, 169–192. New York: Lexington Books.

RICHARD TV. 2018. 'EL TEMA QUE SE LLEVO LA NOCHE CUMBIA PERUANA ((SIN TU AMOR)) SONIDO CARIBE 66 SAN JUAN DE ARAGON.' YouTube, 14 February 2018. www.youtube.com/watch?v=ZR-fY3u6pS4.

RICHARD TV. 2019. '(((LA ULTIMA SAMPUESANA))) DE SONIDO CONDOR EN LA MERCE ANIVERSARIOS 62 ANIVERSARIO EQUIPO RETRO.' YouTube, 25 September 2019. www.youtube.com/watch?v=HpzElRKLENE.

RICHARD TV. 2020a. 'CUMBIA EL FINAL DE NUESTRA HISTORIA TEMA LIMPIO – SONIDO CONDOR.' YouTube, 17 January 2020. www.youtube.com/watch?v=UsiSTlOf7UA.

RICHARD TV. 2020b. 'DESDE LA AZOTEA HACE SUS BAILES Y ASI LE RESPONDEN SUS VECINOS.' YouTube, 6 July 2020. www.youtube.com/watch?v=Es-KMEtUg4s.

RICHARD TV. 2020c. '¿POR QUE HACEN UN BAILE EN PLENA PANDEMIA? ESTO DIJO.' YouTube, 20 October 2020. www.youtube.com/watch?v=zsA5gsm0PoU.

RICHARD TV. 2021. 'LOS DE TEPITO SI SABEN BAILAN SONIDO MEMO MIX ((LA CUMBIA DE MATILDA)) ANV ESPONJADO.' YouTube, 5 July 2021. www.youtube.com/watch?v=3tGcTF_P9eY&list=PUS4H5zxSjH_WoH92q2_xtMg&index=15.

Rietveld, Hillegonda C. 2013. 'Journey to the Light? Immersion, Spectacle and Mediation.' In *DJ Culture in the Mix: Power, Technology, and Social Change in Electronic Dance Music*, edited by Bernardo Attias, Anna Gavanas, and Hillegonda Rietveld, 79–102. London: Bloomsbury.

Rinaudo, Christian. 2019. 'Musique et danse du Sotavento mexicain dans la construction locale et (trans)nationale des appartenances.' *Revue européenne des migrations internationales* 35 (3–4): 175–196. https://doi.org/10.4000/remi.13769.

Roberts, Les and Sara Cohen. 2014. 'Unauthorising Popular Music Heritage: Outline of a Critical Framework.' *International Journal of Heritage Studies* 20 (3): 241–261. https://doi.org/10.1080/13527258.2012.750619.

Román García, L. Elena. 2016. 'Relatos colectivos de espacios no oficiales de artes vivas: tensiones metaforizadas.' *Córima, Revista de Investigación en Gestión Cultural* 2 (2). https://doi:10.32870/cor.a2n2.6082.

Sarno, Giulia. 2018. 'Schermi, mediazioni e autenticità nel concerto popular: due casi di studio.' *Between* 8 (16). https://doi.org/10.13125/2039-6597/3342

Shaw, Robert. 2010. 'Neoliberal Subjectivities and the Development of the Night-Time Economy in British Cities.' *Geography Compass* 4 (7): 893–903. https://doi.org/10.1111/j.1749-8198.2010.00345.x.

Spanu, Michaël. 2022a. '¿Es posible mantener la diversidad musical a flote en la Ciudad de México?' *El País*, 18 January 2022. https://elpais.com/mexico/opinion/2022-01-18/es-posible-mantener-la-diversidad-musical-a-flote-en-la-ciudad-de-mexico.html.

Spanu, Michaël. 2022b. 'La plateforme et le patrimoine: une enquête sur les valeurs des vidéos de concert sur YouTube.' *Canadian Journal of Communication* 47 (1) : 49–78. https://doi.org/10.22230/cjc.2022v47n1a4047.

Spanu, Michaël and Catherine Rudent. Forthcoming. 'La captation et ses paradoxes. Réflexions sur la mise en vidéo des concerts à l'ère de la plateformisation de la musique.' In 'La musique "live" à l'ère des plateformes et de la captation numérique,' edited by G. Guibert and M. Lussier, special issue, *Communiquer*, forthcoming.

Sonidero Latino TV. 2019a. 'ASI SE BAILA UNA CUMBIA EN LOS BARRIOS DE LA CDMX – ((SAMPUESANA SIBONERA)) SIBONEY – LA MERCED.' YouTube, 27 September 2019. www.youtube.com/watch?v=BKmw-kFZhno.

Sonidero Latino TV. 2019b. 'THIS IS WHAT HAPPENS WHEN SHE DANCES CUMBIA.' YouTube, 3 December 2019. www.youtube.com/watch?v=G_AD-qnnAUg.

Wirz, Mirjam. 2013. *Sonidero City*. Zurich: Buzz Maieschi.

10 Sardinian Traditional Music during the COVID-19 Pandemic

Marco Lutzu and Ignazio Macchiarella[1]

As with every aspect of social life, the COVID-19 pandemic had a profound impact on traditional music making. This impact concerns not only (and perhaps not so much) sound outcomes, but also (and perhaps above all) the ways, the value, the meanings, and the reflections on the actuality of musical practice. Certainly, by choice or by necessity, performers and listeners have increasingly oriented themselves towards the possibilities offered by social media and the digital world. The web, after all, is increasingly a site of complex processes concerning music creation, making, listening, circulation among cultures, symbolising processes, and so on, re-elaborating mere mouth to ear transmission. Sometimes the COVID-19 pandemic stimulated the development of processes based on digital media that were already underway, while in some cases it fostered experimentation and the search for an original use for both performance and dissemination, also in the oral tradition domain of music.

At the same time, due to the impossibility of carrying out field research, social media platforms have themselves become an increasingly important tool for interaction within the relationships between scholars and local actors, which are the core of any ethnographic research. Social media offers different strategies of dialogue and interaction beyond ordinary long-distance telephone conversations. So, to maintain relationships in the field, many scholars have established specific strategies for communicating through Facebook or Instagram. This means that, willingly or not, greater attention has been paid towards the methods and perspectives offered by digital ethnography (Pink et al. 2016), or so-called netnography (Kozinets 2020), which are ever more recurrent in musicological studies and in the human sciences. These new approaches offer effective theoretical tools to investigate the influence of the digital world on music making, an aspect that was indispensable even before the pandemic.[2] The following pages aim to introduce some case studies from our research conducted in Sardinia, a context in which new media have a diffusion consolidated within different oral practices (Lutzu 2017). The issues concern musical practices on which we have both been working for years in a dialogical perspective, in the wake of close and frequent interrelations before the pandemic exploded.

Imaginary Polyphonies

On 21 April 2020, a headline released by ANSA (the leading wire service in Italy) read: '*A tenore* singing in the fight against Covid-19. Rearrangement of the Sardinian anthem of the 1794 revolutionary movements.' The article states that 'the arrangement of the Sardinian anthem (known as "Procurade 'e moderare") becomes an invitation to Sardinians to join together to eradicate coronavirus. The power of singing *a tenore* and *a cuncordu* and the magic of the *launeddas* remotely accompany a video produced by the Tenores Sardinia Association (ANSA 2020). The news (with a link to the video) was shared by several media

DOI: 10.4324/9781003200369-13

126 *Marco Lutzu and Ignazio Macchiarella*

in Sardinia and abroad. In actual fact, there is no *cantu a tenore* (or *a cuncordu*) in the video![3] Let us say that it is a sort of virtual presence. In the video, the first seven stanzas of the Sardinian anthem[4] are performed, with each stanza assigned to a different *boghe* (solo singer) from well-known *a tenore* and *a cuncordu* quartets from seven different villages. Each *boghe* sang the text according to the melodic system of the *gosos* or *gòcius* (a form of devotional singing spread throughout the island),[5] and recorded their song with a smartphone: the tracks were assembled over a background performed by a *launeddas*[6] player, also following the *gosos* performance scheme (Macchiarella 2008). This sound was accompanied by aerial images of the inland areas of the north-central part of the island, with images taken from a film by Davide Melis and Sebastiano Pilosu (2019) on the *cantu a tenore*.

From a technical point of view, it is certainly a homemade, and one might even say rudimentary, initiative, using a smartphone for recording and the open source Audacity software for editing. Its promoter Sebastiano Pilosu considered it to be an 'invitation to Sardinians to feel united as a people' (ANSA 2020), an invitation based on the symbolic value of this practice for identity, the 'power of *cantu a tenore*' (pers. comm., July 2020) which is evoked without being present, and works even without sound. The representative strength of choral practice and the ability of choral singing, whether polyphonic or not, to create and reaffirm the sense of community belonging are well known. And it is no coincidence that many of the musical initiatives during the lockdown concerned choral singing.

Media Strategies for Multipart Singing

In many Sardinian villages, the main events of the liturgical year are characterised by local and very particular sound emissions. This is especially true for Holy Week, which includes the most important ritual systems whose time and spaces are qualified by multipart singing, carried out differently from village to village. The most common songs are local elaborations of a singing pattern based on sacred texts (mostly 'Miserere' and 'Stabat Mater,' usually sung in Latin), or religious texts in the Sardinian language like the *gosos*, which deal with crucial moments of Holy Week or other religious events (Macchiarella 2008). Of course, the pandemic interrupted these singing practices and the continuity of the related secular rituals – or better, the asserted continuity, seeing that they have not always been documented year by year.[7] In actual fact, singers are very proud of such (supposed) continuity, and many of them shared the idea that COVID-19 and the connected lockdown have 'been able to do what not even wars have been able to do' – a strong statement from a symbolic point of view. As a result, many singers thought they 'had to do something,' according to Marcello Mallica, *bassu* of a singers' quartet (pers. comm., July 2020), in order to at least sonorise village time and spaces with the appropriated ritual sounds.[8]

Several local initiatives are of special relevance. For instance, in the village of Santu Lussurgiu, Su Cuncordu 'e Santa Rughe (the vocal quartet of the Holy Cross Confraternity – see Macchiarella 2009) took part in the Palm Sunday liturgical celebration on 7 April 2020, almost in the way it is prescribed by *su conotu*, which is to say the local tradition. The celebration was held without the usual processions, inside the Confraternity church, which was actually empty, without any congregation and with only the emblematic presence, apart from the celebrant, of the mayor of the town. In the centre nave, where the faithful usually stand, the singers were placed more than two metres apart – a distance that surely led to a certain standardisation of the vocal result (but, as one of the singers Gianluca told me, normal performance dynamics were of course not taken into account because it was a special occurrence). The service was broadcast in real time to the whole village, both by loudspeakers on the roof of the church and by streaming on the parish's Facebook page.[9] On the same day, the Confraternity posted a six-minute video with black and white pictures showing rituals from previous years (including some historical pictures from 1950), and two non-contextual recordings of the main ritual songs, the 'Miserere' and the 'Novena,' a paraphrase of the 'Stabat Mater' in the Sardinian language. A short text in Italian introduces the video, explaining that 'since we cannot perform the ritual we can only revive the memories of our faith for a few moments' (Cuncordu Santa Rughe 2020).

The following year, in the same ritual scenario, the Cuncordu and the priest offered the Palm Sunday Mass again, allowing all the members of the four village confraternities to participate. The performance was not only broadcast by the church loudspeakers, it was also livestreamed on the Cuncordu channel on YouTube. An excerpt can still be found on the same website, entitled 'Singing a Cuncordu in a Time of Pandemic.' The description is really emblematic:

> A short verse from the Miserere sung during the Holy Tuesday Mass 2021 by Su Cuncordu Santa Rughe; for the first time ever, the singers of the confraternity are wearing masks and practising social distancing which allows them to respect the anti-Covid rules as much as possible.
>
> (Cuncordu Santa Rughe 2021)

The choice to propose a live performance, 'whatever it may be' (as the singers of the Cuncordu told me), with all the limits imposed by COVID-19, is emblematic of the importance given to performance behaviours, rather than to the contents of the performance (Macchiarella 2009). The broadcasting of sound recordings from the ritual performances of previous years has been documented in several villages, serving as an immediate representation of the local community. In some cases, such as in the town of Bosa, the presence of local radio stations, effectively community radio, has favoured the operations of sonorising the reduced or virtual Holy Week rituals. New media, especially Facebook streaming, is however the most widely used instrument.

Special musical attention has been paid to the music quality of these edited recordings that were often selected and mixed by important local actors. In the village of Bortigali, for instance, this was done by Mario, one of the older singers and *contra* of Su Cuncordu Sas Enas. During our telephone conversation over Holy Week 2020, he told me that the idea of putting good recordings into circulation was most commendable. He was very worried about these disruptions to performances because, as he said, when they allow us to start singing again, it will 'take a lot of time and a lot of practice to reach the singing quality lost due to the quarantine' (Mario Carboni, pers. comm., August 2020).

Virtual Multipart Singing

On 10 April 2020, the traditional vocal quartet Cuncordu sos Battor Colonnas from the village of Scanu Montiferro posted a special video entitled 'A cuncordu singing on Good Friday despite Covid-19' on the local parish's content-rich YouTube channel. The video shows a virtual choir performance: a choir whose members do not meet physically but perform by coordinating each other from separate places. Produced with the Acapella app,[10] the video is edited from the recordings of the four voices singing 'Seven Swords of Pain – *gozos* for Holy Week,' one of the most representative songs in the local sound code of Holy Week (see D. Pani 2013). The video immediately reached a large audience in the small world of multipart singing, appreciated as an attempt to pay tribute to tradition and to the island's culture.[11] In actual fact, a previous attempt (dated 19 March) to edit a short virtual choir video appears on the quartet's Facebook page. The Good Friday video received several likes and comments such as 'The passion for singing *a tenore* is stronger than everything. A safe performance for Su Cuncordu "Sos Battor Colonnas,"' and 'Divided but always united! We are at home but do not give up, strength and courage' (Parrocchia 2020). Thus, in some way, due to the pandemic, Scanu Montiferru's group has introduced a technological innovation into the world of multipart singing. Seeing the enthusiasm of certain comments by non-insiders, one might have expected other groups to do something similar. In actual fact (perhaps due to the app's obvious synchronisation problems) the proposal was not followed up, apart from one case in the aforementioned village of Santu Lussurgiu, where the creation of a virtual video became the pretext for an interesting dialogical experience between a young ethnomusicologist and a group of young singers.

Nicola, one of the youngest singers of Su Rosariu, came up with the idea. Having seen several homemade music videos, he thought of creating a video to be transmitted via social media in the days preceding Holy Week. His idea was to contribute to the continuity of the village's multipart practice. Since he was aware of the technical difficulties of the operation, he decided to ask for help from his fellow villager and scholar, Diego Pani, a PhD candidate in ethnomusicology at the Memorial University of Newfoundland in St John's, Canada, who in those days was in Washington, D.C. for his research. Despite the distance, Diego accepted the invitation, which gave life to a remote collaboration, from which he wrote an article for the journal *SEM Student News* (D. Pani 2020). In short, the four singers recorded their singing independently, using their smartphones. They started from the *boghe*, a sort of pivotal voice in the musical form (Macchiarella 2009). Pani, helped by his wife, went on to edit the four tracks together, using complex software that allowed the synchronisation limits of most apps to be avoided. The singers soon posted the file on Facebook, collecting hundreds of comments of gratitude, some of which even stated that the video is a demonstration of how it is possible to live through the pandemic time 'thanks to traditions' (D. Pani 2020).[12] The quartet decided what to perform:

> They chose to sing an *istudiantina*, a song form mostly sung on mixed profane-secular occasions. The lyrics of the chosen version are usually sung by a senior member of the confraternity every year at the end of the Palm Sunday ritual. This particular version of the *istudiantina*, for them, means that Holy Week is about to start.
>
> (D. Pani 2020, 12)

It was a symbolic beginning entrusted to the expressive power of singing in four voices. The video was posted on various Facebook pages (receiving thousands of likes and comments), introduced by a short text in Italian in which they apologise for any mistakes, 'but during quarantine everything is allowed!! Good listening and good Holy Week to everyone!!!!' (Meloni 2020).

The last two cases, briefly described here, have long been a topic of discussion via social media for Sardinian music makers (and beyond). In actual fact, I expected to see a multiplication of such virtual polyphonies by other groups that are present and very active on social media (Facebook, Twitter, YouTube etc.), but no other cases are known. Chatting with some local actors, I gained the impression that they considered the virtual polyphony model to represent something rather didactic,[13] and therefore far from the concreteness of a normal video recording of a performance. In fact, these virtual choirs are based on the idea that singing in a choir is just emitting sounds with the right notes and in time; in truth, singing in a choir is an experience of collective interaction and negotiation between several people, that goes much further than simply emitting sounds while listening to others – listening within the choir takes place not only with the ears, but with the whole body. Virtual polyphony creates an idea of polyphony as a sum of melodic lines and therefore an artificial construction of a community that is perhaps not very interesting for performers of such a refined practice as Sardinian multipart singing.

Cantzonis: Narrative Song in Southern Sardinia

In 1961, the anthropologist Alberto Cirese underlined the lack of narrative songs in Sardinian musical culture in the face of a prevalence of lyric poetry. According to Cirese, this distinctive feature of the Island's popular literature had already been highlighted by Canon Giovanni Spano in the 19th century (Cirese 1961, 43). However, as Paolo Bravi recently pointed out,

> it was clearly a mistake. Probably the research on this subject had been misdirected or had fallen victim to some schemes and stereotypes of Romantic origin related to the so-called popular poetry and had not focused on the ponderous repertoire of popular *cantzonis* which, to a large extent, is just made up of 'stories.'
>
> (Bravi 2012, 127)

Cantzonis (sing. *cantzoni*) is a family of sung poetic forms widespread in southern Sardinia, where the Campidanese Sardinian language is spoken. They consist of strophic compositions with different metrical and rhyme structures depending on the type (*cantzoni a curba, a torrada, longa* and so on). Campidanese *cantzonis* are an interesting example of the merging of oral and written culture. In the 18th century, lyrics started to be written and published on loose sheets. They were occasionally composed by members of the cultured bourgeoisie, as in the case of the lawyer Efisio Pintor Sirigu (1765–1814) who wrote several *cantzonis* that are still known today (Liori 1991). But more often than not, the authors were illiterate or semi-illiterate, and sometimes even blind, so they entrusted scribes with fixing their compositions on paper. In any case, the lyrics of the *cantzonis* were intended to be sung with orally transmitted melodies, usually performed by a single voice accompanied by a musical instrument such as *launeddas* or guitar. In addition to the *cantzonis*, the realm of narrative songs in the Campidanese language also includes profane *gòcius*.

Because of their strophic structure, *cantzonis* and *gòcius* are poetic forms particularly suited to storytelling. Based on the analysis of an archival fond containing numerous loose sheets of *cantzonis* from the 18th and 19th centuries,[14] Paolo Bravi (2012) has identified the main subjects in these poetic genres. *Cantzonis* can deal with various topics such as love, religion, or moral and social issues; moreover, the theme of death is commonly discussed. Several *cantzonis* focus on various aspects related to Sardinia. We can also find autobiographical *cantzonis* and those intended to offend, mock, or defame a person (*cantzonis po mali*). A rich and popular genre of *cantzonis* includes ironic or comic ones, many of which employ metaphors with sexual references and allusions. But the most widespread kind of *cantzonis* is certainly the one that reports on events which, because of their dramatic nature or the notoriety of the people involved, remain impressed in the collective memory. We can find *cantzonis* about the death of Giuseppe Garibaldi, the regicide of Umberto I of Italy in 1900, and local or national crime events. According to Bravi:

> By portraying the protagonists, recalling the dynamics of dramatic events, exalting the innocent, and condemning the guilty on the basis of a conventional and strongly stereotypical moral code, the *cantzonis* not only serve an informative function for the benefit of the 'popular' public, but also provide a basic educational function, and play a cathartic role. They represent a central moment in the process of overcoming personal and

130 *Marco Lutzu and Ignazio Macchiarella*

collective anguish, achieved through the act of reliving the trauma through the filter of a dramatic narration of its salient moments, and through emotional identification with and collective participation in the event.

(Bravi 2012, 128)

During the 20th century, new ways of spreading *cantzonis* gradually replaced the loose sheets. Initially, they were printed in booklets (*libureddus*) released by small local publishers and marketed by street vendors, often combined with the transcription of improvised poetry contests. In the 1960s a local record market came into being, thanks to small labels that published LPs, and later cassette tapes, sold mostly at stalls during religious festivals. Thus, dozens of recordings of *cantzonis* entered the houses of many Sardinians, and began to be broadcast on the radio. Meanwhile, *cantzonis* continued to be performed live mostly on informal occasions or during festivals at the end of improvised poetry contests. In recent decades, however, *cantzonis* have gradually lost their primary communicative function. If in the past, texts were published or *cantzonis* were recorded to chronicle significant news events shortly after they happened, such performative occasions have become increasingly rare. In more recent times, performers tend to always return to the same texts, and there are fewer and fewer authors who compose new texts to tell of contemporary happenings.

I argue that the COVID-19 pandemic, as a traumatic and exceptional circumstance, revitalised the practice of composing *cantzonis* intended to discuss and give an account of events in contemporary life. Moreover, due to the increasingly active presence of authors of *cantzonis* and aficionados of traditional Sardinian music on social networks, these songs have spread in the digital world with previously unimaginable dissemination strategies and results in terms of popularity. In this sense, the most significant example is certainly the *cantzoni* titled 'Su Baballoti,' composed and sung by Antonio Pani.

'Su Baballoti' by Antonio Pani

Antonio Pani, a barber from Quartu Sant'Elena, is one of the most prominent improvising poets in Sardinia and also an author and singer of *cantzonis*. In the past, some of his *cantzonis* celebrating his hometown or describing significant events involving its inhabitants have been released on self-produced cassettes or CDs with a fair amount of success.[15] In 2008 Pani opened his Facebook account and in 2009 he launched his first YouTube channel, using both platforms as a means to showcase his work as a poet and musician. On 12 March 2020, during the first national lockdown in Italy, Pani composed a few stanzas of a profane *gòciu*. That same evening, he sang the song in his home studio, recording it with a smartphone. The next day he posted the video on Facebook and, two days later, uploaded it to his YouTube channel. Pani introduces his performance with these words:

Hello everyone. In these cloistered days, we have time to cultivate our passions a bit. Although I prefer to improvise, I occasionally write something. Last night, all in one go, I wrote four stanzas in ten minutes ... For this undesirable creature, whom I refer to as 'su baballoti', I've written four simple, unpretentious, and unrefined stanzas. I did this to try to keep up the interest in this problem that we hope will soon be resolved.

(A. Pani 2020)

The *baballoti*, literally little bug,[16] is the SARS-CoV-2 coronavirus, which Pani addresses first-hand with his song. Below is the transcription in Campidanese Sardinian and the English translation of the opening quatrain and the first two stanzas:[17]

Sardinian Traditional Music during the COVID-19 Pandemic   131

Table 10.1 'Su Baballoti' lyrics (courtesy of Antonio Pani)

Baballoti ascurtamì a mei	Little bug listen to me
Bessiminci dei custa zona	Get out of this area
Mancai giris cun sa corona	Even if you wear the crown
No ses dìnniu de fai su rei	You are not worthy to be king
Fattu prangi as a sa Cina	You made China cry
E tui fòrtzis ti nd'arrisi	And maybe you laugh at it
Ses piticcu chi mancu ti bisi	You're small, you can't even be seen
Però dda scis fai sa faina	But you know how to do your job
As' a biri ca sa mexina	You will see that medicine
Ti dda feus nosu sei sei	We will find sooner or later
Mancai giris cun sa corona	*Even if you wear the crown*
No ses dìnniu de fai su rei	*You are not worthy to be king*
Cun sa màscara a tipu Zorru	I wear a mask like Zorro
E mi tupu sa bucca e su nasu	and I plug my mouth and nose
A bessiri no est su casu	It's not time to go out
E in su divanu m'acorru	I'm shut in on the sofa
Però candu a bessiri torru	When I can go out again
Mi dda fatzu una cursa a pei	I'll run a foot race
Mancai giris cun sa corona	*Even if you wear the crown*
No ses dìnniu de fai su rei	*You are not worthy to be king*

In 'Su Baballoti,' Pani adopts a metrical form and melodic profiles that come from traditional Sardinian music. The lyric style, which includes irony and metaphorical language, as well as the reference to events that have just happened, follows the tradition of narrative songs from southern Sardinia. At the same time, Pani chooses a vocal style closer to mainstream popular music (as he does for some artistic projects in which he collaborates with rock and world musicians) rather than the one he usually adopts to sing improvised verses during poetic contests. The acoustic and visual dimensions further contribute to distinguishing this performance from what Pani would normally do in a traditional public context. In the video, we can see him surrounded by acoustic pyramid foam panels, wearing casual clothes and accompanying himself with an Irish bouzouki.

To better understand how an unexpected real-life event such as the COVID-19 pandemic deeply affected the practice of composing, singing and disseminating narrative songs in the southern Sardinian tradition, we need to turn to digital ethnography. In so doing, the methodological framework proposed by Pink et al. (2016) for research in digital environments is particularly inspiring. In the book *Digital Ethnography*, the authors suggest focusing on seven key concepts used in social and cultural theory for the design and analysis of ethnographic research, 'possibly reshaped in response to the way in which we encounter digital worlds ethnographically' (Pink et al. 2016, 15).[18]

The pandemic inspired Antonio Pani to write 'Su Baballoti,' which I argue should primarily be considered as a digital object. The song was born as a digital video file recorded with a mobile device. Although Pani was accustomed to sharing recordings of his performances on social networks, this activity intensified during the first lockdown in Italy. As he told me during a private conversation, he was inspired by the phenomenon of 'balcony flashmobs': informal gatherings promoted through social networks which in those days brought thousands of Italians to look out from their balconies playing an instrument or singing a song (pers. comm., December 2020; see Scorza Barcellona 2020). Some days after the first post, and again in the following months, Pani released and published a new audio recording and a video version with embedded lyrics (thus new digital objects) on both Facebook and

YouTube. Furthermore, one of these digital objects found a physical counterpart when Pani decided to release a CD version of 'Su Baballoti.'

As if it were an episodic song, Pani then went to compose, perform, record, and share new profane *gòcius* entitled 'Baballoti 2,' 'Momoti 3,'[19] 'Baballoti 4,' 'Baballoti 5,' and 'Baballoti 6' in the following months. As digital content characterised by a high degree of spreadability (Jenkins, Ford, and Green 2013), 'Su Baballoti' (including its various versions and sequels) circulated beyond its creator's control, passing from social networks to video-sharing platforms, to WhatsApp messages. In a short time, these songs had circulated in the digital world in a way unprecedented for Campidanese *cantzonis*. In total, the entire content garnered hundreds of thousands of views and thousands of likes, comments, and shares.

Although I first suggested considering 'Su Baballoti' as a digital object, I accept one of the principles for a digital ethnography proposed by Pink et al., which is the *non-digital-centric-ness* of research. According to the authors, 'such approaches de-centre media as the focus of media research in order to acknowledge the ways in which media are insepar-able from the other activities, technologies, materialities and feelings through which they are used, experienced and operate' (Pink et al. 2016, 9). Born as a digital video recording to share on the internet, 'Su Baballoti' has had an effective impact on the physical world, affecting the social lives, relationships, experiences, and practices of many people. Moreover, as if in a circle between the physical and digital worlds, many of these experiences have been mediatised and shared on social networks.

Sardinian music aficionados and Antonio Pani's fans from different places started to listen to the recordings. In some cases, they filmed the view from their balconies with the song blaring from the speakers and shared the video, which Pani promptly reposted on Facebook. In some kindergartens in Quartu, the teachers taught 'Su Baballoti' to the children, while various people shared the video of their cover version on Facebook. Once the lockdown was over, whenever he met anyone who asked to listen to his song (such as a bedridden elderly person or a young customer at his barbershop), Pani recorded the performance and shared the video on Facebook. Kristina Jacobsen (2020), an ethnomusi-cologist and linguistic anthropologist who was researching in Sardinia, discusses Pani's case in an article on music on the island during the pandemic. On 28 October 2020, an article published in the most important Sardinian newspaper reported the success of 'Su Baballoti' and that 'people in America were talking about this song' (G. Da. 2020). In summer 2020, the posters promoting concerts by Pani and his band explicitly mentioned 'Su Baballoti.'

The COVID-19 pandemic has been an epochal event that strongly impacted music making. In the case of the narrative song of southern Sardinia, as a dramatic event of collective interest, it stimulated the composition of new *cantzonis* intended to comment on contemporaneity, a practice that had long since fallen into disuse. At the same time, the forced stay at home because of the lockdown stimulated authors and performers of *cantzonis* such as Antonio Pani to increase their presence in the digital world and to seek new creative strategies for spreading their music. Inspired by a global phenomenon such as the pandemic, Pani affirms his local belonging in the delocalised environment of the Net. It is no coincidence that the people of Quartu, his fellow citizens, were those who most proudly shared and commented on 'Su Baballoti.' At the same time, Pani can be seen as a local improvising poet who, by using a local language, metrical form, and musical style, speaks about a global event, since digital technologies have allowed him to reach a hith-erto unimaginable pool of listeners and give new life to a genre such as the Campidanese *cantzonis*.

Conclusions

The case studies presented here confirm how digital ethnography has become an increasingly essential methodological tool for the study of traditional oral music. Doing ethnographic research on the digital world requires us to adapt the procedures and techniques we usually employ in our fieldwork. At the same time, it invites us to take into due consideration the ways in which the musicians we work with use digital media as an ever more important part of their music making.

The COVID-19 pandemic led to a further increase in the presence of traditional Sardinian musicians in the digital world, as a consequence of the limitations on performing music both as entertainment and on occasions in which it serves a ritual function. In this situation, a number of Sardinian musicians creatively exploited the potential offered by digital media and assessed its usefulness for their expressive needs. In the case of multipart singing, the virtual choir experiments made during the first lockdown were soon abandoned, as they were unable to replicate the experience of human interaction that is actually the real essence of polyphonic singing. In the case of the narrative songs of southern Sardinia, the pandemic was the pretext, and social networks the tool, to make this genre once again a popular means of storytelling about contemporaneity. The local use of digital media and the strategies of participation in the digital world by oral traditional musicians will increasingly become a central aspect of ethnomusicologists' research, even when – and soon we hope – the pandemic emergency is over.

Notes

1 The general outline of the chapter, the introduction, and the conclusion were shared between the authors. Ignazio Macchiarella wrote the sections Imaginary Polyphonies, Media Strategies for Multipart Singing, and Virtual Multipart Singing, while Marco Lutzu wrote the paragraphs *Cantzonis*: Narrative Song in Southern Sardinia, and 'Su Baballoti' by Antonio Pani.

2 Theoretical and methodological approaches focusing on the virtual and digital world have also been elaborated from an ethnomusicological perspective. See for example Lange 2001; Lysloff and Gay 2003; Reily 2003; Cooley, Maizel and Syed 2008; Wood 2008; Harvey 2014; Hilder, Stobart, and Tan 2017.

3 *Cantu a tenore* and *cantu a concordu* are the main multipart singing traditions of Sardinia.

4 The poem known as 'Procurade 'e moderare' became the anthem of the Sardinian Region on 24 April 2019 (www.regione.sardegna.it/documenti/1_422_20190909110352.pdf).

5 Employed mostly for religious hymns in honour of the Madonna or of Saints, *gòcius* (sing. *gòciu*) are a kind of strophic sung poetry. *Gosos* is the term used in the Logudorese region, while *gòcius* is used in the Campidanese region.

6 A traditional Sardinian triple pipe clarinet.

7 It is impossible to give either an accurate date or a precise starting point for these kinds of ritual and music practices. Many local histories include cases of significant interruption of the singing practices: for instance, in Bortigali the local priest forbade the singing of 'Miserere' and 'Stabat Mater' during para-liturgies for about 15 years since he had some issues with the singers (Macchiarella 2017). This interruption is considered an episode of local history that is fundamentally different from the pandemic urgency, which is seen as some kind of external imposition of prohibitions (during Holy Week, during a telephone conversation, several singers told me that 'they believed they could sing with the appropriate arrangements,' that 'they did not agree with suspending their sacred tradition due to a virus,' and so on).

8 The same thing has been done in various towns in Italy and elsewhere – several traces are found on YouTube.

9 The video remained online for a short time during Holy week and was then removed. It is now proudly part of the Confraternity multimedia collection.

134   *Marco Lutzu and Ignazio Macchiarella*

10 www.etaleteller.com/acapella-app.
11 For example, it was shared on the same day by the online journal Cronache Nuoresi (www.cron achenuoresi.it/2020/04/10/sas-battor-colonnas-di-scano-montiferro-e-i-canti-della-settimana-santa-ai-tempi-del-coronavirus-video). The same video can be found on the quartet's YouTube channel (https://youtu.be/0YkfvRiuInQ).
12 The video is now on YouTube: https://youtu.be/0A2-a0KNqLQ.
13 In Sardinia several ethnographic museums have adopted installations of this type to illustrate the functioning of music practice and more.
14 The Ballero-Sanjust fonds is kept in Cagliari's Biblioteca Comunale. For further information on the *cantzonis* in this fonds, see Milleddu 2012.
15 The most famous are 'Su scrignu da Campidanu' (The Casket of Campidano), dedicated to Quartu, 'Su Papadori' (The Hungry), an ironic cantzoni which, telling of a glutton's meal, describes the main dishes of the local cuisine, and a cantzoni dedicated to a young man from Quartu who died tragically in a car accident.
16 The term generically means insect or bug, but in some cases can refer specifically to certain species of cockroach.
17 The profane *gòcius* composed by Pani is written in nine-syllable lines. It consists of an initial quatrain with an *xyyx* rhyme scheme, followed by six-line stanzas rhyming *abbaac*. During the performance, each stanza is followed by the last two lines of the initial quatrain, the second of which rhymes with the last line of the stanza ($c = x$).
18 According to the authors, 'these concepts were selected to represent a range of different routes to approaching the social world, that that is: through experiences (what people feel); practices (what people do); things (the objects that are part of our lives); relationships (our intimate social environments); social worlds (the groups and wider social configurations through which people relate to each other); localities (the actual physically shared contexts that we inhabit); and events (the coming together of diverse things in public contexts)' (Pink et al. 2016, 14–15).
19 In the Sardinian language the *momoti* is a frightening imaginary creature invoked to scare children.

References

ANSA. 2020. 'Canto a tenore nella lotta al Covid-19. Rielaborazione inno sardo del 1794 durante i moti rivoluzionari.' 21 April 2020. www.ansa.it/sardegna/notizie/2020/04/17/canto-a-tenore-nella-lotta-al-covid-19_46c677bd-7a1b-48e2-9da3-a1ff220f493b.html.
Bravi, Paolo. 2012. 'Il canto a chitarra nella Sardegna meridionale.' In *Enciclopedia della Musica Sarda*, edited by Francesco Casu and Marco Lutzu, 5, 114–139. Cagliari: L'Unione Sarda.
Cirese, Alberto M. 1961. *Poesia sarda e poesia popolare nella storia degli studi*. Sassari: Gallizzi.
Cooley, Timothy J., Katherine Meizel, and Nasir Syed. 2008. 'Virtual Fieldwork: Three Case Studies.' In *Shadows in the Field: New Perspectives for Fieldwork in Ethnomusicology*, edited by Gregory Barz and Timothy J. Cooley, 90–107. Oxford: Oxford University Press.
Cuncordu Santa Rughe. 2020. 'Su Nazarenu 2020.' YouTube, 7 April 2020. www.youtube.com/watch?v=J5eCcbrlrww.
Cuncordu Santa Rughe. 2021. 'Cantare a Cuncordu in tempo di Pandemia.' YouTube, 30 March 2021. https://youtu.be/GIPb-wnKF-w.
G. Da. 2020. 'Un brano in sardo per esorcizzare la paura.' *L'Unione Sarda*, 27 March 2020, 32.
Harvey, Trevor S. 2014. 'Virtual Worlds: An Ethnomusicological Perspective.' In *The Oxford Handbook of Virtuality*, edited by Mark Grimshaw, 378–391. Oxford: Oxford University Press.
Hilder, Thomas R., Henry Stobart, and Shzr Ee Tan, eds. 2017. *Music, Indigeneity, Digital Media*. Rochester: University of Rochester Press.
Jacobsen, Kristina. 2020. 'When Coronavirus Emptied the Streets, Music Filled Them.' *Sapiens*, 26 March 2020. www.sapiens.org/culture/coronavirus-sardinia-music/.
Jenkins, Henry, Sam Ford, and Joshua Green. 2013. *Spreadable Media: Creating Value and Meaning in a Networked Culture*. New York: New York University Press.

Kozinets, Robert V. 2020. *Netnography: The Essential Guide to Qualitative Social Media,* third edition. London: Sage.

Lange, Barbara Rose. 2001. 'Hypermedia and Ethnomusicology.' *Ethnomusicology* 45, (1) (Winter): 132–149. https://doi.org/10.2307/852637.

Liori, Antonangelo, ed. 1991. *Il meglio della grande poesia campidanese.* Cagliari: Della Torre.

Lutzu, Marco. 2017. 'Media, Virtual Communities, and Musics of Oral Tradition in Contemporary Sardinia.' *Philomusica on-line* 16 (1): 121–136.

Lysloff, René and Leslie Gay. 2003. 'Ethnomusicology in the Twenty-First Century.' In *Music and Technoculture*, edited by René Lysloff and Leslie Gay, 1–23. Middletown: Wesleyan University Press.

Macchiarella, Ignazio. 2008. 'Harmonizing in the Islands: Overview of the Multipart Singing by Chording in Sardinia, Corsica, and Sicily.' In *European Voices I. Multipart Singing in the Balkans and in the Mediterranean,* edited by Ardian Ahmedaja and Gerlinde Haid, 103–158. Vienna: Böhlau Verlag.

Macchiarella, Ignazio. 2009. *Cantare a cuncordu: uno studio a più voci.* Udine: Nota.

Macchiarella, Ignazio. 2017. 'Confraternity Multipart Singing: Contemporary Practice and Hypothetical Scenarios for the Early Modern Era.' In *Listening to Early Modern Catholicism: Perspectives from Musicology,* edited by Daniele V. Filippi and Michael Noone, 276–300. Leiden: Brill.

Melis, Davide, dir., and Sebastiano Pilosu. 2019. *A Bolu. Il canto a Tenore in Sardegna.* Cagliari: Karel.

Meloni, Antonio. 2020. 'Salve, Paesani and Non …' Facebook, 7 April 2020. www.facebook.com/1000 00427346759/videos/3177996858891194/.

Milleddu, Roberto. 2012. 'Il fondo Ballero/Sanjust.' In *Enciclopedia della Musica Sarda,* edited by Francesco Casu and Marco Lutzu, vol. 5, 144–145. Cagliari: L'Unione Sarda.

Pani, Antonio. 2020. 'Su Baballoti (Coronavirus).' YouTube, 15 March 2020. https://youtu.be/snrg Q87SjcY.

Pani, Diego. 2013. 'Il canto a cuncordu a Scanu Montiferru.' Compact disc. Track on Cuncordu Sas Bator Colonnas De Iscanu, *Antigos Trazos.* Tronos.

Pani, Diego. 2020. 'A Cuncordu Song in the Age of Quarantine.' *SEM Student News* 16 (1): 11–12.

Parrocchia San Pietro Apostolo Scano di Montiferro. 2020. 'Cantu a Cuncordu il Venerdì Santo nonostante il Covid-19.' YouTube, 10 April 2020. https://youtu.be/99LloHa79WE.

Pink, Sarah, Heather Horst, John Postill, Larissa Hjorth, and Tania Lewis. 2016. *Digital Ethnography: Principles and Practice.* London: Sage.

Reily, Suzel Ana. 2003. 'Ethnomusicology and the Internet.' *Yearbook for Traditional Music* 35: 187–192. https://doi.org/10.2307/4149330.

Scorza Barcellona, Gaia. 2020. 'Coronavirus, l'Italia sul balcone: canzoni contro la paura.' *La Repubblica*, 13 March 2020. www.repubblica.it/cronaca/2020/03/13/news/coronavirus_italia_al_ balcone_canzoni_contro_la_paura-251221289/.

Wood, Abigail. 2008. 'E-Fieldwork: A Paradigm for the Twenty-first Century?' In *The New (Ethno) Musicologies*, edited by Henry Stobart, 170–187. Lanham: Scarecrow Press.

11 Becoming Visible

Proud Roma and Sinti Musicians in Italy during the Pandemic

Antonella Dicuonzo

The presence of both historical and more recently arrived Sinti and Roma communities in Italy remains a hot topic in the national public debate. In those areas where Roma and Sinti[1] family groups have moved and progressively settled, relations with the majority society are often marked by conflict, mutual mistrust, and also by a general lack of awareness on the part of the *gagé*[2] of the specific cultural values of these groups, which still embody an image of otherness despite their longtime coexistence. According to Leonardo Piasere, Italy is part of the so-called 'third Gypsy Europe' due to the rather small percentage of Roma and Sinti present on the territory in relation to the total population (2009, 8–9). However, according to a statistic compiled in 2019 by the Pew Research Center, the country ranks first among those in central and eastern Europe for widespread anti-Gypsyism – a phenomenon also highlighted by national press surveys conducted by the National Observatory on Anti-Gypsyism (Wike et al. 2019; see also ONA 2021).

Since their arrival in the territories that later defined the borders of Europe, Roma and Sinti have been able to implement strategies of *immersion* and *dispersion* (Piasere 2009, 69–76) in the world of the *gagé* in order to survive: strategies that have been redefined from time to time over the course of history in response to the policies of exile, forced assimilation, and extermination to which they have been subjected (Rizzin 2020), and that today allow us to view these groups as a 'people-resistance' (Piasere 2009, 89–90). In the wake of the international political and cultural movements for social redemption,[3] associations run by Roma and Sinti began to spread in Italy at the end of the 1980s and beginning of the1990s,[4] as did a proliferation of internal initiatives driven by demands for recognition. Roma and Sinti associations have therefore played a significant role in the progressive process that has led various communities in Italy to emerge from invisibility and make their voices heard: a process that has seen activist-mediators become spokespeople for their family groups, and artists and musicians seeking to re-evaluate their own culture through their works.

In this chapter we will see how the period of the COVID-19 pandemic, characterised by an enforced distance that weighed heavily on the Sinti and Roma kinship-based structure of social organisation, represented an important moment of self-reflection, during which a series of individual and collective initiatives condensed in continuity with the processes mentioned above. These initiatives, which involved a number of musicians from the current Italian Roma and Sinti scene, testify on the one hand to the presence of forms of creativity linked to a feeling of pride in one's own cultural identity, and on the other to the internal perception of the health emergency in response to which strategies aimed at maintaining family ties were introduced. It is precisely the musicians, in fact, who seem to have positioned themselves in the pandemic as the ideal connectors of the many diverse communities present in our country, representing, at the same time, the main channels of dialogue that the Sinti and Roma universe opens to the world of *gagé*.

DOI: 10.4324/9781003200369-14

Becoming Visible   137

First Wave – Stay Home

When Prime Minister Giuseppe Conte announced the national lockdown live on television on 8 March 2020, a series of initiatives involving more and less famous musicians immediately began multiplying: they invited people to stay at home by posting short videos of musical performances on social networks accompanied by the hashtag #iorestoacasa (I stay home). Even some Sinti and Roma musicians joined the campaign to raise awareness in those communities severely affected by the institution of social distancing. A private virtual group that brought together many Roma from central and southern Italy was *La me rom ćas ku kher* (We Roma stay at home).[5] Created on the initiative of Maria De Rosa, a *romni*[6] singer from Abruzzo, the Facebook group was set up with the intention of keeping each other company and maintaining those ties that were being undermined by the health crisis: some of the most popular singers among the Roma in central and southern Italy, such as Rocco Gitano and Max Spada, took part in the initiative by sending short video contributions with extracts from songs and messages of hope addressed to users (Giulia Di Rocco, pers. comm., 9 November 2021).[7] These performances were shared on the singers' Facebook accounts and relaunched on groups also open to *gagé*: this contributed to their greater visibility and dissemination.

The Roma singer from Campania, Rocco Buccino, also known as Rocco Gitano, was perhaps the most active personality of the current Italian Roma music scene during the pandemic: there were numerous live broadcasts from his Facebook channels, during which he performed his repertoire of songs in Neapolitan dialect and in *Romani*. This social media dynamism, with which Buccino had already experimented before the pandemic, allowed him to reconnect with his many followers at a time when his professional activity was suddenly interrupted.

His first pandemic performance on 12 March was a preview of a track that would be included on his forthcoming album. The short musical performance was preceded by the following message:

> To fight the coronavirus, of course I too, Rocco Gitano, am keeping you company, like so many artist friends. So all together, globally, let's sing, let's play, because everything will pass. Fifteen days go by, don't worry: stay at home, please, as I do.
>
> (Buccino 2020a)

Again, on 28 March, another short video from the private group *La me rom ćas ku kher* was shared on the public Facebook group *Friends of Cefferino Gimenez Malla, known as 'EL PELÉ,'* run by the Abruzzo Roma activist Giulia Di Rocco: Rocco Gitano's greeting was addressed to the leaders of the two groups and to 'all the Roma in Italy,' to whom he recommends staying home, referring to the name of the group itself, and to whom one of his most famous songs in *Romani* is dedicated (Di Rocco 2020a).

Similar initiatives were also launched in Trentino-Alto Adige, where perhaps the largest community of Sinti musicians in Italy is concentrated. On 15 March 2020, the Facebook account *Sinti Nel Mondo*, managed by musician Robert Gabrielli, announced a live musical event planned for that afternoon, closing with the hashtag #meciaukhere (I'm staying home). Hosted by Robert (also known as Sereno) together with his cousin Lahi Colombo Gabrielli, it opened with the words:

> *Laćo díves!* Good day! We are here on the balcony of our house: let's stay home! *Me ćau khere!* We are going to do something with the guitar and the violin: we, Sinti, want

to launch this challenge too. Everyone is singing on the balcony: we also want to tell everyone that they can do something with their guitar, with their violin. Or maybe, whoever can sing, sing: something cultural, something Sinti.

(Sinti Nel Mondo 2020)

After informing listeners about their evangelical faith, the two musicians explained that they would first play some 'spiritual' songs and then some 'cultural' Sinti music. The performance lasted about 43 minutes and received various reactions in real time, including comments from acquaintances and relatives, 'amens,' and blessings: during the first part, it quickly turned into a collective prayer. 'Glory to God for these canticles,' said Robert,

unfortunately, as you know, for the regulations ... there are no religious services, but the Lord has found other ways for us to be able to listen to the Word of God through these social networks, Facebook, Instagram, and there's one Instagram profile in particular to follow and that's the adi_mez_off profile. I encourage you, always follow it in large numbers because there our servants of our Zigana Evangelical Mission have the opportunity to be able to bring the Word of God to everybody – Glory to God – to all those who want to hear the Word at home.

(Sinti Nel Mondo 2020)

The health crisis was pressing and deaths from Covid in Italy were rising exponentially every day. The lockdown, contrary to what was initially announced, was extended until May 2020. Scen Gabrielli, a 24-year-old singer from the same musical family in Trentino-Alto Adige, also decided to establish contact with the wider Sinti community by composing a song about the Coronavirus. The song was released on a private Facebook profile on 8 April 2020, International Roma Day: the video montage features a series of photographs of various moments in family life, playing music and being together, alternating with images from events in which Sinti people have participated in recent years, marching with their musical instruments. As Scen recounted, his song says: 'Let's stay home, because the Coronavirus wants to kill us; let's show that if we stay home we can be the ones responsible for the fact that this virus can't do anything to us.' And regarding his creative process:

I came up with it [the song] on the fly, in five minutes. I just stood there and sang it: what came into my head, I sang. But I recorded it on my phone. I thought about what we could get into with this virus ... when it was in the early days and we didn't know it well. If we stay home we can avoid a lot of things, not only getting sick ourselves, but infecting others ... That's why I did it: not just for me, but for everyone ... A lot of people like you, Italians,[8] wrote me and I dictated the text to them; they told me *bravo* for it, I got a few compliments.

(Pers. comm., 14 November 2021)

His cousin Matthew adds that this is how music is invented, 'on the fly': of course, 'you have to listen to yourself, record yourself, because out of nowhere things come out that you can't imagine.' Loris, another relative, adds: 'but then, you know what it is? Music comes from the heart.' And Matthew concludes: 'so that's how you invent a new, special song: and if you didn't record it in that moment, it doesn't exist!' (Matthew Gabrielli, pers. comm., 14 November 2021).

It can be said that social media actually function as an archive for such performances (Macchiarella 2017), which have increased during the pandemic; nevertheless, precisely because of their nature (accounts, pages, and groups can be closed, blocked, or deleted from one day to the next, as can uploaded content), they guarantee neither permanent preservation nor the future possibility of accessing shared materials. And indeed, the video of the song invented 'on the fly' by Scen is now no longer available on the web, because the account from which it was shared no longer exists. Scen also relied on a social medium to transmit the exact lyrics of his song to me: the words came to me sung in *Sinti*[9] and verse by verse translated into *Gagio*,[10] via WhatsApp by voice message. *Sinti* is in fact an orally transmitted language that Scen, by his own admission, is not sure how to put in writing. But, also, an ad hoc recording remains evidently the most appropriate means of conveying the meaning of a song born 'on the fly,' and 'from the heart,' which – if not recorded in that instant – 'doesn't exist' and is lost in the uncertainty of memory.

On 25 April, Italy's Liberation Day from Nazi-Fascism, a number of Roma and Sinti musicians joined the national celebrations online. Santino Spinelli, also known as Alexian, an internationally famous Roma musician from Abruzzo, shared a solo performance on his Facebook channels and on relatively popular WhatsApp groups:[11] it was a very personal arrangement, played on the accordion, of the well-known song 'Bella Ciao.' In the video, behind him, there is a Romani flag and a piano on which the numerous awards he received during his career are arranged. For some time now, the tune has crossed national borders[12] and, especially for Sinti and Roma, is a source of particular pride because it recalls the history of the *brothers* who took part in the Resistance.[13]

Gennaro Spinelli, a violinist and Santino's son, also shared a personal version of 'Bella Ciao': wearing gloves and a mask, and with the Italian flag hanging from his violin, he played the tune over an audio track of the song. The performance is accompanied by subtitles that follow the lyrics, and the post has the following caption: 'Liberation, Resistance, Resilience in times of #Covid19 That's how we respond #BellaCiao! Dedicated to all those who united "RESIST"!' (Spinelli 2020).

The Facebook page of *Movimento Kethane – Rom e Sinti per l'Italia* shares a performance of 'Bella Ciao' by Damiano Cavazza from Lucca, with the caption: 'Today we make BELLA CIAO resound, the song of resistance from all over Italy. From all our communities, fighting together for a better Italy! … Now and always, resistance!' (Movimento Kethane 2020). On the same page, a video montage appears later, accompanied by a version of the song in Italian and *Romani* by the Milanese children's choir Kethane, led by the musician Eliana Gintoli.

If Gennaro Spinelli and Cavazza himself lent their faces to several so-called pandemic performances, Santino Spinelli shared such content only once during lockdown. The musician – who in his career has proudly 'played for three popes,' participated with his sons in the most famous international festivals of Romani music, and appeared on important stages in Italy and abroad – was held back by the health crisis, choosing silence, or a form of memory evocation through sharing videos of past concerts. On the subject of making music online, Spinelli says:

> It's horrible … because it destroys the emotionality and the presence of the audience, which becomes part of the concert. Emotionality can only come from being there – in a theatre, in a suggestive place, with special acoustics – and live: experiencing it. *Live* means *living*: you have to live the music, you have to breathe it, you have to be moved by it.

On the situation of the Roma during the pandemic in Italy, Spinelli adds:

> They were even accused of being *plague-spreaders*, as if the Roma had brought the pandemic when it already existed. In some Italian cities this happened, with the media

140   *Antonella Dicuonzo*

clamour … The media make the difference, as Malcolm X used to say: they have the power to make a perpetrator out to be a victim and a victim out to be a perpetrator.

(Pers. comm., 5 October 2020, emphasis author's)

In May 2020, a scandal hit a Roma community in Campobasso, the capital of Molise, following the funeral of one of their elders, which was attended by many people. About two weeks after the funeral, an outbreak developed in Campobasso, with numerous cases among the Roma (Mancinelli 2020). This event led to a strong conflict with the city's *gagé*, creating a national case that the media exploited for several weeks. The singer Carmine Bevilacqua, also known as Mino Vastano, who had left his home town to attend the funeral of his father-in-law in Campobasso, was also caught up in the scandal after a spike in contagions occurred in his apartment building after he returned to Vasto, Abruzzo. Roma funerals, which are highly attended events and real moments of social cohesion, have always been the subject of misunderstanding[14] and became, in times of pandemic, a further reason for division. Mino Vastano, who publicly apologised to the townspeople for what had happened, immediately received numerous video messages of support from various musicians and singers, including Rocco Gitano, who commented: 'I would like to send a warm hug to my cousin Mino Vastano … I would also like to announce that there will be a surprise, as soon as everything is over … we will make a duet … a big surprise for our people, for those who love us, for our fans, of whom there are so many' (Buccino 2020b).

On 8 April 2020, International Roma Day, a large virtual contest was launched across Europe. The initiative, entitled Stay Home and Make Music, was started by ERIAC, the European Roma Institute for Arts and Culture. It was aimed at families of Roma musicians, who were invited to send short videos in which they play and/or sing from home (ERIAC 2020a). Cash prizes of 1,500, 1,000, and 700 euros were up for grabs for the top three. In just over a month, 160 videos by 365 musicians from 20 European countries were received and published on ERIAC's Facebook page. From Italy, two young activists of the Kethane Movement took part: Damiano from Tuscany, already featured in the musical performances that appeared on the movement's Facebook page, and Consuelo from Calabria. They decided to participate together, albeit from a distance, under the encouragement of the movement's leader Dijana Pavlovic. Their song was entitled 'Odio' (Hatred) (ERIAC 2020b) and, as Damiano explained, was a song he had composed years earlier:

It goes back to more or less five or six years ago … It is very much linked to current affairs … the State, the Government, [as the refrain goes] *the troubles of the world* … The evil of this society, we know, is that there is a large percentage of people who are not even aware that they belong to life itself.

(Pers. comm., 11 November 2021)

The song, therefore, does not speak specifically about the condition of the Roma/Sinti but is a sort of denunciation of the indifference in which a large part of society lives. It is interesting that in the textual concatenation of verse-bridge-refrain – built respectively on words in Italian, English, and Spanish – a short phrase in *Sinti* is placed at the end: 'In the *Sinti* part I say: *asuném, ciavále*, listen, people, *t'aven báxtale*, be lucky. You know this expression, I think: may luck be with you, may it accompany you' (ibid.).[15]

The song has a funk flavour; Consuelo, the drummer and saxophonist, inserted some instrumental interludes, and the audio-video editing was put together thanks to Roberto, her partner, who owns a recording studio. Although the performance did not reach the winners' podium, it won over 500 reactions and numerous comments, reaching over 7,000

Becoming Visible 141

views. Damiano, whose musical experience began with the neomelodica genre and then moved on to international pop, rock, and funk, was no stranger to composing songs in *Sinti*:

> You know the risk you run, don't you? There's a risk that [the song] won't even be heard. But it's OK: it seems that up to now, anyway, something has been done with this idea of singing in *Sinti*; and then it's obvious that you have to sing in *Sinti*. ... it's a great opportunity for those who can do it ... [The language] changes over time, one develops other concepts ... one is enriched, one can also use other expressions ... It is important, it is beautiful.
>
> (ibid.)

The topic of preserving the language as a means of carrying forward one's own culture soon led to talking about the general cultural impoverishment that the Roma and Sinti communities have been experiencing for some time. According to Consuelo,

> Down in Calabria, the communities have completely lost the *Romani* language ... and only the elderly, they just say a few words, and this is a very bad thing. I must say that down there they don't even know about the *Porrajmos*,[16] they don't know anything at all, they don't even know about our flag.
>
> (Pers. comm., 11 November 2021)

Culture, identity, and belonging are words that recurred in the dialogue I had with the two young musician-activists, and which immediately went beyond the song object with which we had started. Consuelo told me that

> there hasn't been a serious awareness process: in the communities, very superficial work has been done. You know that stuff about dependency culture, don't you? I give you a bag of groceries, you stay good and well: you stay there, on the margins of society, so you are not part of anything, you are nobody. All this has led to a degradation that unfortunately continues.
>
> (ibid.)

And when asked how music can help in this situation, Consuelo replied:

> Music can save a lot of kids. We need an organisation to go to the camps, to the neighbourhoods, to see these kids, because there are so many talents. And then you have to work with those talents. ... in Milan, there is Eliana Gintoli who has created orchestras of Roma children: she follows them, makes them play ... But see, it is a *gaǵi* who has taken responsibility.
>
> (ibid.)

Damiano added: 'it is useless to bring up the little competition that Consuelo and I participated in, with all due respect for ERIAC ... but if there were situations in which there is a real investment in talent ... then there could be development.' And Consuelo concluded:

> Eliana gives [Roma children] immediate gratification. So, parents see that there is something there, and they invest in it. But invest in what way? You know what it means for our children to be sent there for an hour of music: to get out of those realities and go

142 *Antonella Dicuonzo*

and make music is fundamental. ... Because music is also about respecting others, as Sennett said.[17] I read it some time ago, he says that music already gives you an education because, in any case, you have to listen to others. ... For our children, listening to others is already a lot.

(ibid.)

Towards the Second Wave and Beyond

With summer, the contagion curve began to fall. The network of social relationships set up during the lockdown period paved the way for a series of initiatives that finally saw the light of day. One of the most important is the creation of a political party that can provide representation for the Italian Roma and Sinti. As Giulia Di Rocco, one of the founders, explains, the idea was born during the confinement period: 'The pandemic ... led to a greater confrontation between our associations.' The founders involved singer Rocco Gitano, who immediately agreed to compose the Party's anthem, whose title 'Mistipe' encapsulates a cultural concept that represents respect and mutual love for the Roma (Giulia Di Rocco, pers. comm., 9 November 2021). For Giulia, Rocco is a musician involved in 'popular' music, the kind that 'is closest to the Roma world.' She appreciates the singer's humanity: talking to me about one of the recent events concerning the presentation of the Party organised in Isernia, Molise, she told me that Rocco had come from Campania to participate, 'bringing his audio equipment with him, because he believed in the cause. ... They call him from North to South: he started out as a neomelodica singer but then he really took a stand by singing in *Romani*' (ibid.). Rocco, for his part, dedicated numerous posts to the project, which aspires to reunite the Roma and Sinti communities on Italian soil. His anthem, like Scen's song, took shape immediately: the musician 'felt inspired,' said Giulia (ibid.), and composed it all in one go, text and music, accompanying himself on guitar. However, a preview of 'Mistipe' was only published on Facebook on 31 August 2021 (Di Rocco 2021b).

Another singer who, during the pandemic, set out to enhance her cultural identity was Eva Pevarello: a young Sinti from Veneto, known nationwide for participating in the television programme 'X Factor' and the Sanremo Festival, Eva announced the release of a new single on her Facebook page in autumn 2020. The caption of the photo, depicting her among the caravans of an amusement park, read: 'I am a gypsy, daughter of the wind, I bring my origins on a golden plate. Who is poor, being loved?' (Pevarello 2020a). And the single arrived, like a gift to her audience, on Christmas night, with this explanation:

'Tine mal.' A song in *Sinti*, my language of origin. Written by my father, he accompanies me on guitar. We recorded this song live ... but I didn't want to touch anything on this track ... because I hope I can convey some of the emotion I felt that night. As soon as we finished recording, I burst into a valley of tears, I hadn't been so moved by singing for a long time.

(Pevarello 2020b)

'Tine mal' (Little Friends) is a song that celebrates the value of friendship, as Eva explained. The tune is actually well known among the Sinti of northern Italy – probably a song whose origin dates back to less recent times.[18] The video, set in the Fantasiland amusement park in Torvaianica, Pomezia, run by Sinti Cesare Calzolai, was filmed during the photo shoot that saw Eva among those featured in volume four of *Survivors Rhapsody*, published by Flewid magazine. Born out of the lockdown experience, 'Tine mal' represents a pivotal point in the singer's musical output: although she had never made a secret of her origins, these now seem to be starting to trace a path to be followed, both from a creative point of view (a song

Becoming Visible 143

released in December 2021 by XXXXL and Eva also includes a refrain in *Sinti*)[19] and from an image-related one. After this period, Eva decided to support the cause of Sinti and Roma activists in the country, creating a short video contribution for International Roma Day 2021 that was launched on the Facebook pages of the Kethane Movement (2021) and Men Sinti (2021), an organisation that brings together many Sinti in northern Italy.

A final mention – in relation to this swarming of projects involving Sinti and Roma musicians during the pandemic – is warranted by the collaboration of Rocco Gitano with the Italian-French rapper Speranza: in autumn 2020, during what would turn out to be the second wave of the virus, the single 'Camminante' was released, as part of the album *L'ultimo a morire* (Sugar Music, 2020). Speranza, the stage name of Ugo Scicolone, grew up in a dormitory neighbourhood in Behren-lès-Forbach, in France's Grand Est, and returned to Caserta, Italy, his father's hometown, after his schooling: his music resonates with the harsh existence of those who live on the margins. As the rapper said during an interview with *Rolling Stone* on the eve of the album's release:

> I grew up around a lot of ethnic groups, and once I was in Italy I started hanging out with the Roma often. By dint of hearing them speak, I started to delve into the language: it fascinated me a lot, so I wanted to do this madness of trying to use it to rap … Roma are the classic easy target, a common enemy who has never had a voice. For things to change, we have to fight degradation with education and inclusion.
>
> (Tripodi 2020)

Speranza's collaboration with Rocco Gitano takes as its point of departure the song that marked the Roma singer's debut as a child, "O Zingariello' (the Little Gypsy):[20] its refrain is placed between the rapped verses of 'Camminante.' The new song's title already condenses a shift in meaning from the degrading condition of gypsy to that of an individual in perpetual movement, who no longer begs to be considered like everyone else but shows off his power to the extreme. The song proved to be a real success, and by 2022 had over 250,000 views on Speranza's official YouTube channel.[21]

Conclusions

What happened, then, in the Italian Sinti and Roma musical world during the pandemic? We have seen, by scrolling through a few examples of musical performances and songs published in this period, how, despite the diversity of genres and styles, these musicians all move towards an ideal of pride in their cultural origins, origins that become a personal brand (Rocco Gitano); that reconnect deep ties beyond the distance imposed by the pandemic (Robert, Lahi, Scen, and Matthew Gabrielli); that are constructed through paths of activism and knowledge (Consuelo and Damiano); that are projected on a national and European dimension (Gennaro and Santino Spinelli); and shown to the world 'on a golden plate' (Eva Pevarello). These paths go beyond awareness – a sentiment that has probably already matured in the souls of these musicians – and appear to be oriented by the determination not to remain silent in the face of the emergency. At the same time, these paths show others – whether in the *small* family and social universe of Sinti and Roma or in the boundless world of the web – their own essence, which is no longer a reason for concealment. It is fair to say that these paths did not emerge from the pandemic, but they were certainly accelerated and at times inspired by it: the desire to reconnect at a distance, to maintain strong social ties, to sensitise families to stay at home and, at the same time, to shake up society so that it considers and recognises a multifaceted cultural world, which lives *within* our own but which, all too often, is placed on a parallel track.

144   *Antonella Dicuonzo*

Perhaps the most extreme example of this desire for visibility, and at the same time of the importance of reconnecting family ties across distance, is a number of funerals that were live streamed, both by local broadcasters and private accounts. For example, that of Santino's father Gennaro Spinelli, one of the last witnesses of the Nazi-Fascist deportations, who died after contracting COVID-19 in February 2021: the funeral rite was broadcast live by *abruzzolive.tv* and was physically attended by only a small number of people, with the police guarding the space in front of the Lanciano Cathedral, after the events in Campobasso and Vasto. Santino and his sons, in quarantine, followed the funeral from a distance, as did the family's wider network of relatives and numerous friends and acquaintances.

And before that, in January 2021, the funeral of little Tristan, a Sinti child just four years old and suffering from a serious form of leukaemia, for whom the entire Sinti community of northern Italy, supported by the Kethane Movement, mobilised. This funeral procession, live streamed from a relative's Facebook account, was very well attended. A group of men wearing masks carried the small white coffin on their shoulders towards the cemetery; behind them, a long line of people wound its way through the rain. A few cars opened and closed the procession, proceeding at a walking pace and rhythmically honking their horns. The man filming on the other side of the phone said 'we are here to accompany our cousin Tristan. Even the sky is crying with us.' This is the story, from an inside perspective, of a crucial moment for the community: despite the prescriptions and possible judgements of the *gagé*, what mattered was to reconnect the community in the widest possible way, using means that guaranteed greater visibility. In front of the threshold of the cemetery some children were waiting, ready to release doves and make white balloons fly into the sky. There was a big round of applause, and the commentator said movingly, 'Goodbye, cousin!' The crowd sang the song 'Alzo gli occhi al cielo' (I Lift My Eyes to the Sky), which accompanied the little coffin from its entry into the cemetery to the niche, where a last song was sung in *Sinti*, and the child was bid farewell with a long round of applause.

I have been able to observe these dynamics from a privileged position: being in contact with some of the people mentioned in this article allowed me to access documents that are not publicly shared on the web, and to be informed, almost always in real time, about the musicians' live broadcasts, in addition to being able to watch performances shared among smaller groups (through WhatsApp, for instance). At the same time, my personal relationships with some of the musicians featured here helped me to have a fuller view of the phenomena observed on the web: discussing those same performances with their performers, even after the pandemic, was essential for understanding points of view and ideas about music and identity that inevitably go beyond that moment in time and space.

It remains difficult, however, to assess the extent to which the gradual unveiling of these musicians has an effective resonance beyond the social and cultural universe they come from: if better known personalities, such as Eva Pevarello and Santino Spinelli, reach a greater audience at the national level, other musicians seem for now to remain anchored to a local and/or internal dimension in which Sinti and Roma listeners are the most faithful supporters of their music. And yet, I have good reason to believe that the possibility of becoming visible as a people in the world of *gagé* is also in the hands of these lesser-known musicians, who – with their art – work to consolidate the internal relations within the Sinti and Roma universe, and convey an image of its truth and complexity.

Notes

1 These are self-designations that define groups settled in different areas of the Country. Leonardo Piasere (2009, 15–17) traces this distinction, whereby Sinti are more present in central-northern Italy and Roma in central-southern Italy, to a broader European distribution of these

Becoming Visible 145

groups: communities that call themselves Roma settled mainly to the east of a line that connects Rome to Helsinki via Vienna and Prague, while to the west of it Roma are in the minority compared to other groups that call themselves *Sinti*, *Manuś*, *Kale*, and *Romanićels*. Roma and Sinti arrived in Italy through several waves of migration from the 15th century onwards: the first account of their arrival in the city of Bologna, contained in two contemporary or partly contemporary chronicles, dates back to 18 July 1422 (Piasere 2011, 41–66). The permanence of these family groups gave rise to communities 'of ancient settlement' (ibid.), with Italian surnames and situations of greater or lesser integration in the social fabric of the inhabited cities. However, groups that arrived in Italy during the most recent migrations from eastern Europe will not be considered here.

2 A term by which Sinti and Roma define non-Sinti and non-Roma (masculine *gagó*, feminine *gagí*, masculine and feminine plural *gagé*), a word that condenses 'the expression of otherness that individual Roma communities have constructed over time' (Piasere 2009, 27).

3 An important moment was marked by the First World Roma Congress held in London in 1971, which 'gave the Roma nationalistic symbols that will remain: a flag (blue and green with a red wheel in the middle), an anthem ("*Ǵelem, ǵelem*," adapted by Jarko Jovanović from a Yugoslav Roma folk song), and a date to celebrate (8 April, the first day of the congress, will be considered as the *Romano dive*, Roma Day)' (Piasere 2009, 116–117).

4 To name but a few: *Nevo Drom* (Bolzano and Trento), *Sucar drom* (Mantua), *Upre Roma* (Milan), *Thèm Romanó* (Lanciano and Reggio Emilia), *Amici di Zefferino* (Lanciano), *Rom in progress* (Isernia), *Lav romanò* (Cosenza).

5 For virtual group names and hashtags in *Romani* (the term used to define the orally transmitted language spoken by Roma and Sinti) I will use, when available, the transcription adopted by the members of the communities themselves. Alternatively, a phonetic transcription will be provided (as in the case of the group *La me rom ćas ku kher*), taking into account that the internally diverse linguistic variants spoken by the Roma in central-southern Italy are different from those spoken by the Sinti in central-northern Italy and that, in both cases, *Romani* is strongly influenced by regional Italian and local dialects.

6 Roma woman.

7 As this is a closed group for Roma in the Centre-South of Italy, I was not a member and was unable to find any links on Facebook. It is possible that the group has special restrictions that prevent Facebook users from tracking it, or that it no longer exists. However, both Giulia Di Rocco and her uncle, the musician Alexian Santino Spinelli, told me about it on different occasions.

8 Although my Sinti and Roma interlocutors are also Italian, in my experience, this term is sometimes used instead of *gagé*, mainly to make oneself understood (the *gagó* generally does not speak *Romani* and does not know what *gagó* means), but probably also not to offend an interlocutor who knows that term, as the word *gagé* can have an oppositional, and often negative, connotation.

9 That is, in the *Romani* spoken by the Sinti.

10 That is, in the language of the *gagé*, in this case Italian.

11 Together with Santino Spinelli, I am in several such WhatsApp groups.

12 One thinks, for example, of its recent use in the cult Netflix series *La casa de papel*; but Roma in Spain used it in 2018 to show solidarity with Italian Roma and Sinti after the then Interior Minister Matteo Salvini planned to activate a reconnaissance procedure for Roma in Italy that brought to mind Nazi-Fascist ethnic profiling (see Vega de los Reyes 2018).

13 It is worth mentioning the partisan messenger Erasma 'Vicenzina' Pevarello, singer Eva Pevarello's grandmother, whose story is told by Rui (2012).

14 A case that aroused great indignation in public opinion was the funeral of the organised crime boss Vittorio Casamonica in August 2015 in Rome, which nevertheless featured some elements traditionally present in Roma funerals, especially of the elderly: the use of a carriage to transport the coffin, the use of a band in the funeral procession, and the large number of relatives.

15 *T'avés baxtalò* (may you be lucky) is a form of transnational greeting and wish that is common among both Roma and Sinti.

16 The term *Porrajmos* (Great Devouring) is used to define the genocide of Roma and Sinti during the Second World War (see Bravi 2011). In Italy, Santino Spinelli promoted the replacement

146 *Antonella Dicuonzo*

of the term (which in *Romani* has an etymology that refers to the sexual sphere) with the word *Samudaripen*, literally All Dead.

17 She is referring to the work of sociologist Richard Sennett.

18 I found an earlier version of the song sung in Piedmontese by a Sinti performer on the net: www. youtube.com/watch?v=_SOP4omAR5U.

19 www.youtube.com/watch?v=4aPwp39rj4Y.

20 www.youtube.com/watch?v=9XNilEClZ_U.

21 www.youtube.com/watch?v=o3brHCgCsZE.

References

Bravi, Luca. 2011. 'La persecuzione e lo sterminio dei rom e dei sinti nel nazifascismo.' *VOCI* 7–8: 2–367.

Buccino, Rocco (@RoccoBuccino). 2020a. '(Live) … Zingara do mare …' Facebook, 12 March 2020. www.facebook.com/roccogitanoofficial/videos/1379943038842787.

Buccino, Rocco (@RoccoBuccino). 2020b. 'Video messaggio per mio cugino.' Facebook, 22 May 2020. www.facebook.com/roccogitanoofficial/videos/1441111606059263.

Di Rocco, Giulia (@friends of Cefferino Gimenez Malla aka 'EL PELÉ'). 2021a. "Grazieeee Rocco Gitano." Facebook, 28 March 2020. www.facebook.com/groups/1792151691101418/posts/2527434100906503/.

Di Rocco, Giulia (@friends of Cefferino Gimenez Malla aka 'EL PELÉ'). 2021b. 'Un anteprima dell'inno del Partito Nazionale Rom e Sinti.' Facebook, 31 August 2021. www.facebook.com/groups/1792151691101418/posts/2953540281629214/.

European Roma Institute for Arts and Culture – ERIAC (@EuropeanRomaInstituteERIAC). 2020a. 'ERIAC is Calling All Roma Musicians.' Facebook, 8 April 2020. www.facebook.com/EuropeanRomaInstituteERIAC/posts/2660197467563997.

European Roma Institute for Arts and Culture – ERIAC (@EuropeanRomaInstituteERIAC). 2020b. '#ERIACMusicContest Presents Damy and Consu Band.' Facebook, 1 May 2020. www.facebook.com/EuropeanRomaInstituteERIAC/posts/2680355952214815.

Macchiarella, Ignazio. 2017. 'Making Music in the Time of YouTube.' *Philomusica on-line* 16 (1): 1–11. http://dx.doi.org/10.13132/1826-9001/16.1873.

Mancinelli, Silvia. 2020. 'Boom di positivi dopo funerale rom, Toma: "Ma il Molise è ancora virtuoso."' *Adnkronos*, 13 May 2020. www.adnkronos.com/boom-di-positivi-dopo-funerale-rom-toma-ma-il-molise-e-ancora-virtuoso_5QAVPAc2XYwlECQqh646wf?refresh_ce.

Men Sinti (@mensinti). 2021. 'Eva (cantante).' Facebook, 8 April 2021. www.facebook.com/mensinti/videos/1478645489149411.

Movimento Kethane (@KethaneRomeSinti). 2020. 'Oggi facciamo risuonare BELLA CIAO.' Facebook, 25 April 2020. www.facebook.com/watch/?v=563020001030016.

Movimento Kethane (@KethaneRomeSinti). 2021. 'Eva Pevarello (cantante).' Facebook, 8 April 2021. www.facebook.com/KethaneRomeSinti/posts/2873578299637059.

ONA (Osservatorio Nazionale sull'Antizaganismo). 2021. https://sites.dsu.univr.it/creaa/osservatorio/.

Pevarello, Eva (@EvaPevarello). 2020a. 'Sono una gitana.' Facebook, 7 September 2020. www.facebook.com/permalink.php?story_fbid=1017654842031835&id=102766853520643

Pevarello, Eva (@EvaPevarello). 2020b. '"Tine mal" Una canzone in Sinto.' Facebook, 24 December 2020. www.facebook.com/eva.pevarello/posts/1803734793109042.

Piasere, Leonardo. 2009. *I rom d'Europa. Una storia moderna.* Roma-Bari: Laterza.

Piasere, Leonardo. 2011. *La stirpe di Cus.* Roma: CISU.

Rizzin, Eva, ed. 2020. *Attraversare Auschwitz. Storie di rom e sinti: identità, memorie e antiziganismo.* Roma: Gangemi.

Rui, Irene. 2012. *Erasma, Vicenzina, Pevarello, storia di una sinta italiana.* Vicenza: VampaEdizioni.

Sinti Nel Mondo (@SintiNelMondo). 2020. 'Laćo díves!' Facebook, 15 March 2020. www.facebook.com/SintiNelMondo/videos/3425993717418028.

Spinelli, Gennaro (@Gennaro Spinelli ArtPage). 2020. 'La Liberazione, La Resistenza, La Resilienza.' Facebook, 24 April 2020. www.facebook.com/GennaroSpinelliGS/videos/316479146008402.

Tripodi, Marta Blumi. 2020. 'Speranza non è il rapper di strada che vi aspettate.' *Rolling Stone*, 15 October 2020. www.rollingstone.it/musica/interviste-musica/speranza-non-e-il-rapper-di-strada-che-vi-aspettate/535035/#Part1.

Vega de los Reyes, José (@JoséVegadelosReyes). 2018. 'Matteo Salvini recuerda.' Facebook, 26 June 2018. www.facebook.com/josevegadelosreyes/videos/2142812379079735/.

Wike, Richard, Jacob Poushter, Laura Silver, Kat Devlin, Janell Fetterolf, Alexandro Castillo, and Christine Huang. 2019. 'Minority Groups.' In *European Public Opinion Three Decades after the Fall of Communism*. Pew Research Center. www.pewresearch.org/global/2019/10/14/minority-groups/.

12 Rethinking Intermedia Practices during the Pandemic

Staging and Conception of Alexander Schubert's Virtual Reality Video Game *Genesis*

Luca Befera

Everything is encoded
and decoded.
We are parametric.
Everything parametric can be altered
parameter by parameter.
That is the definition of such a model.
Normally we don't see these sliders,
these values,
these adjustment dials,
but they can move into our consciousness
through illness
through hallucination
through drugs
through psychotic states
or through computation processes.

Alexander Schubert, Convergence *(2021),*
text excerpt projected during the performance

Introduction

Since the beginning of 2020, the emergence of the pandemic dramatically changed living conditions, implying isolation and an ever-growing need for human interaction. Digital communication consequently thrived, leading to the hegemony of computer mediation and the redefinition of relationships in everyday life. People employed telecommunications to supply social closeness (Aron, Mashek, and Aron 2013) and limit the perceived separation (Nimrod 2020; Roberts and David 2019). Leisure activities were consequently shifted to the digital sphere, with a preference for media that enabled synchronous interaction (Meier, Noel, and Kaspar 2021). For example, instant messaging applications, social media, and video game usage spread dramatically around the world (Steinert 2020; Vuorre et al. 2021). Also, the digital surge shifted working conditions towards internet dominance and home employment (De', Pandeyb, and Pal 2020).

Even if they were growing significantly, these trends were not new. The so-called human-computer interaction (HCI) has been a salient factor of social communication over the course of recent decades. Scholars have described the current period as the HCI 'third generation' (Norman and Kirakowski 2018, 3), in which it is not only possible to connect with people through the World Wide Web and make use of sophisticated screen technology, but digital aspects have also become pervasive in everyday life. Augmented and virtual realities are

DOI: 10.4324/9781003200369-15

Rethinking Intermedia Practices during the Pandemic 149

rooted in this dimension, reflecting an interface that is everywhere (ubiquitous and mobile), visual (watching and showing), conversational (talking and listening), and smart (ibid.). The pandemic also confirmed pre-existing digital inequalities based on socioeconomic status, age, level of education, and internet experience (Nguyen, Hargittai, and Marler 2021; Schumacher and Kent 2020), thus enhancing the informatic habits of specific social groups.

Genesis (2020) by Alexander Schubert was closely related to the aforementioned issues, encompassing in its very concept the attempt of setting up a virtual artwork to be attended from home (Barber-Kersovan and Kirchberg 2021; Velvick 2021).[1] Compared to similar computer-based theatrical works such as those made by Blast Theory or BeAnotherLab (Dixon 2007; Giannachi 2004; Jarvis 2017), *Genesis* stood as an unprecedented attempt to provide an immersive performance over an extended period and within a dynamic of control over human beings.[2] Indeed, the work was a web-based video game conceived as an experiment, which took place continuously over seven days – from 00:00, 27 April to 24:00, 3 May. Anonymous home gamers from all around the world controlled four avatars impersonated by human actors in one-hour slots. Connecting on a dedicated website, users gave verbal commands to performers. The experiment started in an empty industrial hall of 40 metres by 25 metres, whose borders could not be crossed. Avatars wore VR-resembling glasses equipped with microphones and cameras, through which gamers saw and heard from the first-person perspective of the avatars. Also, a graphic interface augmented the screen view (see Figure 12.1), recalling the appearance of video games through four bars in the top-right corner, indicating the avatar's attributes (fatigue, hunger, thirst, and temperature); a chat window on the bottom-right corner to communicate with other gamers; and the inventory on the bottom-left corner, to select up to ten items from about 2,500 objects. Avatars interacted with the closed environment, the items, and the other avatars, also replying or manifesting their needs with essential pre-set pop-up messages sent by wristband device buttons. Items remained inside the hall for the whole performance once they entered. Thus, gamers interacted within an ever-changing configuration of space. The performance was also streamed on YouTube for its entire duration.

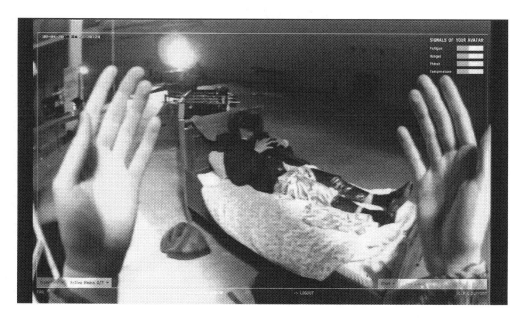

Figure 12.1 *Genesis* game interface (courtesy of Alexander Schubert).

150   *Luca Befera*

Two prominent topics emerged from *Genesis* in relation to the pandemic trends: first, it mainly involved people linked to the German composer's environment, and therefore also interested in the application of new technologies;[3] and second, it fostered a world-wide interaction between human beings through the internet. In both cases, digital devices became tools to establish and frame relationships, filtering them through specific settings and syntax. Participants thus established a connection with performers or other gamers through the interface, despite lacking sensory stimuli and physical sociality.

The performance was designed before the pandemic occurrence and the final setting stuck by the original idea despite the emerging limitations. Therefore, I posit here an already existing relationship between the pandemic informatics hegemony and the most recent works by Schubert. I aim at inquiring this assumption through the analysis and contextualisation of *Genesis* preliminary stages and final enactment, assuming the preordained immersive and digitally mediated environment as significant evidence in itself.

Theoretical Background

In an earlier publication, I analysed *Genesis* from the standpoints of performance evolution, participants' involvement, and communities' development (Befera 2021). Here, instead, I will study its enactment, by relying on Schubert's written texts and video documentation; recorded material including pictures and videos made before, during, and after the performance; and a dedicated online interview.[4] More specifically, I will focus on Schubert's strategies for stimulating the participants' will to believe (in game accuracy, author aesthetics, and human relationships). The concept of 'will to believe' (James [1896] 2005, 95) belongs to radical empiricism, where it is assumed that any belief depends on the many interconnected and experienced factors that together justify the statement of its truth. Richard Grusin further related this assumption to the mediation process, which 'does not stand between a preexistent subject and object, or prevent immediate experience or relations, but rather transduces or generates immediate experiences and relations' (2015, 138). Therefore, I am referring here to the will to believe not only about contextual elements, but also concerning their experienced mediation. This process relies

> on intermediality as a form of identity construction that goes beyond any particular media; on the basis of what connects all the ideas related to a specific performance, it builds itself into a broader picture that is subjective and strongly related to the individual knowledge of the listener/spectator.
>
> (Bratus 2019, 14–15)

Being medially crafted, the performance's reality depended on specific predetermined suggestions and rules orienting the audience's perception and active participation. Therefore, I will study the symbolic and mediated universe through which *Genesis* was defined, which not only fostered the perceived closeness of avatars, items, and space, but also involved the authentication of Schubert's aesthetic itself.

Two points of Alessandro Bratus's tripartite framework of authentication will be considered:

- the *performative concept*, implying the pre-existing 'complex of ideas, opinions, anecdotes, and conventions regarding the social uses and interpretations potentially mobilized by a particular cultural object' (2019, 54);
- the *performative event*, 'expressed as the witnessing of an originating circumstance, confirming the presence of an action authenticated through the bodily presence of the performer' (ibid.).[5]

Hence, the authentication process can be summarised in the following stances: *after* media, considering the role generally played by information technologies in the contextualisation of the performance within the author's poetics; and *through* media, regarding environmental aspects pre-set by Schubert to foster specific possibilities of communication within the piece. The comparison between these aspects and the pandemic influence will show to what extent *Genesis* reflected or rejected the emerging digital instances, highlighting some focal points and evolving patterns of Schubert's aesthetics.

After Media

Mapping Schubert's Latest Production

Digital-mediated communication between human beings has been a focal point of Schubert's entire artistic production. Especially in recent years, the German composer has aimed at creating immersive settings by mixing electronic objects and analogue devices, where 'the digital should always shimmer through, through a tangible, warm, human surface' (Schubert 2021d, 14). While pieces before 2016 dealt with a canonical staged representation, his most recent artistic research has involved digitally mediated 'experimental setups with an open outcome … to learn, together with the audience, something about the changing modes of perception and interaction in the post-digital age' (ibid., 15). This renewed approach has included pieces composed both before and after *Genesis*. Figure 12.2 maps the salient factors of Schubert's production over the last four years:[6] virtual realities, artificial intelligence,

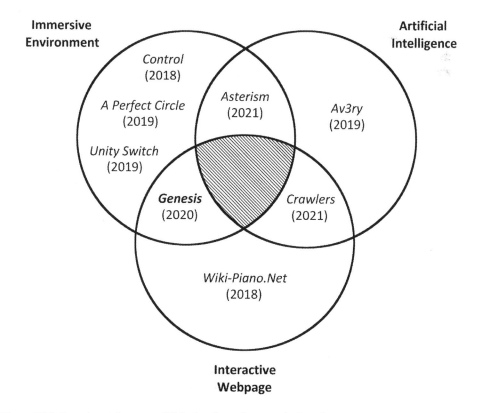

Figure 12.2 Prominent features of Schubert's works over the last four years.

152　*Luca Befera*

and internet-mediated communication stand as cornerstones for concepts' articulation and formal management.[7]

Control (2018) was a participative installation in which, similarly to *Genesis*, users commanded avatars by joining a dedicated setting.[8] *Unity Switch* (2019) and *A Perfect Circle* (2019) were also based on a similar embodiment through VR glasses: the former provided for audience members and actors to see from the perspective of each other while both performing actions in front of a table; the latter involved a therapeutic group session directed by a voice-over. Even if providing an immersive experience, these pieces were not entirely tailored to 'virtual reality' as a 'synchronous, persistent network of people, represented as avatars, facilitated by networked computers' (Bell 2008, 2–3): users related to embodied avatars and received immediate (synchronous) feedback, but computer-mediated environments were not 'persistent,' meaning they lacked the property of existing independently from participants. Furthermore, the installations took place in dedicated locations, addressing 'the viewer directly as a literal presence in the space' similarly to installation artworks (Bishop 2005, 6). Therefore, a website through which relating was not strictly necessary, even where interaction was computer-mediated. Instead, *Wiki-Piano.Net* (2018) is an ongoing piece that allows users to edit an online score by inserting musical notes, texts, videos, and images. A pianist occasionally interprets the result. Even if the piece does not make use of avatars, people are remotely located and interact on the same webpage. Therefore, it allows the emergence of a persistent online community where the environment and social interaction are influenced by the piece's mechanics, interface, and users' choices (Cheng, Farnham, and Stone 2002, 105; Williams et al. 2006).

Over the latest years, the German composer has also been interested in artificial intelligence, presumably due to the relevance of bioinformatics in his training (Schubert 2021a; 2021d). *Av3ry* (2019) is an artificial persona that creates tracks, poems, and images. People might chat with the bot on Telegram, making requests and obtaining a digital outcome. *Crawlers* (2021) and *Asterism* (2021) also embed this algorithmic process on different levels: the former implies a collective of anonymous bots establishing a parallel social network through users' Facebook data; the latter regards an augmented reality installation where an AI voice-over communicates with the audience according to automatically reproduced text sequences. Therefore, Schubert has gravitated around the same macro-categories, progressively mixing them towards specific aesthetic needs. These poles have become connotative of his symbolic universe, also reflecting significative trends by which the audience authenticates his production. *Genesis* merged the web-based interactivity of *Wiki-Piano.Net*, the intimate embodiment of *Unity Switch* and *A Perfect Circle*, and the hierarchical dynamics of *Control*, focusing on the emerging events within the industrial hall. The piece stood as an application of a broader and well-defined technological horizon, where digital mediation became distinctive not only through specific processes, but in its intrinsic nature.

Aesthetic References

Some specific theoretical references and sociological reflections have brought Schubert to define his production towards recognisable aesthetics. First, he reflected on the increasing blending of the technological and human dimensions of the post-digital era (Negroponte 1998), as transcending novelty and integrating into everyday life:

> digitality disappears from our field of vision: for example, because it becomes invisible, or because we no longer see it, it being omnipresent . . . If we are to understand the permeation and invisibility of digital content as a deception, the process of

Rethinking Intermedia Practices during the Pandemic 153

disillusionment can be seen as a tool that directs the gaze back to digital content in everyday analogue life.

(Schubert 2021d, 31–32)

The concept of technology disenchantment discussed here, as a moral reflection on the mentioned digital issues, also means rendering visible. Considering the performative point of view implied in the interactive settings, this process might be further conceived as a 're-enchantment,' where human beings are encouraged 'to enter into a new relationship with themselves and the world' (Fischer-Lichte 2008, 207). Insofar as a *mise en scène* evolves in real-time and concerns the audience's active participation, it triggers liminal experiences: 'the body of the actor and the objects appear and show themselves to the spectators in their own ephemeral presence, (revealing) the "intrinsic meaning" of man and things' (Fischer-Lichte 2008, 186). According to a post-digital perspective, this process involves not only humans and things, but also their merging through digital mediation (Parker-Starbuck 2011). Thus, the human-computer coexistence implies technology as a mirroring tool, 'something that reflects back how we all are' and 'exposes the way how our identities might be constructed' (Schubert 2021c).

At the same time, the construction of auto-poietic systems dependent on the participants' and machines' interactions refers to stylistic choices originally made by the author, which implicitly orients the audience. This consciousness is stimulated by graphic and textual materials, such as those embedded in the web homepages presenting some of Schubert's works (see Figure 12.3). In the case of *Control* and *Genesis*, the embodiment and immersivity principles are immediately clear, showing the VR glasses or the avatar's first-person perspective. On the other hand, web-based pieces such as *Genesis*, *Wiki-Piano.Net*, and *Crawlers* recall the ubiquitous nature of the interface, fostering users' interactivity and implying the persistence of virtual worlds. *Av3ry* and *Crawlers* graphics, instead, imply the depiction of anthropomorphic figures and evoke the liminal conception of artificial and human intelligence.

Moreover, the composer always entailed conceptual issues in uncanny dimensions rendered through visual means. In this regard, he referred to theories like those of 'lost futures,' as the slow cancellation of the future 'interred behind a superficial frenzy of "newness," of perpetual movement' (Fisher 2014, 6); 'render ghosts,' as 'people who live inside our imaginations, in the liminal space between the present and the future, the real and the virtual, the physical and the digital' (Bridle 2014); and 'hyperreality,' as a simulation to restore the escaping real, 'translated by the hallucinatory resemblance of the real to itself' (Baudrillard 1994, 23).

The sense of mystery is also inherent in the programming procedure, which conceals the black box underlying the structure of the pieces. Digitality always represents something other than its source code, and the preordained representation or transformation automatically leads to a media performance (Boast 2017, 165). The encoding (or 'transcoding,' as follows) becomes a stylistic matrix when explicitly aimed to reconceptualise experienced relations 'on the level of meaning and/or language, by new ones that derive from the computer's ontology, epistemology, and pragmatics' (Manovich 2002, 46).

Therefore, Schubert's virtual environments are neither anchored in nostalgia nor striving for a replication of the real, but rather are projected into a renewed conception. Being theoretical references organically (re)integrated within specific aesthetics and sociocultural dynamics, his fictitious worlds are not entirely 'non-places' – namely, without identity, relations, or history (Augé 1995, 87). Rather, they have created a bridge between real and virtual experiences, maintaining the ontological relationship between the two dimensions.

154 *Luca Befera*

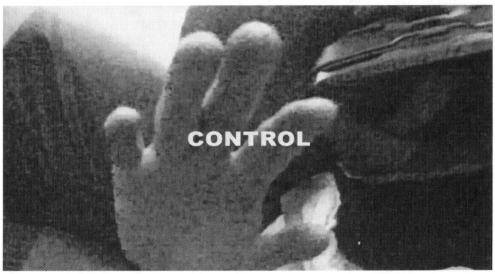

Figure 12.3a-e Homepages of selected pieces by Schubert, in order of appearance: *Genesis, Control, Wiki-Piano.Net, Av3ry*, and *Crawlers* (courtesy of Alexander Schubert).

Through Media

Institutional Aspects

Except for the streamed opening concert, *Genesis* was the only event confirmed within the 2020 Hamburg International Music Festival (Elbphilarmonie 2020a). Even if Germany substantially contributed to sustaining the art scene with emergency funds, the economic loss due to the pandemic has been huge, calling for a quick adaptation (Barber-Kersovan and Kirchberg 2021). Presumably to maintain a minimal continuity – which involved the online

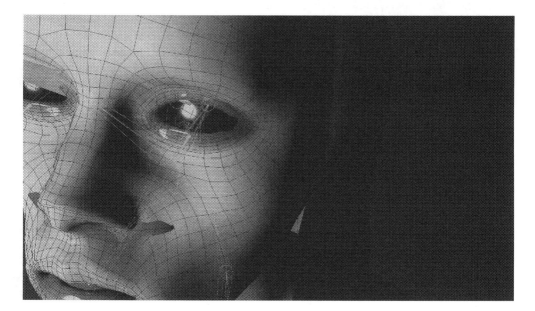

Figure 12.3a-e Continued

transfer of concert venues during 2021 – the Elbphilarmonie persuaded the audience to join the virtual reality game, as it responded 'to our crippled social life with a visionary approach: users [could] create their own world together, live and seen through the eyes of avatars – without leaving the house' (Elbphilarmonie 2020b).[9]

156 Luca Befera

Figure 12.3a-e Continued

As Schubert acknowledged, the final home setting implied some significant changes (A. Schubert, pers. comm.). Besides logistical problems related to staff reductions and extended working periods, one of the most relevant redefinitions regarded the technical aspects. Indeed, the interaction had been originally designed to take place in gaming stations around Hamburg, with the audience reaching physical rooms to join the computer game through tickets. The local connection for in-person participants turned into a worldwide interplay for anonymous users, enhancing the complex relationship between physical and digitally mediated beings. The development of technical components became crucial, requiring data transmission on a global scale and a detailed encoding of additional features concerning the online browser interface programming.

These aspects highlight how 'something that was on an organisational level in the first moment difficult, turned out to be conceptually coherent' (A. Schubert, pers. comm.). Indeed, the original plan would have been more tailored to an installation artwork – as *Control*, *A Perfect Circle* and *Unity Switch* – in which the audience would have physically reached an expressly set physical location. Making the performance available worldwide through the internet enhanced the ubiquitous facet of the media and extended its implied presence to every laptop able to access a browser. It also implied a more relevant role for social media and online streaming, where the composer used these platforms more than usual because of the crucial importance to the project of having people 'there, playing with the avatars, communicate, and watch it' (A. Schubert, pers. comm.). Consequently, *Genesis* finally benefitted from the pandemic adaptation in terms of digital influence, sticking more evidently to online sociality, video game dynamics, and immersive experience as a whole.

Into the Unknown

The symbolic horizon of *Genesis* involves not only Schubert's aesthetic references, such as immersive environments and digital ubiquity, but also specific elements expressly portrayed. First, the performance conception was situated within the festival's main theme, which was faith. The Elbphilarmonie webpage titled the program's descriptive article as ' "Believe" and Music: It is Only Belief that Makes us Human' (Matuschek 2020). Insofar as this 'belief'

Rethinking Intermedia Practices during the Pandemic 157

is here considered as socially constructed, *Genesis* involved no religious faith, but rather the empirical purpose of 'a social experiment on forms of digital communication, interaction, and embodiment' (A. Schubert, pers. comm.). The religious aura was still relevant for enhancing the 'creation' concept and thus focusing on the idea of an empty universe to be defined.[10] The ancestral space was well-connoted in a theatrically mediated environment where 'the self [was] in its own fiction, suspended between the real and the virtual, transcending its physicality, striving to become other' (Giannachi 2004, 94).

The aseptic setting that Schubert progressively defined was strictly related to the experimental purpose (Befera 2021, 15–20). In this regard, the artist provided the absence of predetermined areas and an almost dark stage, to establish a static dimension out of time; also, the audience *in loco* was not allowed, to limit the perception of a standard enactment.[11] Regarding avatars, speech was forbidden and any factor related to identity or gender concealed, thus circumscribing their activity and fostering a virtual persona to be further customised. Reducing any dramaturgical, performative, or theatrical parts drastically limited predetermined knowledge. On the other side, the hall presented an industrial appearance linked to the author's aesthetics and Hamburg urban social movement (Novy and Colomb 2013). Also, *Genesis* was not a scientific experiment in its prerequisites, lacking a restricted hypothesis and a clear analytical methodology. To this extent, it still referred to a theatrical transformative principle (Fischer-Lichte 2008) rooted in speculative fiction rather than academic research: the concept being tested recalled a dystopian future where 'digital spaces are used as a means for conducting experiments and controlling the social order' (Cirucci 2018, 17), constantly questioning the roles of watcher and watched (Blackwell 2018).

Digital Trespassing

As stated, the performance goal was related strictly to the game mediated perception, filtered through specific conditions and aesthetic references. The game interface (see Figure 12.1) assumed a crucial role in this dynamic: it framed the environment and defined interplays, becoming the junction point between the composer's will and the audience's beliefs. The first-person view, the representation of objects' reality, and the immediate reaction to stimuli facilitated the feeling of 'telepresence' as 'the experience of presence in an environment using a communication medium' (Steuer 1992, 76). This perception enhanced the 'imagination' principle pertaining to virtual realities – namely 'the mind's capacity to perceive non-existent things' (Burdea and Coiffet 2003, 3) – while still holding clear references to the physical environment.

Indeed, the interface offered specific and easily accessible insights, recalling a video game appearance, and fostering the players' embodiment. The displayed attributes stimulated users to consider (or not) the avatar's physical condition. On the other side, actors were not only obeying characters but could instantaneously respond to players through the wristband device (e.g., saying yes, no, rejecting an order, or thanking the gamer). Also, the extended working period – involving limited breaks of five minutes every 55 minutes in each slot, one hour five minutes in the middle of the daily schedule, and six to seven hours to sleep – exposed them to an extreme psychological state, which occasionally made their personal needs emerge. The inventory showed the wide range of items grouped into 13 categories. Their pictures were realised by shooting the real object from above on a white canvas, then edited with Photoshop to render the background as white and uniform as possible (Figure 12.4).[12]

Physical and virtual realities were therefore always explicitly interwoven. Still, the virtual experience changed significantly between gamer and avatar. As highlighted in Figure 12.5, the former joined the game through the web-based interface; the latter – and the items as

158 *Luca Befera*

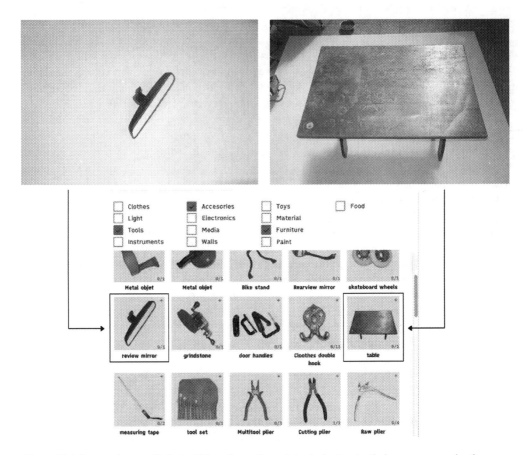

Figure 12.4 Items pictures digital editing, from the original photos to their appearance in the game inventory (courtesy of Alexander Schubert).

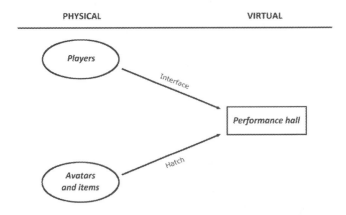

Figure 12.5 Different ways of joining *Genesis* virtual reality regarding gamers, avatars, and items.

well – physically entered the hall through a hatch, the threshold that separated the virtual world from the room where items were stored, and avatars' VR glasses were kept charged and ready to be wear. Also, the avatar underwent the digital setting and referred only to the gamer's verbal commands. This acknowledged unbalanced relationship fostered the many possibilities reported by the artist: surveillance, alienation, emerging empathy, cooperation, and abuse of power (Schubert 2020). On the other hand, the consciousness of the avatars' physical condition and the intimate one-to-one communication with the artificial character made the bodily perception even more relevant than a theatrical performance or an ordinary computer-based interaction. Not so paradoxically, the perception of closeness resided precisely in the distance implied by digital mediation, as expressly oriented by the artist's will.

Conclusions

To summarise, the pandemic enhanced the expressive potential of *Genesis*, amplifying the relevance of the digital means and the authentication process before and during the performance. The problems that emerged certainly affected Schubert's approach. However, as I tried to demonstrate, this influence did not imply a radical redefinition of his expressive means, being as incisive in the performance as marginal in a more general aesthetic conception. Indeed, the analysis of the latest works of the German composer showed how certain post-digital elements were already an integral part of his pieces prior to *Genesis*, and have continued to be so equally afterwards. Though contrasting with everyday life telecommunication, the game's hierarchical dynamic made it even more tailored to the human need for closeness, and also fostered a wider reflection on digital influence. Technological mediation, hence, persisted as a tool for investigating human possibilities and providing renewed insights, even enhancing the role of the composer in re-conceptualising technology and making its social implications clear.

Notes

1 Hamburg Elbphilharmonie commissioned the performance. Also, Internationales Musikfest Hamburg, Decoder Ensemble, Kraftwerk Bille, Hochschule für Musik und Theater Hamburg, and A MAZE (Media Partner) contributed to its realisation. The project was defined along with Heinrich Horwitz and Carl-John Hoffmann; nevertheless, Schubert will be hereafter considered as the author because he is mentioned in *Genesis* credits as the 'artistic head' (Schubert 2020).
2 Significant examples of similar projects were *Desert Rain* (1999) by Blast Theory, providing a theatrically unified VR experience for six players pursuing their human targets through a digital desert landscape, and *Embodied Narratives* (2015) by BeAnotherLab, where a far person enacts a personal story from a first-person perspective while the spectator experiences it through a VR headset.
3 Being that the users are anonymous, the relationship of many of them to Schubert's context has been deduced by cross-referencing data retrieved from the questionnaire on users' nationality and website statistics on where they accessed the game with testimonies from the author and performers who were aware of some users' identities (Befera 2021, 30–32). Maintaining anonymity also implied insufficient data to delve into the pandemic influence during the performance, scarcely mentioned even in the collected testimonies.
4 The interview was held on 13 June 2021.
5 The tripartite framework also involves the 'performative object,' namely the recorded artefact that simulates the viewers proximity to the performance. From the author's perspective, the crossed analysis of these three tenets aims to inquire about the alleged authenticity of mass-mediated products within the socioeconomic dynamics of the popular music industry. Bratus adopted his

160 *Luca Befera*

    matrix from Allan Moore (2002), who first discussed authenticity in terms of mechanisms of attribution, even if not strictly focused on the mediation process and artefacts.

6 Even if they provide similar aspects, the scheme does not involve staged performances such as *Acceptance* (2018) and *Convergence* (2021) because they are not strictly related to digital means as essential for real-time interaction.

7 Indeed, asides from the sandbox video games *Minecraft* and *The Sims*, Schubert stated his own virtual reality projects as *Genesis*' main references (Schubert 2021b).

8 For information about all the reported pieces, see Schubert's official website (Schubert 2021a).

9 This quotation was evidently realised for the pandemic constraints, *Genesis* having been planned since before its emergence.

10 The performance duration of one week also recalled the biblical Genesis.

11 Even if the pandemic occurrence made the audience's absence inevitable, the authors had previously discussed whether to allow viewers inside the performance hall. Therefore, the final YouTube streaming manifested the will of an entirely virtual setting that also equalised the access of the player and the viewer, moving towards a more fluid process of interchanging roles. *Genesis* remained accessible online even after its end, with only the final setting observable through pictures.

12 This information refers to my own participant observation, which also involved shooting items and categorisation. Instead, the reported pictures come from *Genesis* archive by Alexander Schubert, who kindly allowed the use of the material within this text.

References

Aron, Arthur P., Debra J. Mashek, and Elaine N. Aron. 2013. 'Closeness as Including Other in the Self.' In *Handbook of Closeness and Intimacy*, edited by Debra J. Mashek and Arthur P. Aron, 27–42. Mahwah: Lawrence Erlbaum Associates.

Augé, Marc. 1995. *Non-Places: Introduction to an Anthropology of Supermodernity*. London: Verso.

Barber-Kersovan, Alenka and Volker Kirchberg. 2021. 'The Show Must Go On(Line): Music in Quarantine.' *The European Sociologist* 46 (2). www.europeansociologist.org/issue-46-pandemic-impossibilities-vol-2/covid-arts-show-must-go-online-music-quarantine.

Baudrillard, Jean. 1994. *Simulacra and Simulation*. Translated by Sheila Glaser. Ann Arbor: University of Michigan Press.

Befera, Luca. 2021. *Rise of a Transcoded World: Field Study and Performance Analysis of Alexander Schubert's 'Genesis.'* www.alexanderschubert.net/on/Luca-Befera-Genesis-Book.pdf.

Bell, Mark W. 2008. 'Toward a Definition of "Virtual Worlds."' *Journal of Virtual Worlds Research* 1 (1): 2–5. https://doi.org/10.4101/jvwr.v1i1.283.

Bishop, Claire. 2005. *Installation Art: A Critical History*. London: Tate Publishing.

Blackwell, Derek R. 2018. 'All Eyes on Me: Surveillance and the Digital Archive in "The Entire History of You."' In *'Black Mirror' and Critical Media Theory*, edited by Angela M. Cirucci and Barry Vacker, 55–68. Lanham: Lexington Books.

Boast, Robin. 2017. *The Machine in the Ghost: Digitality and Its Consequences*. London: Reaktion Books.

Bratus, Alessandro. 2019. *Mediatization in Popular Music Recorded Artifacts: Performance on Record and on Screen*. Lanham: Lexington Books.

Bridle, James. 2014. 'The Render Ghosts.' *Reading Design*. www.readingdesign.org/render-ghosts.

Burdea, Grigore C. and Philippe Coiffet. 2003. *Virtual Reality Technology*. Hoboken: John Wiley & Sons.

Cheng, Lili, Shelly Farnham, and Linda Stone. 2002. 'Lessons Learned: Building and Deploying Shared Virtual Environments.' In *The Social Life of Avatars: Presence and Interaction in Shared Virtual Environments*, edited by Ralph Schroeder, 90–111. London: Springer-Verlag.

Cirucci, Angela M. 2018. 'Digitally Natural: Gender Norms in "Black Mirror."' In *'Black Mirror' and Critical Media Theory*, edited by Angela M. Cirucci and Barry Vacker, 15–26. Lanham: Lexington Books.

De', Rahul, Neena Pandeyb, and Abhipsa Pal. 2020. 'Impact of Digital Surge during Covid-19 Pandemic: A Viewpoint on Research and Practice.' *International Journal of Information Management* 55 (102171). https://doi.org/10.1016/j.ijinfomgt.2020.102171.

Rethinking Intermedia Practices during the Pandemic 161

Dixon, Steve. 2007. *Digital Performance: A History of New Media in Theater, Dance, Performance Art and Installation*. Cambridge: MIT Press.

Elbphilharmonie. 2020a. 'Cancelled: Hamburg International Music Festival.' www.elbphilharmonie.de/en/festivals/cancelled-hamburg-international-music-festival/499.

Elbphilharmonie. 2020b. 'Genesis: A Computer Game Becomes Reality.' www.elbphilharmonie.de/en/mediatheque/genesis-a-computer-game-becomes-reality/331.

Fischer-Lichte, Erika. 2008. *The Transformative Power of Performance: A New Aesthetics*. London: Routledge.

Fisher, Mark. 2014. *Ghosts of My Life: Writings on Depression, Hauntology and Lost Futures*. Hants: Zero Books.

Giannachi, Gabriella. 2004. *Virtual Theatres: An Introduction*. London: Routledge.

Grusin, Richard. 2015. 'Radical Mediation.' *Critical Inquiry*, 42 (1): 124–148. https://doi.org/10.1086/682998.

James, William. (1896) 2005. 'The Will to Believe.' In *James and Dewey on Belief and Experience*, edited by John M. Capps and Donald Capps, 95–110. Urbana: University of Illinois Press.

Jarvis, Liam. 2017. 'The Ethics of Mislocalized Selfhood: Proprioceptive Drifting towards the Virtual Other.' *Performance Research* 22 (3): 30–37. https://doi.org/10.1080/13528165.2017.1348587.

Manovich, Lev. 2002. *The Language of New Media*. Cambridge: MIT Press.

Matuschek, Clemens. 2020. '"Believe" and Music: It Is Only Belief That Makes Us Human.' Elbphilarmonie. www.elbphilharmonie.de/en/mediatheque/believe-and-music/290.

Meier, Jennifer V., Josephine A. Noel, and Kai Kaspar. 2021. 'Alone Together: Computer-Mediated Communication in Leisure Time during and after the COVID-19 Pandemic.' *Frontiers in Psychology* 12. https://doi.org/10.3389/fpsyg.2021.666655.

Moore, Allan. 2002. 'Authenticity as Authentication.' *Popular Music* 21 (2): 209–223.

Nguyen, Minh Hao, Eszter Hargittai, and Will Marler. 2021. 'Digital Inequality in Communication during a Time of Physical Distancing: The Case of COVID-19.' *Computers in Human Behavior* 120. https://doi.org/10.1016/j.chb.2021.106717.

Nimrod, Galit. 2020. 'Changes in Internet Use When Coping with Stress: Older Adults During the COVID-19 Pandemic.' *The American Journal of Geriatric Psychiatry* 28 (10): 1020–1024. https://doi.org/10.1016/j.jagp.2020.07.010.

Negroponte, Nicholas. 1998. 'Beyond Digital.' *Wired* 6 (12). https://web.media.mit.edu/~nicholas/Wired/WIRED6-12.html.

Norman, Kent L. and Jurek Kirakowski, eds. 2018. *The Wiley Handbook of Human Computer Interaction*. Hoboken: John Wiley.

Novy, Johannes and Claire Colomb. 2013. 'Struggling for the Right to the (Creative) City in Berlin and Hamburg: New Urban Social Movements, New "Spaces of Hope"?' *International Journal of Urban and Regional Research* 37 (5): 1816–1838. https://doi.org/10.1111/j.1468-2427.2012.01115.x.

Parker-Starbuck, Jennifer. 2011. *Cyborg Theatre: Corporeal/Technological Intersections in Multimedia Performance*. Houndmills: Palgrave Macmillan.

Roberts, James A. and Meredith E. David. 2019. 'The Social Media Party: Fear of Missing Out (FoMO), Social Media Intensity, Connection, and Well-Being.' *International Journal of Human–Computer Interaction* 36 (4): 386–392. https://doi.org/10.1080/10447318.2019.1646517.

Schubert, Alexander. 2020. 'Genesis: About.' www.virtual-genesis.net/about.

Schubert, Alexander. 2021a. 'Alexander Schubert Official Website.' www.alexanderschubert.net/works_date.php.

Schubert, Alexander. 2021b. 'Genesis Lecture @ Graz University.' www.youtube.com/watch?v=bizdiXyV4ng.

Schubert, Alexander. 2021c. 'Lecture "Projected and Fluid Identities" [Darmstadt 2021].' www.youtube.com/watch?v=YMWRsR2KOXE.

Schubert, Alexander. 2021d. 'Switching Worlds.' Wolke Verlag. www.wolke-verlag.de/wp-content/uploads/2021/02/SwitchingWorlds_DIGITAL_englisch_210222.pdf.

Schumacher, Shannon and Nicholas Kent. 2020. 'Eight Charts on Internet Use around the World as Countries Grapple with COVID-19.' *Pew Research Center*. https://pewrsr.ch/2wOyAYy.

Steinert, Steffen. 2020. 'Corona and Value Change: The Role of Social Media and Emotional Contagion.' *Ethics and Information Technology* 23: 59–68. https://doi.org/10.1007/s10676-020-09545-z.

Steuer, Jonathan. 1992. 'Defining Virtual Reality: Dimensions Determining Telepresence.' *Journal of Communication* 42 (4): 73–93. https://doi.org/10.1111/j.1460-2466.1992.tb00812.x.

Velvick, Lauren. 2021. 'Adapting to the Pandemic.' *Art Monthly*, 443 (February): 39–40.

Vuorre, Matti, David Zendle, Elena Petrovskaya, Nick Ballou, and Andrew K. Przybylski. 2021. 'A Large-Scale Study of Changes to the Quantity, Quality, and Distribution of Video Game Play During the COVID-19 Pandemic.' *PsyArXiv*. https://doi.org/10.31234/osf.io/8me6p.

Williams, Dmitri, Nicolas Ducheneaut, Li Xiong, Yuanyuan Zhang, Nick Yee, and Eric Nickell. 2006. 'From Tree House to Barracks: The Social Life of Guilds in World of Warcraft.' *Games and Culture* 1 (4): 338–361. https://doi.org/10.1177/1555412006292616.

13 Musicians in the Brazilian Pandemic

Facing COVID-19 during the Bolsonaro Regime and the Aldir Blanc Emergency Bill

Suzel Ana Reily and Regina Machado

In the first months of 2020, people across the planet came to the realization that the novel coronavirus first detected in Wuhan, China, SARS-CoV-2, was spreading across the globe. COVID-19, as the disease caused by the virus would come to be known, would soon be declared a pandemic by the World Health Organization, affecting populations the world over in very different ways, depending on how health authorities and governments chose to respond to the crisis. Closed borders and lockdowns of different degrees of severity became the norm. Masks, social distancing, and sanitary gels entered everyday practices all around the world.

The lockdowns had serious negative impacts on the economies of all nations while, within nations, the effects of the pandemic varied significantly, with poorer communities being hit especially hard. Across the globe, the sectors most affected by the economic consequences of the pandemic were those associated with public culture, including restaurants and bars, tourism, sports, and the arts, especially the performing arts such as theatre, dance, and music, that depend on the presence of an audience. As many performing artists have repeatedly noted, they were the first to shut down and the last to reopen (Muniagurria 2021).

In Brazil, the onset of the lockdown to halt the spread of COVID-19 was declared on 17 March 2020, and, as elsewhere, only essential services such as pharmacies, supermarkets, and gas stations were allowed to remain open. Where possible, people were encouraged to work from home through virtual platforms, but even so, many businesses went bankrupt and job losses were widespread, particularly in the most vulnerable sectors and among intermittent and seasonal workers. Despite some measures to contain the virus, the impact of the pandemic on the country was calamitous: over 60 percent of the population experienced a decline in income and over 50 percent of Brazilian households were living in conditions of food insecurity (Rede Penssan 2020). While an emergency funding bill to assist the struggling population was eventually signed on 14 May 2020 (Auxílio Emergencial Lei N° 13.998), it excluded many categories of vulnerable workers, including musicians, artists, and workers in the entertainment industries generally, consistent with the broader neoliberal stance guiding public policy. It would take several months of intense struggle before a bill directing funds to minimize the impact of the pandemic specifically to these workers – the Aldir Blanc Emergency Bill (Lei Emergencial Aldir Blanc – LAB) – would be signed.

Alongside these worrying matters, the country registered a particularly devastating death toll: the second highest in the world, at over 660,000 people as of April 2022. All spheres of the diverse music worlds of Brazil were impacted in this regard. There were numerous victims among the prominent names in Brazilian popular music, such as Aldir Blanc and Agnaldo Timóteo, and other well-known musicians, such as Paulinho of the pop band Roupa Nova, and Izael Caldeira, a former member of Demônios da Garoa, the São Paulo

DOI: 10.4324/9781003200369-16

164 *Suzel Ana Reily and Regina Machado*

based samba ensemble associated with the prominent composer Adoniran Barbosa. Deaths also occurred in the world of art music; following a rehearsal by the choir of the São Paulo Municipal Theatre at the very start of the pandemic, both choir directors – Naomi Munakata and Martinho Lutero de Oliveira – were infected and sadly both died on the same day, 26 March 2020. Countless lesser known musicians also lost their lives, impacting local music scenes across the country: deaths like that of Claudio Augusto Motta, a button accordionist from Ibirubá, Rio Grande do Sul, who was active in the gaucho nativist movement; or of Benedito Sebastião dos Santos, also known as Ditinho, who made his living by playing in bars in Ribeirão Preto, São Paulo; or of São Paulo's street singer Jonathan Neves, known as the Tim Maia of Avenida Paulista, where he performed every Sunday when the street was closed off for pedestrians.

Traditional music spheres were also seriously impacted. Raimundo Aniceto, the oldest living member of the Banda Cabaçal dos Irmãos Aniceto, an ensemble that still performs on hand crafted fifes, died a victim of the pandemic. Tragically, the Quilombola Community of the Arturos, that safeguards the *congado*[1] tradition in Contagem, in Greater Belo Horizonte, lost six members to COVID-19, including Mestre Mário Braz da Luz, the last remaining son of Artur Camilo Silvério, who secured the land on which the *quilombo*[2] sits; his wife, Maria Auxiliadora da Luz; and one of their daughters, Maria Antônia Vieira, all in the space of two weeks. Numerous Indigenous leaders crucial to the preservation of Indigenous languages and cultures, which of course include music and dance, became victims of the coronavirus. Among the most prominent were Aritana Yawalapiti and Paulinho Paiakan, but one can also remember the Guarani shaman Gregório Venega, and chiefs Warini Surui, Acelino Dace, Artemínio Antônio Kaingáng, Elizer Tolentino Puruborá, Puraké Assuniri, and João Sõzê Xerente, among others. In June 2020, recognised Macuxi cultural master Vovó Bernaldina, died from coronavirus related illness. According to Ananda Machado, coordinator of the Roraima Program for the Valorization of Indigenous Languages and Culture, 'Vovó Bernaldina … was always singing and dancing, teaching the girls how to make traditional crafts, and representing her ethnic group at cultural events. An unreplaceable loss' (Raquel 2021).

The way the federal government handled – or mishandled – the health crisis is a major factor contributing to the alarming figures registered in Brazil. Throughout the pandemic, President Bolsonaro, in Trumpian fashion, openly dismissed science and the recommendations of the medical community; promoted medications that had been proven to be ineffective against the virus; cut funds to research and sustained attacks on universities; neglected the Ministry of Health, which remained without a proper minister for several months; organised mass gatherings and rallies with his supporters; refused to wear a mask and to be vaccinated; purposefully delayed the purchase of vaccines and the organization of mass vaccination; and continually opposed lockdowns and social distancing for all but the elderly.

While the president's stance initially fuelled the political divide that had taken hold of the country following the impeachment of left-leaning President Dilma Rouseff in 2016, the general pro-vaccine orientation of the Brazilian population played a critical role in reducing the death toll once the vaccines had been (belatedly) made available, thanks to the fierce negotiations sustained by civil society, the medical and scientific communities, and a pro-science political opposition, which found ways around the obstacles set up by the government. Ultimately, Congress was led to instate a Parliamentary Inquiry Commission (CPI) to look into the Government's handling of the pandemic, which identified numerous instances of crimes against humanity perpetrated by Bolsonaro, some of his immediate relatives, and other politicians that supported him.[3] These revelations and the continuous loss of income exacerbated by high rates of inflation have led to a steady decline in the President's popularity, but not, as yet, to his impeachment.

It is against this backdrop that we look at how the Brazilian music world (see Becker 1982) and, more specifically, professional musicians in the Greater São Paulo area confronted the pandemic, noting that the same disdain the President demonstrated towards science was mirrored in his stance on the arts and culture. By focusing on musicians, one of the sectors most affected by the economic crisis engendered by the pandemic, one can assess the precarity that circumscribes this highly specialized labour force that, ironically, played an important role in assisting isolated individuals to survive the lockdown by contributing to their tool kits for managing loneliness and depression, hardly a task for so-called non-essential workers![4]

As elsewhere, the shutdown of public culture and the cancellation of live performances in Brazil meant that from one moment to the next, musicians, actors, artists, technicians, costume and set designers, and other artistic professionals were no longer able to work and thereby guarantee their own survival and that of their dependents. Estimates suggest that more than eight thousand major music events were cancelled or delayed in Brazil in 2020, affecting eight million people at a loss of around R$483 million (ca. US$900,000) (Rodrigues, Kamlot, and Carvalho 2021, 3).

If at first it seemed the situation would be resolved in a few months, it was soon obvious that this would not to be the case, as the number of infections kept rising. The prospect of reopening concert halls, theatres, and music venues became ever more distant. As conditions worsened, countless workers in all of the performing arts found themselves facing a profound crisis, seriously aggravated by the absence of any prospect of state support. Many musicians, like their colleagues in other art worlds, had to rely on family and friends, crowd funding, teaching, occasional tenders for projects sponsored by local governments or other institutions, commercial sponsorships for 'lives,'[5] and meagre social media royalties, among other alternatives. Within this scenario, the announcement of the R$3 billion (ca. US$550 million) Aldir Blanc Emergency Bill, named after the composer Aldir Blanc[6] who died from COVID-19 on 4 May 2020, marked a sea change for many artists, particularly those in the most vulnerable positions, even though this was hardly sufficient to cover the needs of the art world.

Nearly two years on, live performances have resumed, but numerous restrictions limit audience numbers and new variants continue to cause concern. As such, many professional musicians still experience feelings of financial uncertainly and conditions of precarity. We will now turn our attention to the Emergency Bill itself, looking at how it was negotiated in relation to prior funding practices in music and the arts; the current government's stances on culture; the circumstances of the art world; and the demands of the wider population.

Negotiating the Emergency Bill

In Brazil, as perhaps throughout Latin America, public policies to promote culture and the arts have tended to be fragile and limited at best. However, in the period between 2004 and 2016, when Brazil was administered by the Workers Party, or Partido dos Trabalhadores (PT), the governments of Presidents Luiz Inácio da Silva ('Lula') and Dilma Rousseff made significant advances towards the establishment of policies based on intense dialogue between government officials and organized sectors of society. Even though these policies may not have become as consolidated as one might have hoped, this dialogic orientation to policymaking in the arts is evident in the speech made by Juca Ferreira, the Minister of Culture during the Rousseff government, at the ceremony to launch the National Arts Policy in 2015:

> We are here in the name of dialogue and participation, convinced that we still need a vigorous policy for the arts in Brazil. This is the certainty and method that motivate us to instate a process of listening and discussion, like the one we are initiating today.
>
> We are not starting from scratch, we know that. We already have accumulated many joint reflections; we have a history; we have some good starting points. We have the debates and propositions that emerged within the Conferences on Culture, the Sectorial Colleges, which were organized around the National Council of Culture, as well as during the entire process of the construction and assembly of a National System. Today, we have a National Plan for Culture (Plano Nacional de Cultura) in place.
>
> (Juca Ferreira in Machado 2017, 125–126)

The National Plan for Culture is best seen within the broader movement initiated within the Ministry of Culture (MinC) under the management of the popular musician Gilberto Gil and, later, Juca Ferreira. A key protagonist in this process, however, was the historian, cultural activist, and administrator Célio Turino. Between 2004 and 2010, Turino was MinC's Secretary of Cultural Citizenship, working under the command of the Minister at the time, Gilberto Gil. During this period, he collaborated in the creation of MinC's Art, Education, and Citizenship Program, better known as Living Cultures (Cultura Viva), an offering that marked a radical change in the paradigm for elaborating public policies in Brazil, precisely because it relied on dialogue and the participation of civil society and the art world in the construction and implementation of its projects, particularly the so-called cultural points (*pontos de cultura*).[7] Through this model, the administrator of Cultura Viva claims to have created more than 2,500 *pontos de cultura* within more than a thousand municipalities across the country, directly and indirectly reaching eight million people, and creating around 30,000 jobs.

According to Lia Calabre and Deborah Rebello Lima, Cultura Viva instated:

> A fundamental experience for reflection on such issues as: the notion of the participation of civil society in the management and production of culture; the idea that the exercise of citizenship also allows for civil and political rights and guarantees cultural rights; the challenges of a form of management grounded in networks, in which the State and social groups seek to jointly define the directions to be taken, and the construction of new conceptions regarding development.
>
> (2014, 6)

In effect, they argue that Cultura Viva established a modus operandi that promoted links between government and civil society, through a 'cooperative posture' and 'the sharing of responsibilities' (ibid., 6). Célio Turino refers to this mode of policymaking as 'shared administration,' in which, 'instead of imposing a cultural program or calling cultural groups to say what they want (or need), we ask them how they wish [to proceed]. Instead of understanding culture as a product, it is recognized as a process' (2010, 23).

The cooperative posture that guided Cultura Viva would be adopted by other programs implemented by the MinC, such as the National Plan for Culture. As set out in its preliminary proposal, the National Plan contains 53 stated aims encompassing the breadth of the plan,[8] which was structured around the contemplation of the nation's diversity, social inclusion, and sustainability, and meant to reach the entire country. The Plan would promote tourism across the country; increase the number of cultural workers in all spheres;

fund university programs related to the promotion of culture; support existing performance groups, from the fine arts to popular and traditional arenas; and provide internet access to all, among a range of other ambitious objectives. It was to be funded by a significant increase in federal moneys, highlighting the centrality the PT attributed to culture.

The original plan was to run for a period of ten years, which, on paper at least, it did, even being extended for a further two years by the current government. However, very little of it has been implemented since Rousseff was overthrown in 2016, and the idea of shared administration has been systematically downplayed. In effect, the policies advanced by former Vice-President Michel Temer upon taking the Presidency exemplify this, as he quickly took steps to begin dismantling the previous government's initiatives in the field of culture by merging the Ministry of Culture with the Ministry of Education, weakening the portfolios of both agencies. Swift action on the part of the art world was able to reverse the measure, though as soon as Jair Bolsonaro came into office the Ministry of Culture was extinguished and rendered a secretariat within the Ministry of Tourism, where it has remained since then. This act draws attention to an ever more common orientation toward the arts: viewing them in terms of their potential for economic return, and shifting the focus outlined in the Plan. As Luisa Marques Barreto has noted,

> the trajectory of public cultural policies over the past ten years has been tremendously erratic, with advances and sudden ruptures, growth and contraction in the concept of culture and a still indefinite amalgam around the relations between culture, creativity, urban and territorial development.

(2020, 33)

Even as the post-impeachment administration worked to weaken the policies implemented by the PT, the dialogic and participatory culture that guided MinC in its various programs established an ethos of discussion within the country's non-commercial and alternative art worlds, as well as strong networks within and between them. For many musicians and artists, these networks proved critical in ensuring some degree of security during the period of uncertainly brought on by cancellations of work opportunities, and they were mobilised to negotiate state financial support, most notably through the establishment of the Aldir Blanc Emergency Bill.

It is worth noting that pressure was also forthcoming from the wider society. The same pandemic that locked musicians and other artists in their homes, blocking access to traditional performance venues, created a context in which large sectors of the general public began reassessing their relationship with the arts. Unable to leave their homes, people turned to music, movies, books, television, internet, and other forms of entertainment to pass the time and deal with the emotional strain of isolation, lockdown, and the fear of infection. The demand for culture and artistic activity increased.

In effect, then, these circumstances were critical to allow for the mobilization of civil society, the artistic classes, and some parliamentarians, who joined forces to address the plight of the nation's culture workers. Even within the current political climate so averse to the arts, MinC culture still had sufficient strength to mobilize opinions in the political sphere and achieve the approval of a bill – the Aldir Blanc Emergency Bill – that injected much needed funds into the sector. Although the Bill was originally passed on 29 June 2020, it was not until 1 September that funds began to be transferred to states and municipalities, where they would still need to be transferred to musicians, artists, and others involved in cultural activities, according to the procedures to be defined by each state and town.

One of the figures spearheading the negotiations, though no longer linked to government, was Célio Turino. Following well-rehearsed practices of policymaking, he joined forces with

168 *Suzel Ana Reily and Regina Machado*

cultural administrators, culture advisers, community leaders, artists and representatives of civil society, and after months of concerted discussion, the group produced a draft outlining their needs. In dialogue with sympathetic parliamentary bases, the demands were incorporated into the draft of a bill presented by Rio de Janeiro parliamentarian Benedita da Silva (PT); parliamentarian Jandira Feghali (PCdoB),[9] also from the state of Rio de Janeiro, acted as rapporteur for the bill in the Federal Chamber. The role of rapporteur involves convincing other members of the Chamber to support the bill they are representing, and Feghali was particularly effective in this role, given that LAB passed in both houses of parliament with almost 100 percent of the votes.[10]

The next step was to achieve the sanctioning of the Bill by the President, which occurred on 29 June 2020. The R\$3 billion allocated to the program started being distributed across states and municipalities in September, so that the funds could reach needy art world professionals. It is of special significance that the bill decreed that funding would be administered directly by state and municipal governments, rather than through federal agencies, ensuring it reached local musicians and artists as quickly as possible. This meant of course that states and municipalities would need to define how they would organize the transfers of funds. Needless to say, there was considerable diversity in the local policies regarding the use of these funds.[11]

The bill itself proposed three domains for the distribution of funding: Category I, emergency aid paid in three monthly instalments to artists of all the different art worlds who had lost their income; Category II, monthly instalments to sustain small businesses, such as music schools, cooperatives, and community cultural centres whose activities had been interrupted by the pandemic; and Category III, the promotion of artistic activity through tendering, in which artists were invited to submit projects to be undertaken individually or in collectives.

An important issue to highlight is the extraordinarily broad scope that these categories were able to encompass. Monthly stipends to artists encompassed professionals, as well as traditional artists typically excluded from such programs. In Santa Helena de Minas, Minas Gerais, for instance, several members of the local traditional mummers groups, the *folias de reis*, were awarded stipends, as were participants in the *batuque* dance and percussion group, capoeira players, and performers linked to a *quilombo* community (Souza and Araújo 2021). Moreover, the concept of small cultural business could be quite elastic, encompassing registered institutions as well as informal ones, such as *folias de reis*, *congados*, capoeira groups and so on. Similarly, with the term artistic activity, projects submitted to Category III could involve finished artistic productions as well as workshops, the production of teaching material, and payment of licensing rights and copyrights, among other possibilities.

In the state of São Paulo, the municipalities were allocated funding to be distributed in Categories I (focusing on local artists) and II, while the state government allocated funding in Categories I (privileging artists with broader remits) and III. With respect to the public competitions for funding, four spheres of activity were proposed: production, licensing,[12] fruition,[13] and awards.[14] The state's call for proposals was encompassed by the São Paulo Programme for Cultural Action (PROAC), which ran a range of parallel calls.

Having discussed the general conditions of the Brazilian music and art worlds during the pandemic, the historical backdrop in public policymaking around cultural promotion in Brazil, the negotiations leading to the implementation of the Aldir Blanc Bill, and the remit of the Bill, we now turn our attention to the way the Bill has been used by a few non-mainstream professional musicians in São Paulo who found themselves and colleagues in conditions of vulnerability and uncertainty with the onset of the pandemic. Drawing on these cases, we assess the impact of the Bill on these musicians and their peers.

LAB Projects and Their Impacts

The musicians with whom we spoke could all be classed as professionals, involved in what can be loosely called popular music, albeit in alternative, independent, non-mainstream genres. Even before the pandemic, many of them relied on a range of activities beyond musical performance to generate income. While the degree to which each of the musicians was financially affected by the pandemic varied, all of them indicated a level of insecurity brought on by the almost complete cancellation of their performance agendas. The submission of proposals to obtain funding from LAB, and from other sources, was critical to provide them with some financial support during the pandemic. It is noteworthy that the preferred category for their bids involved Category III, which encompassed musical productions of various types.

Singer and producer Jeanne de Castro participated in the conception and production of four projects submitted for LAB funding proposed by other participating partners, and one of her own involving a video production on cultural production. She highlighted the importance of these funds for her, as these projects provided the only income she received over a period of a year and a half. Izabel Padovani, a singer and teacher of the Alexander Technique, noted that the requirements for submitting proposals in São Paulo were complex, as proposals had to be accompanied by extensive documentation of the artistic careers of the candidates. Nonetheless, for her it was important to receive financial support since, besides the cancellations of her live performances, she lost many students who had themselves experienced a decline in income and could no longer afford to pay her.

While most of the interviewees submitted projects to Category III, Mônica Marsola, a guitarist, composer, and singer, presented her project within Category II, in order to keep her music school open. The counterpart for this project entailed the production of a series of videos aimed mainly at music teachers.[15] Marsola claimed that the support received for the school was important, but it arrived far too late, as she had already taken out a loan from the bank to keep her institution afloat. Nevertheless, she noted that she was happy with the videos, as they allowed her to share some of her knowledge with other people, something that probably would not have happened had she not received the funding. Now that the stipend has ended, she is again struggling, and is back in the red.

Musician and arranger Mario Manga and singer and instrumentalist Ana Deriggi have been working together for some years as the duo O Bom e Velho. When they were prevented from performing live shows, they submitted a bid for LAB funding (Category III) to produce a season of an online program that could show the influence of the Beatles on the Tropicália Movement.[16] They aimed to generate pairs of songs, one from the Beatles and one from Tropicália: 'A Day in the Life' by The Beatles, for example, could be paired with Gilberto Gil's 'Domingo no Parque.' According to Manga, the repertoire was chosen based on parallels in song construction, that is, the ways in which the lyrics, melody, and arrangement of a song were structured, as well as in terms of song themes.

The recordings for the shows involved techniques that minimized direct contact between the musicians, who only met in person twice during the entire process, as neither of them had been vaccinated. The first meeting involved testing the tuning, tonality, and the viability of some of their ideas. Having sorted these matters out, Ana Deriggi recorded her parts at her own home and then sent them to Mario Manga. They each recorded several parts to each song: Ana recorded vocals and guitar; Manga, guitar, cello, and vocals. Once all the parts had been joined and mastered by Manga, the duo met again, now at Ana's house, to create the video by dubbing over the recording. Ana's husband, Pipo Gialluisi, was responsible for the video recording and editing, done in chroma key with a green background,

170 *Suzel Ana Reily and Regina Machado*

so that other images could be superimposed on the video recordings. The final show was entitled *O Bom e Velho – A Day in Tropicália*, and it premiered online on 28 April 2021.[17]

Guitarist, singer, and songwriter Badi Assad, who is accustomed to an intense international schedule, was on a tour around Brazil as part of the Sounds of Brazil Project (Sonora Brasil) sponsored by Sesc Nacional, the national branch of a non-profit sports and cultural institution, when the pandemic broke out. Despite her national and international profile, she claimed that the interruption in her performance activities brought feelings of insecurity and fear; but she also noted that the circumstances gave her strength and courage to take on new challenges in order to continue producing artistically, despite the crisis.

When the PROAC call for proposals within the LAB funding was made, Assad submitted a project within Category III entitled *Around the World with Guests* (*Volta ao Mundo com Convidados*), linked to a long-standing project that culminated in the publication of her book *Volta ao mundo em 80 artistas* (2018). In the book, Assad presents personal views on the work of 80 musicians from different parts of the world, and she planned a series of shows based around musicians profiled therein. The arrival of the pandemic made this project impossible: it would have to be adapted to a format without a live audience. The call for proposals provided an opportunity to attempt a new format. So for her LAB project, she proposed the production of a series of six 45-minute videos drawing on profiles included in the book. The first video is solo: she presents her project and then discusses a few of the musicians profiled in her book, including Duo Assad, Clarice Assad, Tori Amos, Mumford and Sons, and Uakti. Each is then represented through a song or two. The other five videos begin with a discussion of one or more of the musicians she profiled and a few solo performances of their work; she then introduces her guest, with whom she proceeds to discuss a few more profiles, before she and her guest provide a joint performance of a song by the musician profiled. For her first video, for instance, she begins by profiling Brazilian popular musician Zélia Duncan and blues singer Roy Rodgers. She is then joined by guitarist Swami Jr., and together they perform Bola de Nieve's iconic song 'Drume Negrita.' Then, profiling another Brazilian popular musician, Zizi Possi, they perform 'Bom dia,' a song Possi interpreted, but that was originally composed by Paulo Freire and her guest, Swami Jr. The guests in the other videos are Fernandinho Beatbox, Carlinhos Antunes, Livia Mattos, and Marcelo Pretto. Five of the songs recorded as part of the project were also released by Tratore as an EP with the same name and available on all platforms.[18] Besides providing financial support for herself, the project included payment for her guests, the production crew involved in the filming and editing, and the production costs.

Cristiano Cunha – a singer, percussionist, and composer from Belo Horizonte, Minas Gerais, now living in São Paulo – received LAB funding to cover the licensing fees of a production entitled *48 Minutes*. He noted that, besides the much-needed support for his own subsistence in the middle of the pandemic, the project allowed him to collaborate with other close artists, for whom the funds were also very welcome. The concern in constructing projects that could integrate many people was frequently noted by musicians and artists in their proposals, and as we have seen in previous cases, some musicians were involved in projects they did not themselves submit. This was a central concern for theatre director Regina Galdino, a member of the Movimento Artigo Quinto.[19] She submitted two projects based on past productions that would now be produced in virtual formats: *Operilda na Orquestra Amazônica Online*, a musical that tells the story of classical music for children; and *Memórias Póstumas de Brás Cubas*, in which music has a strong presence.

Both productions involved large crews. *Operilda* had a crew of 17 people, including one actress; six musicians; one director; one music director and arranger; two lighting technicians; one sound technician for capturing, mixing and post-mixing; one stage technician; one cameraman and video editor; one production director; one executive producer;

and one administrator. Further payments were made to a sound leasing company; a light leasing company; the use of theatre space for a day and a half of filming; and transportation for scenery, lighting and sound. *Memórias Póstumas* engaged 16 professionals as well as a few small businesses. Galdino noted that she made a point of trying to help as many members of the production chain as she could, from fellow musicians and artists to costume designers and technicians. Indeed, she claimed the structure of the Aldir Blanc Bill itself encouraged this form of inclusion. Moreover, Galdino helped many other candidates to write and submit proposals for funding from LAB and other sources.

As a final example, we discuss the various proposals put forward by Magda Pucci, a singer and arranger with a consolidated career in São Paulo, nationally and internationally. She is the director of the São Paulo-based research and performance company Mawaca, composed of 12 musicians: six female singers accompanied by a six-piece band. At the onset of the pandemic, Mawaca had been performing an average of two shows per month, in addition to their participations in several international festivals. For 2020, they were looking forward to an intense schedule: they planned to record a new album celebrating their 25th anniversary; they were to participate in a festival in Malaysia in which they would present music of the native peoples of Brazil; and they aimed to produce video recordings of a series of events to be held at Mawaca Studio, a cultural space located in the city of São Paulo and directed by Pucci, which hosts various artistic activities in addition to those carried out by the group. Furthermore, Magda Pucci was on tour with SESC's Sonora Brasil project with a group called Wiyae, composed of herself, her Mawaca colleague Gabriel Levy, Indigenous singer Djuena Tikuna, and percussionist Diego Janatã; the quartet was presenting a program based on songs from native peoples of Brazil in several cities in the north and northeast of the country. This entire agenda was cancelled, leaving all the music workers involved in these projects, including singers, musicians, and technicians, without the income upon which they had relied.

Pucci claimed that the arrival of the Aldir Blanc Bill brought a sense of hope to the group. Taking advantage of her vast experience and production dynamics within Mawaca, Pucci and her team made bids for four LAB grants: one within the Maintenance of Cultural Spaces category and three within the remit of Production.[20] With the funding received for the Maintenance of Cultural Spaces, the company was able to cover some of the expenses related to the upkeep of Estúdio Mawaca as well as organize a few workshops with guests, such as percussionist Vitor da Trindade and Guarani rapper MC Kunumi. These activities were recorded and have been made available free of charge through social networks and digital media.

There was an award of approximately R$5,000 (US$ 930) among the projects linked to Production, for which some members of the group offered workshops related to their specialties, which were filmed in digital formats. The production of these videos was extremely laborious: they involved a team of technicians to carry out the filming and editing, and throughout the year of 2021 they were broadcast free of charge through the group's social media channels. Despite the intense work involved, Pucci claimed that the project boosted morale within the company, as it drew out the ways in which different members of the group understood their experiences within the company. Financially, however, the sum they received only brought a temporary respite, hardly constituting a resolution to the problems related to the significant decline in income the company members experienced in the absence of regular performances.

A second Production project presented by the company involved the licensing rights for a digital project, in which Mawaca recorded a composition by guitarist Carlinhos Antunes. Each musician involved in the project recorded their part at home and, later, these videos were edited into a single video, and disseminated on various social media.[21] A third project

172 *Suzel Ana Reily and Regina Machado*

in the Production category involved the filming of classes on the music of the native peoples of Brazil. Magda Pucci taught the classes, but the funding allowed for payment to two technicians and a producer in addition to herself. She noted that the production of the videos provided her with an opportunity to contemplate the work she has been doing for some years in creating material on Indigenous music for use in the classroom.

Like other artists and musicians who submitted projects to LAB bids, Pucci was also concerned about designing her projects in a way that included other musicians and workers involved in musical production, such as technicians and producers, considering that many of them were also confronting economic difficulties from lack of work. In effect, she noted how the health crisis drew special attention to the collaborative nature of the music world, generating feelings of mutual responsibility.

Final Considerations

The funds made available by the Aldir Blanc Bill were the largest directed toward the cultural sector from a single source during the pandemic, and they injected much needed assistance into this social sphere. Given the continental dimensions of the country, however, these funds were far from sufficient to meet the needs of the country's art worlds. Yet, besides bringing some respite to very needy artists and ensuring many cultural spaces could survive, these monies also had an important impact on cultural activity during the pandemic, providing many musicians and artists with new skills, particularly in the digital realm; they created opportunities for projects to be reimagined in new formats; they heightened collaborative links within art worlds; and they created a huge national archive on social media of artistic endeavours in a range of artistic fields.[22]

Furthermore, this injection of funding demonstrated how in need of financial support this sector found itself in during the pandemic, such that a range of other funding options emerged, both within the public and the private sectors. It is noteworthy that when PROAC launched a bid for cultural projects in the state of São Paulo in May 2021, approximately 42,000 projects were submitted, of which only a fraction could be funded. In effect, the mobilisation of artists in all spheres continued, with a view toward negotiating further programs to support cultural endeavours across the country. In September 2021, for instance, the so-called Paulo Gustavo Bill was passed, which aims to inject R$ 4.2 billion (over US$ 780,000) into the audiovisual sector by the end of 2022; many bids for these funds are envisaged to involve musicians. Currently, the Aldir Blanc Bill No. 2 is moving through Congress, which like the original bill is directed at all artistic fields. The city of São Paulo runs the Programa para a Valorização de Iniciativas Culturais (Programme for the Valorization of Cultural Initiatives), better known as the Programa VAI. This program existed before the pandemic, but the emergence of COVID-19 gave it greater visibility. Its objective is to provide financial support for artistic and cultural activities involving groups or collectives of low-income young people living in areas with limited cultural opportunities. Projects can last up to eight months and the maximum funding for a project is R$94,000 (US$ 17,500), a sum that presupposes these funds will be widely distributed.

It is also worth pointing out that the assistance musicians and artists received helped give them a degree of stability as they identified new sources of livelihood, such as teaching and the monetisation of online productions, including 'lives' and recordings of shows, music video clips, workshops, and masterclasses. Now, with the country's high vaccination levels, live events have of course begun to take place again, although it cannot be said that the pandemic has come to an end. Indeed, numerous restrictions on audience-based indoor activities remain in place, just as the possibility of new variants threaten to impose further closures. Income security in the art world therefore remains fragile. The pandemic only

Musicians in the Brazilian Pandemic 173

deepened the vulnerability of a large class of workers in Brazil, highlighting the limitations in the country's public and private policies directed towards an egalitarian support of arts and culture.

Facing the COVID-19 crisis was, without doubt, difficult for artists all over the world. In Brazil, where the challenge involved circumventing the federal government and a society strongly divided along political lines, it was primarily thanks to the strong links of solidarity shared by many musicians and a broad range of cultural workers engaged in music worlds, across their artistic worlds and critical sectors of civil society, that they have, for better or worse, weathered this storm. Musicians and artists generally were critical for assisting many people to cope with the anxieties and fears brought on by the pandemic. It is now fundamental that there be an extensive reflection on ensuring this essential workforce receive the support they need in times of crisis.

Notes

1 *Congados* are music and dance troupes associated with African-Brazilian communities. Such groups are common throughout the state of Minas Gerais as well as in other parts of southeast Brazil.
2 While the term *quilombo* once referred to a community of runaway slaves, it is now understood as a space of resistance for communities of African descendants. The 1988 Constitution of Brazil recognises the rights to the land of those living in these communities, and their right to preserve their own culture.
3 The full Senate report can be viewed at https://static.poder360.com.br/2021/10/relatorio-final-renan-calheiros-cpi.pdf.
4 Across the globe, researchers have highlighted the role of music in coping with the tensions brought on by the pandemic. See, for instance Cabedo-Mas, Arriaga-Sanz, and Moliner-Miravet 2021; Krause et al. 2021; Ziv and Hollander-Shabtai 2022.
5 During the pandemic live shows, typically broadcast from the musician's own home, became common and these events were referred to as 'lives' (in English).
6 Aldir Blanc is best known as the lyricist of over 600 compositions, and his most important partnerships were with João Bosco, Guinga, and Cristovão Bastos.
7 In line with MinC's orientation, it should not be the role of government to produce culture, but rather to create conditions that assist groups and individuals to engage in cultural activities. Using the analogy of do-in, Minister Gilberto Gil suggested administrators should massage the points capable of liberating blocked energies (Calabre and Lima 2014, 11). Thus, local cultural spaces came to be understood as 'cultural points,' and their members were invited to register their points in order to receive funding from the Cultura Viva program. Generally speaking, *pontos de cultura* were based in low-income neighbourhoods, such that the program targeted precisely those communities typically excluded from private forms of culture sponsorship.
8 The full National Plan for Culture can be found at: www.ipea.gov.br/participacao/images/pdfs/confe rencias/IIICNCultura/metas-do-plano-nacional-de-cultura.pdf; the list of aims is on pages 10–11.
9 Partido Comunista do Brasil (Communist Party of Brazil).
10 Celio Turino wrote the first document, with the idea that it would come to be the Aldir Blanc Bill. He then participated in the Cultural Convergence, the working group that helped write the Bill. Moreover, he was active on the YouTube channel called Cultural Emergency (www.youtube. com/c/Emerg%C3%AAnciaCultural), together with Alexandre Santini, Marcelo das Histórias, and parliamentarian Jandira Feghali. As a member of Fifth Article, a group of artists constituted to confront the growing number of cases of censorship occurring in the country following the onset of Jair Bolsonaro administration and of which this chapter's co-author Regina Machado is also a member, Turino continuously informed the group on how negotiations for the bill were proceeding.
11 While the system worked remarkably well in many places, some have criticized the way the funds were distributed (see Melo 2021), while others have noted that not all municipalities understood

174 *Suzel Ana Reily and Regina Machado*

how the funds could be obtained and how they could be used (Silva Junior 2021, 357). It is beyond the scope of this essay to engage with this issue.

12 This sub-category covered expenses connected to licensing rights and the use of copyrighted material.

13 Fruition refers to expenses linked to the dissemination of existing artistic or cultural material, such as the construction of virtual environments to display material held by a museum, for instance.

14 To view in full, see www.proac.sp.gov.br/proac-editais-editais-expresso-lab/.

15 The videos produced within the project include: (1) How choral singing works (https://toutu.be/i83fKmZvnws), (2) Musicalisation for children from ages six to nine (https://youtu,be/XwZc mtmPdW0), (3) Corner for teachers of early childhood and elementary education I (https://youtu.br/T7Ksp225A4), (4) Troubadourism and RAP (https://youtu.be/qhLPD6mjdqw), (5) Yellow Submarine – a story sung and played (https://youtu.be/LCSnlbCKF4).

16 In the late 1960s the Tropicália Movement emerged in Brazilian popular music, spearheaded by such musicians as Caetano Veloso, Gilberto Gil, and others. Its experimental style united Brazilian musical elements with elements associated with international pop articulating what these musicians and artists understood to be Brazilian culture at the time.

17 To view the show, go to: www.youtube.com/watch?v=IgaUn5XXT8I.

18 All the videos produced in the *Around the World with Guests* project are available on YouTube at the following links: (1) Badi solo (https://youtu.be/HZIjDPJ5gRA), (2) Badi com participação de Swami Jr. (https://youtu.be/YNi2Hj3VcEg), (3) Badi com participação de Carlinhos Antunes (https://youtu.be/rjVITuEKouE), (4) Badi com participação de Fernandinho BeatBox (https://youtu.be/Z6khvigdrJw), (5) Badi com participação de Livia Mattos (https://youtu.be/9YnqJ7-hA1k), (6) Badi com participação de Marcelo Pretto (https://youtu.be/pXAmRuYG_R0).

19 See note 10.

20 Other members of Mawaca presented LAB bids, and many of those that generated videos can be found on Mawaca's YouTube channel: www.youtube.com/channel/UCSaHrceApVYv3L3J6s9T rXg. See for instance projects produced by Gabriel Levy.

21 See www.youtube.com/watch?v=PMreHHXVgAs.

22 It is worth noting, however, that such critics as João Roque da Silva Junior (2021, 359) have alerted to the way in which part of the funding from state-sponsored bids is ending up in the hands of Big Tech like YouTube, Facebook, Instagram, Spotify, Vimeo, and others.

References

Assad, Badi. 2018. *Volta ao mundo em 80 artistas*. São Paulo: Pólen.

Barreto, Luisa Marques. 2020. 'Lei Aldir Blanc de Emergência e o fim do Plano Nacional de Cultura (2010–2020).' *Boletim de Políticas Públicas* 7: 29–42.

Becker, Howard. 1982. *Art Worlds*. Berkeley: University of California Press.

Cabedo-Mas, Alberto, Cristina Arriaga-Sanz, and Lidon Moliner-Miravet. 2021. 'Uses and Perceptions of Music in Times of COVID-19: A Spanish Population Survey.' *Frontiers in Psychology* 11. https://doi.org/10.3389/fpsyg.2020.606180.

Calabre, Lia and Deborah Rebello Lima. 2014. 'Do do-in antropológico à política de base comunitária – 10 anos do Programa Cultura Viva: uma trajetória da relação entre Estado e sociedade.' *Políticas Culturais em Revista* 2 (7): 6–25. https://doi.org/10.9771/1983-3717pcr.v7i2.12867.

Krause, Amanda E., James Dimmock, Amanda L. Rebar, and Ben Jackson. 2021. 'Music Listening Predicted Improved Life Satisfaction in University Students during Early Stages of the COVID-19 Pandemic.' *Frontiers in Psychology* 11. https://doi.org/10.3389/fpsyg.2020.631033.

Machado, Cacá. 2017. 'Música e ação política, Brasil 2003/2006.' *Políticas Culturais em Revista* 10: 119–147. https://doi.org/10.9771/pcr.v10i2.24319.

Melo, Túllio. 2021. 'A desigualdade na distribuição dos recursos da Lei Aldir Blanc entre os estados e o Distrito Federal.' *Jusbrasil* (March). https://advtulliomelo.jusbrasil.com.br/artigos/880019600/a-desigualdade-na-distribuicao-dos-recursos-da-lei-aldir-blanc-entre-os-estados-e-df.

Muniagurria, Lorena Avellar de. 2021. ' "Os primeiros a para e os últimos a voltar": trabalhadores da cultura no Brasil em tempos de Covid-19.' *Revista Música e Cultura* 12: 232–243.

Raquel, Martha. 2021. 'Brasil ultrapassa marca de mil indígenas mortos em decorrência da Covid-19.' *Brasil de Fato*, 13 March 2021. www.brasildefato.com.br/2021/03/13/brasil-ultrapassa-marca-de-mil-indigenas-mortos-em-decorrencia-da-covid-19.

Rede Penssan. 2020. *Food Insecurity and Covid-19 in Brazil*. http://olheparaafome.com.br/VIGISAN_AF_National_Survey_of_Food_Insecurity.pdf.

Rodrigues, Marco Aurelio de Souza, Daniel Kamlot, and Anita Vasconcelos de Carvalho. 2021. 'Pandemia, samba e as *lives* de Diogo Nogueira: desafios de gestão de plataformas.' *Revista de Administração Contemporânea* 25 (Special Issue). https://doi.org/10.1590/1982-7849rac2021200225.en.

Silva Junior, João Roque da. 2021. 'Os desafios dos setores criativo e cultural brasileiros durante e depois da pandemia covid-19.' *Extraprensa* 14 (2): 344–363. https://doi.org/10.11606/extraprensa2021.188687.

Souza, André Luis Santos de and André Luiz Ribeiro de Araújo. 2021. 'Lei Aldir Blanc, política cultural imaterial e folia de reis em Santa Helena de Minas (MG).' *Em tempos de Histórias* 1 (39): 191–210. https://doi.org/10.26512/emtempos.v1i39.39557.

Turino, Célio. 2010. 'Ponto de Cultura: a construção de uma política pública.' *Cadernos Cenpec* 7: 23–31. http://dx.doi.org/10.18676/cadernoscenpec.v5i7.61.

Ziv, Naomi and Revital Hollander-Shabtai. 2022. 'Music and COVID-19: Changes in Uses and Emotional Reaction to Music under Stay-At-Home Restrictions.' *Psychology of Music* 50 (2): 1–17. https://doi.org/10.1177/03057356211003326.

14 Musical Performance during and after the COVID-19 Pandemic

Days of Future Passed?

Alessandro Bratus, Alessandro Caliandro, Fulvia Caruso, Flavio Antonio Ceravolo and Michela Garda[1]

All of the dimensions related to music performance have experienced deep transformations, crises, and reinventions during the COVID-19 pandemic. The scope of this *multivocal piece* – composed by five scholars from various academic backgrounds, such as musicology, ethnomusicology, sociology, and popular music studies – is mapping different ways of attributing meaning to the enactment of a wide range of musical practices that proposed a rearticulation of the reception and distribution of live musical performance. The act of music making has been, since then, digitally remembered, represented, reimagined, relocated, and remediated in a variety of forms. At the same time, the performance of music – especially since it became impossible, as its practical organisation seems incompatible with any effective form of physical distancing – has become an ideal referent and arena. A great deal of the identity claims related to music, as a form of personal and collective self-recognition and construction of cultural values, rely on the (now impossible) act of live performance. These dynamics are observed through the lens of a number of Italian case studies related to traditional religious rituals, operatic performance, discourses around live clubs, and the circulation of musical performance through social media. We discuss the extent to which they reflect our current understanding of the cultural activities generally referred to as music production and consumption, and how much they foreshadow future developments in these sectors.

The Great (Social) Narration of the Pandemic

From a sociological point of view, the initial period of the lockdown (that is, from the first days of March to the end of May 2020) can be divided into three different phases (Anzivino, Ceravolo, and Rostan 2020). The first phase was characterised by the spread of a general awareness of the impending danger, and by the development of a new common sense. This common sense was the outcome of the widespread feelings of fear concerning both the health risks and the possible consequences of the lockdown policies on the economic situation. In other words, people were concerned both for themselves and for the general situation within the country. This increasingly common sense of fear provoked an overwhelming social disorientation (Affuso, Giap Parini, and Santambrogio 2020; Colombo and Rebughini 2020). People were continually connected to different devices (mainly television), looking for information on the developments of the pandemic and the government's decisions (Anzivino, Ceravolo, and Rostan 2020). The gradual restriction of the individual freedom to meet other people augmented the sense of isolation, and consequently the fear felt by individuals, in a sort of vicious circle. The news provided by media and institutional actors fed the general apprehension and uncertainty. In this very first period, the role of

DOI: 10.4324/9781003200369-17

music was not yet so recognizable or notable. Everyone's attention in Italy was directed to news about the pandemic (Affuso, Agodi, and Ceravolo 2020).

The second phase can be identified as the first collective reaction to face the declining atmosphere. In this part of the chapter, we focus our attention mainly on this phase and the subsequent one. This social reaction phase began during the very first days of the general lockdown restriction (around 10 March). During these days a common 'great narration' started emerging, everywhere on social and traditional media (Affuso, Agodi, and Ceravolo 2020; see also Giorgino 2020). Different actors, belonging to different social, political, and economic backgrounds, contributed to create a sort of rhetoric landscape, which was based on a few collective values shared by the national community:

- the necessity of common responsibility to protect oneself and the others, represented primarily by the hashtag *#iorestoacasa* (I stay at home);
- the idea of a common national community. One of the key-points in the fight against the virus was respecting the severe social distancing rules imposed by the government. The social and political sustainability of these rules was based on the sense of *belonging to the same community*. Renewed national pride functioned to sustain social compliance to the measures;
- the vision of the future to fight the present fear. The other hashtag which was shared the most during this period was *#andràtuttobene* (everything will be alright). The collective optimism towards the future, and the overcoming of the crisis, became one of the most important ways to face the fear of the present. Such general optimism proved to be very important for the public because, during these days, the media constantly reported the daunting situation occurring in the hospitals and the increasing death rate caused by the disease.

The role of music in this great narration was crucial on many levels. All of the actors who contributed to creating this landscape of rhetoric used music as a universal language to communicate feelings and to promote collective values. Very often, institutional communication was introduced and underlined using classical musical pieces from the Italian tradition. Company brands participated in this great narration by including a specific Italian music repertoire (both popular and classical) in their spots and campaigns (Giorgino 2020). The aim was to use their communicative power to both instil a sense of belonging among the members of the Italian community fighting against the virus and, at the same time, to demonstrate their civic engagement to the public. This membership was a crucial strategy of corporate social responsibility during a period of crisis.

Obviously, many artists sustained this great narration too. They often offered online concerts and produced music related to the topic. A clear example was the song 'Rinascerò, rinascerai' (I will be reborn, you will be reborn) written and produced by Roby Facchinetti regarding the dramatic situation of Bergamo. This song became a sort of hymn for national resilience. In any case, famous musicians were not the only ones to take a stand against COVID-19 and/or see their profession reconfigured by the pandemic emergency, this also pertained to ordinary, non-professional musicians. In fact, one of the most important transformations occurred in the relationship between artists and their public, concerning the level of intimacy produced by the situation. The necessary stoppage of all live performances led the artists to look for new ways of communicating with their followers. Before the pandemic, many artists used digital tools to disseminate music or create and maintain their personal community of fans. This was an important part of their self-branding process. However, before the virus crisis, digital tools were only one of the channels in a more complex set of

178 *Alessandro Bratus et al.*

strategies of artistic communication. The pandemic forced the development of new and more direct methods to create online music performances.

The real novelty was the participation of non-professional musicians and ordinary people in creating performances to keep up morale and to overcome the solitude brought by physical distancing. Music became the most common language to create social connections over the necessary physical isolation. Everyone could see the effect of the improvised choruses that people sang from windows and balconies. The repertoire of these grass-roots collective performances was based on traditional Italian music, from traditional regional songs to more well-known tunes written by famous songwriters. The live performances of common people became viral, and so we can observe a sort of union between the physical and the digital world. This was not completely unpredictable, but the real novelty was the amount of shared live music experienced in the digital arena.

The third phase was characterised by the consolidation of social resistance to the virus (Affuso, Agodi, and Ceravolo 2020). We place this phase from the end of March to the lifting of the severe lockdown restrictions at the end of May. During this period, the role of music was mainly related to the necessity of helping people to resist a very difficult situation for a long time. The great narration became a sort of *basso continuo* in people's everyday lives, but they needed to find ways to spend an enormous amount of free time (Giorgino 2020). Therefore, the consumption of digital music increased because of the hyper-connection created between people and (mainly social) media (Santambrogio 2020). This continuous connection can be identified as a compensation for the lack of other relations. Also, the production of music by professional and non-professional musicians increased dramatically, which is a good sign for the future.

Experiencing and Reflecting on Digital Liveness after the Lockdown

The suspension of live events during the pandemic produced serious economic consequences and individual psychological damage. Yet, the pandemic cultural crisis can also be investigated as an unintentional and upsetting experiment about the prospective transformations of our artistic and cultural experiences in the digital age. Before the pandemic, sociologists, psychologists, and media scholars were concerned with the consequences, both positive and negative, of the increasing pervasiveness of digital media in everyday life. The 'Onlife,' the experience of a hyperconnected reality (Floridi 2015), suddenly became the only possible way of life outside the limited boundaries of our homes. It is therefore urgent to investigate how this previously unknown condition impacted our understanding of live and mediatised art experiences, as well as how it produced new forms of creativity.

Over the past decade, the experience of liveness has been conceptualised beyond the binary reduction of live/recorded, from which this term originated, and extended to an experiential and aesthetical quality that encompasses nearly every technologically mediated interaction (Auslander 2008, 2012; Sanden 2013). Auslander proposes two different definitions. The first, which I call the *affective argument*, claims that if virtual entities 'respond us in real-time, they feel live to us' (Auslander 2012, 6); the second one, which I call the *relational argument*, maintains that 'digital liveness emerges as a specific *relation* between self and other, a particular way of "being involved with something"' (Auslander 2012, 10). Although Auslander dismisses the affective argument in favour of the relational one, I think that the reference to time makes the former worthy of further discussion. If virtual entities respond in real-time, they do not only feel live to us metaphorically, they are constructed to share something with us, that is, chronological time. In fact, during the lockdown, the opportunity to be together in real-time, although through digital mediation, was the only possibility to surrogate some live experience. While traditional liveness (experienced in person) and extended liveness (technologically mediated) are not ontologically different, the experiences show different emotional nuances, as Alfred Schütz indicates in a now old-fashioned phenomenological approach to the live/recorded opposition:

> It is of no great importance whether performer and listener share a vivid present in face-to-face relation or whether through the interposition of mechanical devices-as records, only a quasi-simultaneity between the consciousness of the mediator and the listener has been established. The latter case always refers to the former. The difference between the two shows merely that the relationship between performer and audience is subject to all variations of intensity, intimacy, and anonymity.
>
> (Schütz 1951, 93)

Beyond fear, distress and mourning, everyday life during the lockdown could be described as the fading out of the experience of the vivid-present: grey anonymity was perhaps the most widespread feeling. Once the possibility of a live experience in the traditional sense had been suspended, the thrilling quality of *liveness* in a digital involvement slowly evaporated. In this case, what was experienced as an extension of the possibilities of human relations turned out to be a surrogate. *Being involved with something* was accompanied by the feeling of a deep loss of intensity and intimacy. Some form of compensation was needed.

During the first lockdown in Italy, opera theatres, concert halls, and streaming platforms, which usually charged consumers for their services, started offering events and past productions for free, both streaming and on demand. Suddenly, accessing repertories that had been difficult to because of economic, social, and geographical reasons, was available to everyone who had access to the internet. Two interesting aspects emerged in this process. The first one was related to time: although the streamed productions were no longer related to new live events, a playbill similar to that of a theatrical season regulated the access to the videos on demand. While traditional liveness was suspended, extended liveness was enhanced by mimicking the ephemerality of live events. The restriction on the availability counterbalanced the unprecedented richness of the cultural resources at the public's disposal.

Second, the suspension of live performances was a threat to both cultural consumers, who were suddenly deprived of their normal routine and cut off from their communities, and to cultural institutions, whose connections with their regular audiences seemed to be shattered. Theatres and concert halls exploited the new possibilities of digital media in very different ways: to produce narratives about their practices, through interviews with collaborators, video tours of their facilities, backstage footage, and so on, all to benefit from the new forms of public intimacy launched by social media. The most spectacular example of this trend in Italy was the Teatro Donizetti in Bergamo, one of the cities in Italy most affected by the pandemic. Francesco Miceli, the theatre's artistic director, succeeded in turning the newly inaugurated theatre WebTV into a powerful means to consolidate and expand his theatrical community. The Donizetti Festival WebTV tried to provide a surrogate for the live experience of the opera theatre in a style closer to popular entertainment, with a clever mix of free videos and paid streaming. The offer included a digital tour of the renovated theatre that was still closed to visitors, interviews, fictive conversations between Gaetano Donizetti and the festival's guest artists, and a digital parlour with guests commenting on the performances. The TV website advertised a newsletter proposing an introspective and

180 *Alessandro Bratus et al.*

interactive journey designed by opera experts and psychologists to address the tragic experience of pain and loss in the 2020 pandemic, through Donizetti's music (Donizetti.org).

As illustrated by the Bergamo example, mediated intimacy and the construction of a popular narrative around traditional high-culture genres like opera suggest that these practices could have a permanent impact on cultural consumption by increasing audiences for traditionally exclusive genres such as classical concerts and opera. According to research carried out in Italy, 16 per cent of respondents were new cultural consumers, who explored cultural programming on television or on digital platforms for the first time during the lockdown (Paltrinieri 2020).[2] Supposedly this group will forego the emotional intensity related to attending the performance in person, in favour of enjoying the comfort of being at home and having the opportunity to share such events with family. The extended liveness of digital media is an opportunity to soften cultural, social, and economic barriers. We can foresee a future in which live streaming could be part of the regular playbill of theatres and concert halls, as a different mode of live performance that can reach a broader and more diverse audience.

The lockdown experience fuelled experimentation and reflections on technological solutions for overcoming distance while being together in time. *Soundhouse: Intimacy and Distance*, an online installation by Barbican, consists of three rooms where listeners can gather to listen to experimental programs (Barbican 2021).[3] This installation explores forms of communal listening as the opposite of more widespread programs on demand, creatively re-mediating older forms of communal listening, such as radio. For the performers, distance is, of course, an insurmountable obstacle. The experience of the lockdown reignited interest in telematic music, which had previously been limited to experimental projects.[4] During the pandemic, Tempo Reale, the centre for music research, production, and education founded by Luciano Berio in Florence, started the 'remote musical socialization platform' *#HOMEPLAYING*, that 'allows distant musicians to perform together, strictly in a live setting and from the comfort of their homes, actively bridging their physical distance' (Tempo Reale 2020). A more popular enterprise, which received significant media attention, was *The Lost Nutcracker*, featured from 17 to 20 December 2020 at the Tulsa ballet in Oklahoma. The Italian choreographer Luciano Cannito succeeded in directing the ballet streaming via Zoom from Italy, overcoming the international travel ban and inaugurating a new practice of remote directing.

Streaming Liveness in the Time of Pandemic (and after?)

The change brought to the music sector by digital technology in the first years of the twenty-first century caused a complete reshaping of the recording industry as we once knew it. First and foremost the perception of musical products was transformed from physical goods to services (Butler 2018). In this brief report on the consequences of the COVID-19 pandemic on the live music industry in Italy, I will argue that the partial closure of live venues and festivals over the last 18 months favoured the economic interests and diffusion of new business models for the phonographic sector of the live music industry, first and foremost by developing positive discourses around live streaming and virtual events. The chronicle of the Italian situation represents an interesting case study of a glocalised phenomenon that can also be aligned with local specificities in other countries, wherever there are similar conditions involving the absence of a long-term state policy and a fragmentation of the live music scene.

A brief survey on the 2019–2020 programming of seven Italian live clubs, conducted during the time that I was gathering ideas for this chapter, offers some first hints of the ongoing transformation, which has been somehow accelerated by the pandemic.[5] According to my data, the total number of events halved in 2020, with significant differences between small and large clubs, which dealt with the emergence of the pandemic through different approaches and strategies. Large venues, considering the limitations regarding both audience admission and the international circulation of performers and bands, completely cancelled the entire programming season in 2020. Mid-level and small clubs, often local and non-profit enterprises, tried to maintain their activities by focusing on local and/or small or solo acts or by proposing events in unusual locations. These latter venues are also those that generally propose more diverse and curated artistic programs, in terms of genre diversity, styles, and development of local music scenes, while larger clubs tend to be more eager to follow the trends of the phonographic market, in order to maximise profits and sustain their larger organisations.

When comparing the Italian situation with other European live music scenes, the trend just described highlights some characteristics of our national scenario that made the consequences of the COVID-19 pandemic even worse than in other countries.[6] State funding for live music in Italy is almost completely absorbed by classical music and opera theatres and foundations, as the guiding criteria is to privilege those initiatives with established cultural prestige. The absence of governmental initiatives specifically supporting live popular music leaves the entire sector in the hands of private entrepreneurship, either no-profit (56 per cent in Italy, 44 per cent in Europe) or commercial (44 per cent in Italy, 48 per cent in Europe). This makes the professional life of those who work at all levels in the scheduling, planning, organisation, and setting up of the events especially precarious; even before the pandemic 85 per cent of workers in the live music sector made their living from temporary jobs. In the three months from February to April 2020, 75 per cent of the live music clubs reported economic losses exceeding €25,000, and exceeding €100,000 for the rest of the year. Such a situation, an emergency within the emergency, strengthened the requests for renewed forms of state support for live clubs, such as the one presented by KeepOn Live, the Italian association of live clubs and festivals. They proposed a series of parameters for public funding: a fixed percentage would be devoted to funding programming of original music, the organisation of courses, and other initiatives related to the professionalisation of the live music sector; fair working terms and conditions for both temporary and permanent employees; the promotion of social cohesion on the local level; and so on. As of October 2021, the proposal presented to the Senate during the spring of 2021 has not yet been reviewed by the relevant commission, a number of festivals have significantly changed the organisation of their schedules to respect the strict rules of social distancing and the limit of 1,000 spectators per event. No indications have been given about a timeline for the gradual reopening of live clubs after the summer of 2021.

While trying to cope with rules that closed or severely compromised the economic sustainability of live music events, live clubs also tried to implement strategies for sustainable, local-based live streaming initiatives, such as DeLIVEry (a tentative circuit starting from Turin and Brescia),[7] and Domeniche d'Essai (Essai Sundays – Brescia and Bergamo),[8] in which the purchase of access to performances was bundled with home delivery of food and beverage. In December 2020, the staff of Domeniche d'Essai tried a different approach with PluggedPolaMolloy,[9] by commissioning the recording of live performances in places of special significance for each artist or band. The pre-recorded performances could be accessed through a dedicated pay-per-view platform for a limited time and were complemented by an exclusive meet-and-greet with the artists before the beginning of the pay-per-view window. The temporal limitations of the pre-recorded content, while mimicking some of the temporal constraints of a live event and the possible direct connection with the artists, had the practical consequence of restricting the

182 *Alessandro Bratus et al.*

potential audience for live audiovisual shows produced as professional products. This also meant additional costs and technical issues that were not covered by the ticket sales for access to the virtual event.

The combination of failed attempts to establish new practices for live music and the lack of state support led to a spectacular demonstration called *The Last Concert?*: after a great deal of publicity announcing an event to take place on the night of 27 February 2021 across more than 130 venues, each live club uploaded to the collective platform a video of their empty stages and concert halls, with silent musicians as testimonials.[10] While it had the immediate effect of prompting the announcement by the Ministry of Culture of a €50 million package for the live music sector, the deluding of audience expectations raised public awareness of the frustration around the impossible task of reconfiguring an entire production chain in such a short period of time. This is especially so for smaller venues that routinely work with limited budgets, often barely breaking even, also before the pandemic. The modes of production and distribution characteristic of digital content revealed how limited the audience for these products are when they involve emerging and mid-level artistic projects, for whom the scale of the potential audience cannot be increased so easily. This is probably also related to the characteristics of the public for such niche artists, who are followed by a more attentive audience: for them, in particular, the live experience cannot be replaced by vicariously attending a live stream performance or a Zoom virtual aperitif.

From the previous summary on the state of the live music sector in Italy during the worst phase of the COVID-19 pandemic, a deep fracture seems to emerge from the clash between the competing interests represented by the modes of production of the live music sector and the recording industry. The latter was obviously at an advantage when they both became service providers, since the technology was already in place from the advent of digital music a couple of decades earlier. The gap is even more apparent when considering the failed attempts of live clubs to transform their shows into live stream productions, in contrast to the successful attempts by global companies to turn live streaming into a gigantic business for mainstream acts. Not by chance, the rhetoric that drove such services refers also to so-called success stories told by big-business-oriented magazines, commenting on how the post-pandemic scenario is a benefit for the additional revenues and forms of direct connection with the fan base (e.g., Bruner 2021; Schabel 2020; Shapiro 2021). One can easily note what kind of artists and audiences these enthusiastic reports refer to: global stars as famous as Dua Lipa, Sofi Tukker, BTS, Steve Aoki, and Deadmau5 – big names whose established reputations guaranteed a return on the technological investment needed to move towards a total paradigm shift regarding live shows.

But is this radical change real, or just a repackaged version of the mediatised immediacy of concert films and live broadcasts? Virtual backstage passes or exclusive meet and greets with artists do replicate the promise of an intimacy with the artists and the access to perspectives from which a live audience is normally precluded. These further delve on the 'will-to-believe' of the media spectator, as a way to trigger their own involvement with the event perceived from behind a screen (Bratus 2019, 26–28). In economic terms, for what part of the current music industry does the inclusion of live stream performance represent a viable option to be incorporated into their individual business model? Would another algorithm-based form of content distribution work towards the preservation of cultural diversity and access to economically sustainable resources for a larger number of artists and musicians? When live streaming services and recording industries coalesce, these questions cannot be downplayed, and emerge as critical issues that we, as academics and music scholars, can explore to better understand our present and near future.

Suspended Rites and Media Compensation

Since March 2020 I have been developing a virtual ethnography, observing different behaviours on the web within various contexts and collecting repertoires of ethnomusicological value, with a particular focus on some religious rituals that were suspended in 2020 and in 2021. Here I will concentrate on the pilgrimage to the Holy Trinity shrine in Vallepietra, a ritual that I know well from previous fieldwork completed between 1999 and 2009 (Caruso 2008). This allowed me to better understand some emergent behaviours brought about by the pandemic.

The Sanctuary of the Holy Trinity, located between Lazio and Abruzzo on a ledge of Mount Autore, has been a site of pilgrimage for centuries. It is open from the first of May to the first of November, a period during which pilgrims visit the spot, especially for the feast of the Trinity on the Sunday following Pentecost, and for the feast of St. Anna on 26 July. The inhabitants of Vallepietra dedicate a procession to the village on 16 February to celebrate the appearance of the image of the Trinity, and a pilgrimage to the Sanctuary for the opening on the first of May. The rest of the opening period is dedicated to welcoming pilgrims.

The soundscape of the Sanctuary is extraordinary. The various companies of pilgrims use a wide range of sonic possibilities to express their devotion: singing, invocations, and prayers all overlap with the buzz of those who have already made their visit and speak, play, or listen to Mass, all amplified by the convex shape of the rocky wall behind the Sanctuary. The songs of pilgrims stand out above this mass of sounds, sometimes accompanied by various types of instruments. When the companies reach the Sanctuary, they mainly sing the 'Canzonetta in lode alla Santissima Trinità' (Canzonetta in Praise of the Holy Trinity), replaced by the 'Canzone sul miracolo eseguito da S. Anna' (Song of the Miracle Performed by St. Anne) for the July visits. Their departure from the Sanctuary is often accompanied by songs of salutation for the Holy Trinity. Many companies continue to sing religious songs even after visiting the holy cave. The song that never fails to be sung is the 'Canzonetta,' performed by every company in a personalised way through micro-variations. It is through singing, in fact, that the company testifies to its presence and faith.

None of this happened in 2020. The Sanctuary was closed shortly after the usual February procession in Vallepietra, then all of the events were cancelled. Visits resumed in the summer, but with safety rules that discouraged many pilgrims until September, when the rules were significantly loosened. However, the autumn brought a second lockdown, with restrictions that continued until the beginning of 2021. Celebrations of the first of May that year only took place with the Sanctuary's Prior and a few representative pilgrims. How did the pilgrims react? In recent years there has been a flourishing of Facebook pages on the reality of the situation that, in various ways, revolves around the Sanctuary:

- Confraternita della Santissima Trinità di Vallepietra (Brotherhood of the Holy Trinity of Vallepietra), created and administered by Prior Paolo De Santis in 2014;
- Santuario della Santissima Trinità (Holy Trinity Sanctuary), created in 2016 by Don Alberto, rector of the Sanctuary;
- Fede e tradizione al Santuario della Santissima Trinità (Faith and tradition at the Sanctuary of the Holy Trinity), created in 2018 and administered by Filippo Graziosi, active member of the Centre for Studies and Documentation of the Sanctuary of Holy Trinity.

I decided, therefore, to develop a virtual ethnography within these three Facebook pages, to understand if and, if so how, their contents have filled the void created by the closure of

184 *Alessandro Bratus et al.*

the Sanctuary. Soon after the first day of lockdown, on 10 March 2020, a letter from the mayor of Vallepietra and the rector of the Sanctuary was posted on the first two Facebook pages inviting faithful to pray the Holy Trinity Sunday, 15 March at 10.30. In a couple of days a Facebook event was created by the Consulta Giovanile of Vallepietra,[11] inviting all the devotees to the shrine to participate from their homes in a 'flash mob sonoro' (sonic flash mob).[12] This was relaunched by the Brotherhood Facebook page and a proliferation of amateur videos animated the Facebook page of the Brotherhood, echoed by the *Faith and Tradition* page.

In this initial production, we can distinguish two cases: the faithful of Vallepietra and the others. In the first case, the very first videos show the Sanctuary from afar, without ever framing the faithful, and in many of these you can hear the young women of Vallepietra singing 'Pianto delle Zitelle,' a lament about the death of Christ, sung only by the women of Vallepietra at the shrine on the Sunday dedicated to the Holy Trinity.[13] The following videos show the streets of Vallepietra with the faithful looking out from their homes singing the 'Canzonetta.' The purpose is to pay homage to the devotees who do not live near the Sanctuary, allowing them at least to see the holy place and hear the lament.

In the second case, the devotees instead posted amateur videos that show them inside their homes or courtyards, playing the 'Canzonetta' alone.[14] These are only instrumental performances, reflecting an established custom of devotees playing instruments to accompany the company, to express vows or their devotion.

As the days went on, many users made amateur montages to accompany the audio, from live recordings of devotional songs made on previous pilgrimages, with photos of people from both their previous visits and from home, the latter holding drawings of rainbows with the words 'everything will be alright,' a very popular expression during the first lockdown.[15] On Easter Sunday (12 April), Don Alberto decided to broadcast a recording of the 'Pianto delle Zitelle' from the speakers of the church of Vallepietra, sharing a video on the Facebook pages that shows the deserted streets of Vallepietra on 30 April.[16] In the following months, this produced requests from numerous companies to upload to the Brotherhood page amateur video footage of their previous visits. The first lockdown finished on 18 May, however, to avoid gatherings the shrine opened only on 15 June, after the feast of the Trinity that fell on 7 June. Pilgrims slowly began to visit the Sanctuary again, but in compliance with strict rules that severely limited the influx. During this period, the pilgrims increased the demand for selfies and videos attesting to the visit to be uploaded, almost always showing the same scene: the company, in a reduced version, singing as they approach the holy cave of the Sanctuary. The publications reached their peak in September, when less stringent controls enticed many more faithful to go to the Sanctuary. The habit of filming the company during the visit has been well established for a decade, because they like to have a memory of the event. However, this footage was rarely put online.

I continued to monitor the pages in 2021. The February procession of Vallepietra took place, limited by the anti-COVID-19 rules to only a few people from the town. The opening of the Sanctuary, as already mentioned, took place on 1 May, but with a poor presence of devotees, since pilgrimages on foot, gatherings, and overnight stays around the Sanctuary were prohibited due to the pandemic. The rector of the Sanctuary established a calendar of visits with the companies from June to September that allowed for staggered attendance. In 2021 the proliferation of video clips of previous and present visits has thus continued, while the ad hoc execution production from home has hardly occurred. The explosion of an online presence, for this ritual as for others, only occurred during the first lockdown, probably because such an abrupt restriction to home has not occurred again.

We learned from studies on pilgrimage that presence is central (Turner and Turner 1978). The way to express this is strictly codified, and it happens at its best when it is shared among many people. In Vallepietra, the performance of devotional songs dedicated to the Trinity is at the centre of this representation. During the lockdown, the presence of the Vallepietra Brotherhood Facebook page became the means to *be there* somehow, reproducing the core of the ritual: the musical performances. Reopening with a limited number of visitors produced the need to amplify the visit on the same page. No other Facebook page was created: that page is recognised as official by the faithful and a special virtual place, since the Brotherhood holds the shrine.

The presence on the web has provided the numerousness that the pandemic has prevented in person. The way they were packaged is well explained by a devotee who emigrated abroad:

> When I was little, I always followed my mamma to the Sanctuary. Unfortunately, now we can't anymore, but I still have the best memories of being with my mum on the bus. Technology is good, in minutes we all know what we are doing and where we are, but memories are what keep us going on and allow us to remember.
>
> (Confraternita S.S.Trinita' Vallepietra 2020)[17]

The contents are produced within a specific cultural circuit of use, and speak exclusively to those who already know the ritual and can be moved by those images through their memories. In this way the virtual world follows the same behaviours enacted for centuries at the shrine. 'In this sense the virtual is used not only in reference to the recreation of the everyday via technology, but also through memory, dreams and psychic subjectivity' (Trainer 2016, 409).

Re-Inventing Virtual Communities on YouTube during the Lockdown

Media scholars tend to agree that, as a matter of fact, social media are *anti-social* environments (Marres 2017), since their affordances favour connectivity and quick consumption of content, rather than communitarian interactions. This was surely true before the pandemic emergency, but not necessary during it. In fact, we argue, not only that the COVID-19 pandemic reconfigured the way in which music is performed, circulated and consumed, but (more profoundly) it pushed musicians and consumers to find new ways of using social media to recreate a sense of community around music.

In order to observe more closely the changes in music practice and consumption during the pandemic emergency, as well as how people create a sense of community around them, we developed a small digital ethnography on YouTube (see Caliandro 2018). Specifically, we used the open-source software YouTube Data Tool to collect 500 videos featuring the keyword *music performance* on the Italian version of YouTube from 9 March to 3 May 2020. After an initial manual screening, we identified 48 in-topic videos (i.e., we discarded the Spanish posts). While these videos are admittedly quite few, and a more refined search needs to be done in the future, they nonetheless perfectly served the scope of our exploratory qualitative analysis. We analysed the videos through a qualitative content analysis as well as by taking advantage of YouTube quantitative metrics (see Rieder, Matamoros-Fernández, and Coromina 2018), which we repurposed for the scope of our research. Specifically, such analysis of YouTube content allowed us to observe how different kinds of actors (music stars, musicians, YouTubers, labels, and ordinary consumers) aggregate around the topic of music performance, as well as how they reimagined it during the pandemic emergency.

186   *Alessandro Bratus et al.*

Table 14.1 Top 10 most viewed videos

Channel Title	Video Title	Topic	View Count
Warner Music Italy	Annalisa – Houseparty (Official Video)	official video	6,802,768
Jack Savoretti	Jack Savoretti – Andrà Tutto Bene	official video	449,963
Andrea Bocelli	Music For Hope LIVE – April 12th 10am LA	official video	262,430
Arcade Boyz	Come creare una HIT MUSICALE in LIVE!? (WalkThrought PARTE 1)| Arcade Boyz	tutorial	11,771
Quanto Basta	LA BELLEZZA CI UNISCE ("Nessun dorma," Venezia 2020)	official video	9,556
Arcade Boyz	Come creare una HIT MUSICALE in LIVE!? (WalkThrought PARTE 4)| Arcade Boyz	tutorial	9,489
Edo Baroni	Edoardo Baroni – Have You Met Miss Jones?	home live	8,661
Arcade Boyz	Come creare una HIT MUSICALE in LIVE!? (WalkThrought PARTE 2)| Arcade Boyz	tutorial	6,601
Arcade Boyz	Come creare una HIT MUSICALE in LIVE!? (WalkThrought PARTE 3)| Arcade Boyz	tutorial	6,363
TeleMia La Tv	CORONAVIRUS: A GIOIOSA FLASH MOB | IL VIDEO	home live	5,810

Looking at the general statistics generated by the software, we can make some interesting initial observations: the 48 videos retrieved were created by 42 users, and only ten videos are licensed content. This indicates that there are no dominant actors monopolising the creation of content, and that only a portion is composed of professional musicians. If we look at the top ten most viewed videos, we see that this ranking is dominated by professional content creators (see Table 14.1). For example, the most viewed video in the dataset is 'Houseparty,' by famous Italian pop singer Annalisa, with 6,802,768 views (as of 3 May 2020): a song about life in quarantine and, not by chance, shot in computer graphics. The other videos include content of professional singers, like Andrea Bocelli or Jack Savoretti, and well-known youtubers such as Arcade Boyz, with 474,000 subscribers. Only in the tenth position do we find a homemade video: 'CORONAVIRUS: A GIOIOSA FLASH MOB,' a typical example of so-called balcony song, performed by (very) amateur musicians.

If we focus instead on the content of the 48 videos and look at the public reaction to them, a different scenario emerges. Of the 48 videos, 15 are posts that we categorised as Home Live; that is, videos in which mostly amateur and semi-professional musicians perform live music from their homes, for example from bedrooms or living rooms. Sometimes these posts are marked by the hashtag #iorestoacasa, like the following one: *#IoRestoACasaGuitarChallenge* Guitar contest Arduini Luca / DV Mark, Musicoff. We categorised another 15 videos as Talkradio, which are posts in which two or more musicians or music lovers discuss artistic topics. Nine videos are Collective Songs: this is the most peculiar category, since it entails videos in which groups of musicians, home alone but connected and coordinated via digital technologies, perform live songs all together. It is interesting to notice that some of the tunes are ad hoc pieces about the pandemic situation, as in the case of 'UNA NOTA PER IL MOLISE COSA NON TI TORNA (FAI LA DIFFERENZA)': a piece written and performed by an ensemble of music teachers in order to support their own region, Molise. Five are tutorials, related to music techniques or specific pieces of software; and only four are videos uploaded by mainstream musicians.

Which of these categories most engaged the public? To answer this question, we calculated the *intensity* for each video, which we then aggregated per category. As Rieder, Coromina, and Matamoros-Fernández explain, intensity is a metric that allows researchers to identify

Musical Performance during and after the COVID-19 Pandemic 187

'those videos able to generate more engagement per view and corresponds to the sum of likes, dislikes, and comments divided by view count' (2020, 15). The final results are quite interesting; the topic distribution divided by intensity is as follows, with higher values indicating higher engagement: Home Live, 68; Talkradio, 59; Collective Song, 37; Tutorial, 24; Official Video, 6. The kinds of videos that engaged the public the most belong to the categories Talk Radio and Home Video. Also, videos falling in the categories Tutorial and Collective Song score decently in terms of engagement; but the same cannot be said for Official Videos. This suggests that YouTube users most appreciated those pieces of music that they perceived as authentic (Marwick and Boyd 2011), and which pursue a collective goal.

These results allow us to draw some interesting, although very preliminary, conclusions. Italian users were forced by the lockdown to spend much more time than usual on digital media; it seems that this circumstance led them to re-discover older ways of using the internet from before the advent of mainstream social media platforms like Facebook, YouTube, and Twitter, that were more virtual but much more participative (Jenkins 2006). In our case, it seems that musicians and music lovers used YouTube to recreate forms of virtual community to accomplish specific collective goals (see Lévy 1997), rather than passively consuming them or uploading music content simply to fish for likes. Not only did music content creators re-discover this (old) way of using digital media, but the public also seems to like it very much. In conclusion, we can say that the COVID-19 emergency reconfigured the way of conceiving of the digital – which is always contextual and never universal (Costa 2018) – pushing musicians and music listeners to use social media in a more communitarian way, although in a virtual mode, and to then re-imagine the concept and the practice of live performance.

Conclusions

The suspension of public cultural life has fuelled the desire to return as soon as possible to live performance; this inconvenient experience has nevertheless provided a unique opportunity to rethink and creatively recast the extended liveness of digital media. COVID-19, and especially the consequent lockdown, forced music producers and listeners to employ the digital in new and unexpected ways. It seems that they re-invented novel practices of music production, consumption, and circulation that re-elaborate the live performance with the typical characteristics of online interaction (collaboration, community building, participation, inclusiveness), compared to those usually experienced in the pre-pandemic period (scattered, individualistic, commercial, and elitarian). It remains to be seen whether such an opposition is part of a bottom-up global reconfiguration of musical activities, or should be considered an adjustment of pre-existing industrial, ideological, and cultural frameworks to the new conditions brought on by the pandemic. To be sure, the reconfiguration of the digital arena brought about by the emergence of the pandemic has opened new frontiers for music production, performance, and distribution. Now the question is: will these new frontiers be consolidated in the future post-pandemic era, or not?

Notes

1 The general outline of the chapter, the introduction, and the conclusion were shared between the authors. Flavio Antonio Ceravolo wrote 'The Great (Social) Narration of the Pandemic,' Michela Garda wrote 'Experiencing and Reflecting on Digital Liveness after the Lockdown,' Alessandro Bratus wrote 'Streaming Liveness in the Time of Pandemic (and after?),' Fulvia Caruso wrote 'Suspended Rites and Media Compensation,' and Alessandro Caliandro wrote 'Re-Inventing Virtual Communities on YouTube during the Lockdown.'

188 *Alessandro Bratus et al.*

2 For a broader overview see www.confcommercio.it/-/consumi-culturali-italia.

3 Online from 28 November 2020 until 28 February 2021, *Soundhouse* has stopped broadcasting. The website is now offering recordings of the broadcasts, plus a series of specially commissioned writings.

4 See, for instance, the utopic tone in Pauline Oliveros' (2009) account of her experimental career in this field. See also *The Online Orchestra*, a research project funded by AHRC, aiming to give musicians who live in remote communities the same opportunities to play in an orchestra as those who live in more urban areas: www.falmouth.ac.uk/research-programmes/digital-creativity/onl ine-orchestra#outcomes---outputs and http://onlineorchestra.com/.

5 For this preliminary survey I compared the live music programming for the entirety of 2019 and 2020, from a sample of music venues chosen for the sake of diversity in their organisational models, geographical contexts, and dimensions: Binario 69 (Bologna), Diagonal Loft Club (Forlì), Hall (Padova), Latteria Molloy (Brescia), Mercato Nuovo (Taranto), Santeria (Milan), and SmartLab (Trento).

6 The data commented on in this paragraph are extracted from the survey published by LiveDMA (2020) and KeepOn LIVE (2020).

7 The project, although announced and widely advertised in the spring of 2020, never began. A presentation can be seen at http://offtopictorino.it/live-delivery/.

8 See the following for a presentation of the events and some interesting thoughts on the topic by Luca Borsetti, artistic director of Latteria Molloy and one of the few entrepreneurs that tried to experiment with different business models in the live music sector during the COVID-19 pandemic: www.ecodibergamo.it/stories/eppen/cultura/musica/le-domeniche-dessai-la-rete-di-brescia-e-bergamo-per-la-musica-dal-vivo_1357721_11/.

9 www.facebook.com/pluggedpolamolloy/.

10 www.ultimoconcerto.it/.

11 www.facebook.com/events/151025052764574/.

12 www.facebook.com/108154787184253/photos/a.175057887160609/199215184744879/.

13 See for example www.facebook.com/ristorantesimbrivio.aziendafamiliarelauri/videos/3497900950285127.

14 See for example www.facebook.com/1429536853955845/videos/1315687388616359.

15 See for example www.facebook.com/1429536853955845/videos/202372934349382.

16 Behaviour that took place in several places to compensate for the absence of the Good Friday procession. See Caruso 2020 and Gugolati and Klien-Thomas 2021. An interesting compensation for the May celebrations in Accettura was also realised by the inhabitants of the village in collaboration with LEAV (see http://leavlab.com/portfolio/accettura-2020-il-maggio-del-silenzio/). The video is here: www.facebook.com/watch/?ref=search&v=734421453760307&external_log_id=fd94a5d3-8a87-4068-851f-e6874fc3cf7c&q=Comune%20di%20Vallepietra.

17 This comment can be found among the 425 comments on the video showing the shrine.

References

Affuso, Olimpia, Ercole Giap Parini, and Ambrogio Santambrogio. 2020. *Gli italiani in quarantena. Quaderni da un 'carcere' collettivo.* Perugia: Morlacchi Editore.

Affuso, Olimpia, Maria Carmela Agodi, and Flavio Antonio Ceravolo. 2020. 'Scienza, expertise e senso comune: dimensioni simboliche e sociomateriali della pandemia.' *Sociologia Italiana: AIS Journal of Sociology* 16: 57–68. https://doi.org/10.1485/2281-2652-202016-4.

Anzivino, Monia, Flavio Ceravolo, and Michele Rostan. 2020. 'La rivincita della scienza sul senso comune? Gli orientamenti di fiducia degli italiani all'inizio dell'emergenza Covid 19.' *Sociologia Italiana: AIS Journal of Sociology* 16: 121–139. https://doi.org/10.1485/2281-2652-202016-8.

Auslander, Philip. 2008. *Liveness: Performance in a Mediatized Culture.* London: Routledge.

Auslander, Philip. 2012. 'Digital Liveness: A Historico-Philosophical Perspective.' *Journal of Performance and Art* 34 (3): 3–11. https://doi.org/10.1162/PAJJ_a_00106.

Barbican. 2021. 'Soundhouse: Intimacy and Distance.' https://sites.barbican.org.uk/soundhouse/.

Bratus, Alessandro. 2019. *Mediatization in Popular Music Recorded Artifacts: Performance on Record and on Screen*. Lanham: Lexington.

Bruner, Raisa. 2021. 'The Livestream Show Will Go On. How COVID Has Changed Live Music- Forever.' 30 March 2021. https://time.com/5950135/livestream-music-future/.

Butler, Adam. 2018. 'Digital Media, Music Consumption and our Relationship with Music in the 21st Century.' https://blog.yorksj.ac.uk/musicproduction/2018/01/22/digital-media-music-consumption-and-our-relationship-with-music-in-the-21st-century/.

Caliandro, Alessandro. 2018. 'Digital Methods for Ethnography: Analytical Concepts for Ethnographers Exploring Social Media Environments.' *Journal of Contemporary Ethnography* 47 (5): 551–578. https://doi.org/10.1177/0891241617702960.

Caruso, Fulvia. 2008. *Evviva la Santissima Trinità*. Pescara: Carsa.

Caruso, Fulvia. 2020. 'Digital Humanity: musica e riti sospesi al tempo del Coronavirus.' In *Verso una musicologia transculturale. Scritti in onore di Francesco Giannattasio*, edited by Giorgio Adamo and Giovanni Giuriati, 131–140. Roma: Neoclassica.

Colombo, Enzo and Paola Rebughini. 2021. *Acrobati del presente. La vita quotidiana alla prova del lockdown*. Roma: Carocci.

Confraternita S.S.Trinita' Vallepietra. 2020. 'Buon giorno e buona domenica a tutti viva la SS Trinità.' Facebook, 26 April 2020. www.facebook.com/watch/?v=1361379774059869.

Costa, Elisabetta. 2018. 'Affordances-in-Practice: An Ethnographic Critique of Social Media Logic and Context Collapse.' *New Media & Society* 20 (10): 3641–3656. https://doi.org/10.1177/14614 44818756290.

Donizetti.org. 'Dal dolore alla rinascita.' www.donizetti.org/it/dal-dolore-alla-rinascita/?_ga= 2.168614369.1310496778.1607006788-172338417.1593512306.

Floridi, Luciano, ed. 2015. *The Onlife Manifesto. Being Human in a Hyperconnected Era*. Cham: Springer.

Giorgino, Francesco. 2020. 'La pubblicità al tempo del Coronavirus.' *Sociologia Italiana: AIS Journal of Sociology* 16: 207–236. https://doi.org/10.1485/2281-2652-202016-12.

Gugolati, Maica and Hannah Klien-Thomas. 2021. *Carnival in Digitalscapes: The Mediatisation of Public Cultural Events in and beyond the Pandemic*. www.isrf.org/fellows-projects/carnival-in-digita lscapes/.

Jenkins, Henry. 2006. *Convergence Culture: Where Old and New Media Collide*. New York: New York University Press.

KeepON Live. 2020. 'Analisi dell'impatto economico e lavorativo sui live club e festival italiani a seguito delle misure restrittive post emergenza COVID-19.' 25 June 2020. www.senato.it/applicat ion/xmanager/projects/leg18/attachments/documento_evento_procedura_commissione/files/000/ 160/901/Keepon_live.pdf.

Lévy, Pierre. 1997. *Collective Intelligence: Mankind's Emerging World in Cyberspace*. Translated by Robert Bononno. Cambridge: Cambridge University Press.

LiveDMA. 2020. 'Key Numbers – Impact of the Covid-19 Pandemic on 2,600 Live DMA European Music Venues and Clubs in 2020.' September 2020. www.live-dma.eu/wp-content/uploads/2020/10/ KEY-NUMBERS-IMPACT-OF-THE-COVID-19-PANDEMIC-ON-2600-LIVE-DMA-EUROP EAN-MUSIC-VENUES-AND-CLUBS-IN-2020_September-2020.pdf.

Oliveros, Pauline. 2009. 'From Telephone to High Speed Internet: A Brief History of My Tele-Musical Performances.' *Leonardo Music Journal* 19 (online supplement 'Telematic Music: Six Perspectives').

Marres, Noortje. 2017. *Digital Sociology: The Reinvention of Social Research*. London: John Wiley.

Marwick, Alice E. and Danah Boyd. 2011. 'I Tweet Honestly, I Tweet Passionately: Twitter Users, Context Collapse, and the Imagined Audience.' *New Media & Society* 13 (1): 114–133. https://doi. org/10.1177/1461444810365313.

Paltrinieri, Cristina. 2020. 'I consumi culturali degli italiani ai tempi di COVID-19.' https://group. intesasanpaolo.com/it/sala-stampa/approfondimenti/arte-e-cultura/2020/consumi-culturali-in-ita lia-ai-tempi-di-covid.

Rieder, Bernhard, Ariadna Matamoros-Fernández, and Òscar Coromina. 2018. 'From Ranking Algorithms to "Ranking Cultures": Investigating the Modulation of Visibility in YouTube Search Results.' *Convergence* 24 (1): 50–68. https://doi.org/10.1177/1354856517736982.

190 *Alessandro Bratus et al.*

Rieder, Bernhard, Òscar Coromina, and Ariadna Matamoros-Fernández. 2020. 'Mapping YouTube: A Quantitative Exploration of a Platformed Media System.' *First Monday* 25 (8). https://doi.org/10.5210/fm.v25i8.10667.

Sanden, Paul. 2013. *Liveness in Modern Music: Musicians, Technology and the Perception of Performance*. New York: Routledge.

Schütz, Alfred. 1951. 'Making Music Together: A Study in Social Relationship.' *Social Research* 18 (1): 76–97.

Santambrogio, Ambrogio. 2020. *Ecologia sociale. La società dopo la pandemia*. Milano: Mondadori Università.

Schabel, Mike. 2020. 'Why the Music Industry Must Think Big about Livestreaming: "Could an Artist Following in Freddie Mercury's Footsteps Execute the Same Call-and-Response with an Internet Audience?"' Music Business Worldwide, 7 December 2020. www.musicbusinessworldwide.com/why-the-music-industry-must-think-big-about-livestreaming-could-an-artist-following-in-freddie-mercurys-footsteps-execute-the-same-call-and-response-with-an-internet-audience/.

Shapiro, Ariel. 2021. 'Concert Streaming Was a Lifeline for the Music Industry during Covid. This Ex-Salesforce Exec Is Making It Part of the Post-Pandemic Future.' *Forbes*, 18 June 2021. www.forbes.com/sites/arielshapiro/2021/06/18/concert-streaming-was-a-lifeline-for-the-music-industry-during-covid-this-ex-salesforce-exec-is-making-it-part-of-the-post-pandemic-future/?sh=155debb940f3.

Tempo Reale. 2020. '#HOMEPLAYING Remote Music Socialization Platform.' https://temporeale.it/en/production/homeplaying-remote-musical-socialization-platform-2/.

Trainer, Adam. 2016. 'From Hypnagogia to Distroid: Positronic Musical Renderings of Personal Memory.' In *The Oxford Handbook of Music and Virtuality*, edited by Sheila Whiteley and Shara Rambarran, 409–427. New York: Oxford University Press.

Turner, Victor and Edith Turner. 1978. *Image and Pilgrimage in Christian Culture*. New York: Columbia University Press.

Part III

Perspectives

Rethinking Sound and Music against the Backdrop of a Global Crisis

15 Coronamusic(king)

Types, Repertoires, Consolatory Function

Melanie Wald-Fuhrmann

Shortly after lockdown policies against the spread of the new coronavirus had been implemented in European and North American countries, media reports about novel music practices started to spread, and were cherished as a welcome break in the stream of dire news about skyrocketing cases, terrible death tolls, and economic crises. These practices were meant as musical responses to the pandemic and the lockdowns, and the anxieties, hardships, and challenges that came with them. They included singing and playing from balconies, performances in front of homes for the elderly, live streams of concerts from artists' living rooms or private studios, the production of splitscreen performances by virtual ensembles, and the creation of COVID-19-themed parody songs. All of these are forms of 'musicking' (Small 1998) that differ in meaningful ways from standard concert and music industry forms of musicking. Some of them – such as balcony singing – disappeared again soon. Others, however, are still being practiced, and might even have the potential to survive after the pandemic.

This chapter will describe these new musical practices (coronamusicking) and their repertoires (coronamusic) – that is, their relevant forms and contents, and how people used them – from two angles. First, a music-sociologically informed typology of the relevant musicking types will be developed. Second, psychological theories of the functions and effects of music will be used to understand what coronamusicking could offer individuals and collectives during the pandemic, with particular regard to the consolation offered by such practices. Scholarly interest in coronamusicking does not only seem worthwhile because of the novelty, diversity, and aesthetic appeal of many of these practices, but also because these practices play a major role in helping people to cope with the emotional distress caused by the pandemic and social distancing, at both individual and group levels (Fink et al. 2021). Thus, they can provide a general model of music-related social coping with crises and challenges.

Coronamusicking: A Typology

The sources for the following data are mainly media reports, many of which are collected in the coronamusic database (Hansen et al. 2020; Hansen et al. 2021), YouTube videos, and some other music streaming services. Although an attempt was made to represent cases from as many different countries as possible, the majority of the examples come from German- and English-speaking countries. Overall, the varieties of coronamusicking found can be divided into forms of live music performances, performances on the internet, and the creation of specific repertoires.[1]

DOI: 10.4324/9781003200369-19

Live Music Performances

The most prominent form of live music performances were joint performances from balconies, rooftops, or windows, which became known as balcony singing. After first occurring in Wuhan in January 2020 (BBC 2020), it also appeared in Italy in March, and immediately attracted widespread public attention. From there, it soon spread not only across Europe, but also to North America, India, and other countries with particularly strict lockdown regimes, such as Australia and New Zealand. Joint balcony singing was clearly a means to uphold social connections and enable communication despite spatial distancing. It made use of some of music's unique features, namely the acoustic fact that it travels better across space than mere speaking. In this way, balcony singing provided a novel form of inclusive and participatory music making, where there was no distinction between audience and performers, and where performers were not required to be professional musicians.

The songs performed seem to have been primarily national or regional anthems and quasi-anthems, as well as popular songs, which is to say songs that are part of people's collective identity. Through their use in the context of the pandemic, their original meaning was actualised to respond to, and comment on, the current situation. They praised people's communities and homes, wished them well, encouraged people to fight this hardship together, and evoked better times ahead. Thus, the participatory practice, as well as the selected repertoire, reflected the perception of the pandemic as a national crisis that affected everyone, but that could be overcome through a collective effort.

For professional musicians, balconies and similar places also served as alternative stages when other public performances were not allowed. In this second category of live performances, which could be called balcony concerts, the standard form of a presentational performance with functionally separate groups of musicians and audience was maintained, but realised outside typical concert venues in places of everyday living, and without financial remuneration. As far as one can see, the pieces that were performed in these settings were mostly from the musicians' existing repertoire, and therefore only slightly, or not at all, related to the current situation. In addition to the desire to communicate and connect with others across spatial distance, and make them feel less lonely, bored, scared, and sad through music, the musicians' motivation for these performances may have also come from their interest in maintaining the self-image of a performer: having a reason to practice, feeling needed, and receiving positive feedback.[2]

Performances on the Internet

Next to making music, and listening to it live and together from a distance, we find internet performances of soloists and musical ensembles that either already existed, or were formed specifically for this purpose. Here, group music making that aimed to produce a performance-like result typically took the form of a splitscreen performance. Similar to balcony performances, the wish to experience feelings of being close despite spatial distancing and quarantine, but also the wish to comfort and encourage others seemed to underlie the majority of these virtual performances – in addition to passing time and fighting boredom (Daffern, Balmer, and Brereton 2021; Onderdijk, Acar, and Van Dyck 2021).

The songs that choirs chose show some overlap with the balcony singing repertoire, although anthems played a much smaller role here. Songs were either famous folk, pop, and opera pieces, or religious hymns that often speak from the perspective of a collective We and/or address a You. The lyrics of these songs can also easily be heard to make sense in the present situation such as 'You'll Never Walk Alone,' 'What the World Needs Now Is Love,' or the Bach chorale 'Befiehl du deine Wege.' The expressive character of many of these is relatively serious, hymn-like, moving, or solemn. This is also true for many of the instrumental pieces performed by lockdown orchestras.

For soloists (or a limited number of musicians who live together), however, the typical online performance format was a concert live stream via Twitter, Facebook, or YouTube. There were also some instances in which hours-long live-streamed concerts were organised with musicians from many different places, if not countries. In the popular music sector, this was the all-star concert *One World Together at Home*, under the aegis of Lady Gaga on 18 April 2020, and its Pan-Asian complement *One Love Asia* on 25 May 2020.[3] As in splitscreen performances, spoken notes by musicians or a moderator were a typical feature, but these also included short documentary clips about life in the pandemic, health workers, and initiatives that were trying to help. At the same time, audience members could give direct feedback or interact with each other via the chat or comment functions. Regarding the choice of pieces, classical musicians typically just played pieces from their repertoire, while the pop musicians created new songs (most famously, The Rolling Stones' 'Living in a Ghost Town'), or picked those from their repertoire that had some kind of fit based on their content.

Repertoire Creation

Finally, there was the practice of creating topical musical repertoires and putting them on streaming platforms. Three sub-types of this exist: arranging existing pieces into playlists, composing new pieces with concrete reference to the pandemic, or producing COVID-19-themed parody songs. All three types are relatively similar regarding the aspects of the pandemic and the lockdown that they touch upon. There are songs that mainly refer to the SARS-CoV-2 virus, the COVID-19 disease, its transmission, and effective safety measures (social distancing and washing hands in the beginning, wearing masks and vaccination later on), either in a primarily informative and exhortative way, in a personal and emotional way, or ironically. The majority of songs, however, centre on first-person experiences of living during the pandemic and under lockdown. Stockpiling and the shortage of goods; being stuck at home, either alone and bored, or annoyed by one's family members; and issues of childcare, homeschooling, and working remotely were the most frequently represented topics. Another group of songs resembles the balcony singing and virtual choirs' repertoires: they are primarily meant as a collective encouragement. Thoughts about death and survival are also present, as are the individual, societal, and political conflicts that accompanied the pandemic, understanding it, and the fight against it. The types differ, however, in their concrete attitudes towards these topics, particularly in terms of the most prominent emotions represented, and whether or not humour, irony, and even sarcasm play a role.

In contrast to balcony singing or splitscreen performances, corona playlists present not only one, but several pieces in a row or in an album-like fashion, and can thus present a much more differentiated picture of the pandemic, its aspects, and people's ways of dealing with it. Corona playlists have mostly been created and shared by private individuals, for example on Spotify, but also by broadcasting stations, or musicians on their own channels and websites.[4] Selecting and compiling songs in such a way that they carry a certain message, or address a given situation, is the standard way for non-musicians to communicate musically and appropriate the meaning and expression of a piece as their own.[5] Besides the wish to communicate, the desire to connect with others in a situation where physical forms of social interaction are severely restricted seems to be the strongest motivation behind this behaviour. Overall, playlists present a primarily unidirectional form of musically mediated social interaction since, in most cases, feedback from those who listen to a playlist is not technically possible.

196   *Melanie Wald-Fuhrmann*

Corona playlists typically assembled pop songs from a wide stylistic spectrum that mostly seem picked to cheer people up, help them through their day, and make them feel less lonely, anxious, and sad. Apart from the mere gesture of creating a playlist for others, and the often upbeat character of those songs, this is achieved mostly via the humorous effect resulting from the recontextualisation of songs such as MC Hammer's 'U Can't Touch This,' or Daft Punk's 'Digital Love.' Overall, we find songs that allude to COVID-19 symptoms; to the transmission of the virus via touching people and surfaces; to social distancing; to the lockdowns, and the loneliness and boredom that sprang from them; to death and the wish to survive; and to staying in contact with people via phone or online.[6] In addition to showing solidarity and confidence with songs such as Elton John's 'I'm Still Standing' or the Bee Gees' 'Stayin' Alive,' there was also a lot of seriousness, desperation, anxiety, and above all, black humour demonstrated, which was not present in balcony singing and splitscreen performances. Examples include Europe's 'The Final Countdown,' The Doors' 'The End,' R.E.M.'s 'Everybody Hurts,' and London Grammar's 'Wasting My Young Years.'

While playlist creation was predominantly an amateur practice, new songs that dealt with the pandemic were mainly composed by professional musicians. So far, no exhaustive overview exists about the number, breadth, and stylistic or regional distribution of these pieces. There are only some playlists and articles from media outlets that provide a first selection, discussion, and evaluation of the emerging repertoire (e.g., Buß 2020; Curcio 2020; Fekadu 2020; Reuter 2020; Volkmann 2020). Whereas earlier pieces, arranged in a corona playlist, could only allude to the present situation, genuine coronasongs could be much more precise and explicit about the pandemic and related aspects. Also, the large number and diverse stylistic and personal backgrounds of contributing musicians allowed the coronasong repertoire to develop into a multi-faceted mirror of the pandemic and how people experienced it.

Two main types of content emerged: in one type, the song tells about various aspects of living under lockdown and the lyrical subject's perceptions and responses. This type unfolds into several sub-types that differ mainly in regard to the expressed and afforded emotions, which range from sad and depressed, to angry and annoyed, to ironic and humorous. In another, the artists address their audience with, or include themselves in, a collective We in an attempt to comfort, encourage, give hope to, strengthen resilience, and to ask for considerate behaviour and compliance with hygiene measures. Examples of the first type are songs such as The Rolling Stones' 'Living in a Ghost Town,' Mahmood's 'Eternantena,' and Luke Comb's 'Six Feet Apart.' The second type is represented, for example, by Bon Jovi's 'Do What You Can,' Senri Oe's 'Togetherness,' or Silbermond's 'Machen wir das Beste draus.' Overall, the boundaries between the lyrical subject (and the person[s] addressed by it) and the biographic subject of the musicians become blurred, and the songs function on the aesthetic premise that the musicians speak as themselves.

Finally, there are corona-related parody songs, or contrafacta,[7] a particularly creative form of musical response to the pandemic that can be seen as a mix of putting existing pieces into the new context of a corona playlist and of composing entirely new pieces. Such songs typically represent a reworking of an existing piece of vocal music, mainly by substituting its text, but the original music can also be modified regarding its arrangement, style, and other features. Corona contrafacta have been created and performed by professional artists, but also by hobby musicians and musical families. The most common form is a one-stop production where one person or group writes the new lyrics, performs the song (either in the form of singing to a playback of the original or performing an entire arrangement of it), shoots or edits a video, and puts it online on an individual channel, often also with additional written content that explains who the musicians are and why and how they did this parody.

Creators of parody songs primarily use songs as models that can be assumed to be extremely popular and well known. The classics and evergreens of popular music stand alongside recent top hits and songs from popular movies. While first-person experiences of living under lockdown form the most frequent sub-type in terms of content, another prominent sub-type features songs that provide information on the disease and necessary hygiene measures, and ask people to follow them. The most popular example of a health information parody worldwide is 'Ghen Cô Vy' with more than 100 million views so far, commissioned by the Vietnamese Ministry of Health at the very start of the pandemic.[8] This song is based on the V-Pop hit 'Ghen' ('Jealous') originally released in 2017, and sung by the Vietnamese artists Erik and Min, who also perform in the contrafactum.[9] The animated video illustrates the information given in the lyrics, but in a funny and engaging way that is meant not to scare people too much.

Similar to playlists and original coronasong repertoires, the spectrum of emotional expressions and communicative goals of corona contrafacta seems to be broader than in live or splitscreen performances of existing pieces. Optimism, perseverance, and cheering people up can be found, along with showing empathy and compassion, offering comfort or a sense of belonging and togetherness, and calling for people's considerate behaviour and solidarity with others. Many of these parodies are also humorous, from simply funny, to ironic, or even sarcastic. We can, however, also find examples in which negative emotions such as boredom, frustration, anger, waning patience, anxiety, depression, loneliness, and even going mad are expressed. Chris Mann's 'Hello from the Inside,' a reworking of Adele's 'Hello,' and 'The 12 Days of Quarantine';[10] Raúl Irabién's 'Coronavirus Rhapsody' after Queen's 'Bohemian Rhapsody';[11] or some of the songs created by the Marsh family, such as 'One More Day' and 'Totally Fixed Where We Are' may serve as examples. Also, corona contrafacta seem to be the only type that also addresses conflicts related to the pandemic and criticises people and groups who trivialise the disease, give false information, or do not follow hygiene measures, as in Randy Rainbow's 'A Spoonful of Clorox' or the 'Hamilton Mask-Up Parody Medley' by the Holderness family.

Taken together, the forms of coronamusicking described so far are characterised by the following essentially social parameters: (1) a strong aspect of participation and direct or digitally mediated interaction within and between performers and audiences, which makes it possible to come together as a group and share a (musical) experience despite social distancing measures; (2) a notion of providing personal attention in the form of giving musical gifts to others, and (3) a selection or creation of musical pieces whose meaning can be directly linked to the pandemic and thus serve as a means of communication. Thus, the social meaning of these kinds of musicking lies in the combination of the concrete piece and the specific way it is performed, shared, and received. They can be interpreted as essentially social acts, by means of which communities share their feelings and experiences and try not only to come to an understanding about the subject matter of the COVID-19 pandemic and living under lockdown, but also seek to jointly cope with it and provide each other emotional support, consolation, and inspiration under the circumstances of spatial distancing.

Coronamusicking as 'Umgangsmusik'

The above presented description and typology of coronamusicking has been informed by theories developed by the German music historian Heinrich Besseler and the American ethnomusicologist Thomas Turino. Both tried to conceptualise observable differences in musical performances, and also music recordings in Turino's case, with what could be called an anthropological and sociological emphasis. In two seminal publications, Besseler differentiated between *Umgangsmusik* and *Darbietungsmusik* ([1925] 1978, [1959] 1978).

198   *Melanie Wald-Fuhrmann*

While Darbietungsmusik is easily translatable into presentational music, or music making, and meant to refer to the classical concert with its two functionally distinct roles of typically professional performers, and a more or less passive and anonymous audience, the other half of the pair presents greater difficulty. In his earlier article, Besseler still used the more common *Gebrauchsmusik*, that is, functional or use music, such as dance music, work songs, liturgical music, lullabies, and the like. In contrast, the neologism Umgangsmusik alludes to *Umgangssprache*, the most common German composite noun with *Umgang*, which means common or colloquial speech, but also to Umgang in the sense of social contact and intercourse. In this sense, Umgangsmusik refers to music that provides a form and means for social intercourse.

Besseler describes Umgangsmusik as participatory music making and a social activity, not an object. It is a social form of musicking of a 'true community of like-minded [literally: similarly tuned] individuals who approach the music in an active attitude and expectation' (Besseler [1925] 1978, 38). Participatory forms of making music do not entail a separate audience role, but everyone present can and will participate in the music making in one way or another, by singing, playing, clapping, or dancing. Furthermore, Umgangsmusik are instances of music and music making that are embedded into the everyday life of people, their work, responsibilities, worries, news, recreation, and so on. Here, a piece of music responds to a real request and will 'emerge from the moment for the moment' (ibid., 39). It is a medium for an everyday behaviour or social act and thus meant to do something (non-musical) to the participants, and at the same time it is a way in which participants do something. This is clearly what we find in balcony singing to fight the pandemic.

Independently of Besseler, Turino developed relatively similar categories. He sees the difference between performances and recordings as the difference between music being understood primarily as a form of a social activity or as an object (Turino 2008). He distinguishes between two main types of performances and recordings: participatory and presentational performances on the one hand, and high fidelity and studio audio art recordings on the other. The participatory-presentational distinction comes very close to Besseler's ideas, although the aspect of dealing collectively with what is affecting the community at present is much less pronounced in his definition of participatory music making. Instead, Turino characterises participatory music and music making as primarily a form of 'heightened social interaction' (Turino 2008, 28), the quality of which 'is ultimately judged on the level of participation achieved' and 'by how participants *feel* during the activity' (ibid., 29). This clearly provides a good explanation of the attractiveness these forms had early in the lockdown. He also stresses that 'participatory music-dance is … a strong force for social bonding' (ibid.).

The high fidelity recording type, in Turino's definition (2008, 67–68), is characterised by the attempt to represent a live music performance as faithfully as possible. Concert live streams would fall under this category. Studio audio art, on the other hand, results in a musical product that could never have been performed live, but relies heavily on music production technology. This would be the case for splitscreen performances of virtual choirs and orchestras, but also for many corona parody songs.

While the four categories of participatory, presentational, high fidelity, and studio audio art help to describe and categorise the various types of coronamusicking, it is Besseler's notion of Umgangsmusik as a form of social communication, interaction, and intercourse that emerges from the moment and addresses the present situation of the participants that has also explanatory value. It points to the fact that it is not just their only partly participatory forms, but their being motivated by a unique situation and their social and communicative function in this situation that makes them stand out. Even concerts on balconies, in courtyards, and in front of residential homes are not purely presentational, but *umgangsmäßig*, that is, they happen in the form of social interaction. Although there are performers and audiences, their relationship is not organised economically and aesthetically around the presentation of autonomous music. Rather, the musicians have a personal motivation to bring their music to the neighbourhood or to the elderly to ease their loneliness and bring them some joy, to communicate with them musically in the here and now of the pandemic situation. Playing music thus becomes a situation-specific means to interact and communicate in and about the pandemic.

Also, whereas music recordings and presentational forms of live performances do not usually attempt to establish personal relationships between the musicians and the audience, or if so, only in quite indirect and not very obvious ways, neighbourhood performances are precisely this: a means to establish or maintain personal relationships at a distance, via the giving and receiving of musical sounds as a form of social attention. Even the online forms of musicking demonstrate the wish to overcome anonymity, they do this either by the musicians letting the audience virtually into their homes and addressing them personally, or via the chat and comment functions, whereby audience members get a voice. The character of coronamusicking as Umgangsmusik becomes particularly clear for the specific repertoires: coronasongs and corona contrafacta establish their own strands of collective discourse about the pandemic and living in lockdown, and thus provide spatially separated communities a possibility to keep in touch and communicate their experiences, thoughts, and feelings via the musical medium. They only make sense artistically because of and during the pandemic.

From this, it is clear where the attractiveness and coping potential of these forms of musicking lie: they allow performers to continue to practice their profession, to do something meaningful, to feel like they have agency, and to contribute their share towards coping with the pandemic, as well as to stay in some form of contact with their audiences. For non-professional musicians and audiences, on the other hand, the attractiveness of coronamusicking consists in maintaining social and emotional exchange via the medium of music, being encouraged by the creative ways others find to stay active and maintain agency despite the lockdown, but also in the somewhat voyeuristic pleasure of being invited into the homes of artists and musicians they admire and catching a glimpse of their private lives. All of this transcends how social relationships are organised in today's music business and renders them more personal and immediate.

Coronamusicking as a Practice of Mediated Consolation

In this section, the classification of coronamusicking as Umgangsmusik will be used to further analyse its specific socio-emotional coping potential, which has been demonstrated in a previous empirical study (Fink et al. 2021; see also Hansen, Davidson, and Wald-Fuhrmann 2021). I argue that basically all of the social and emotional benefits people reported gaining from their interactions with coronamusic can be subsumed under the concept of consolation. That people needed consolation and encouragement during the COVID-19 pandemic goes without saying, but a number of psychological studies, with participants from around the world, have established that it increased levels of depression, anxiety, loneliness, and other negative and potentially detrimental states (e.g., Ahrens et al. 2021; Pieh et al. 2021; Rossi et al. 2020). At the same time, containment measures such as spatial distancing and lockdowns have not only increased the need for consolation, but also made it much more difficult to give and receive. Consequently, other forms of consolation that did not rely on direct and physical personal interaction, but that could function under pandemic conditions were needed, that is to say, media of consolation.

Consolation at Large: Theories, Practices, and Media

Discussed and theorised in Europe since antiquity, consolation is defined as social practices that help people to deal with or overcome emotional distress, most importantly sadness, desperation, sorrow, and hopelessness due to illness, death, loss, failure, or the *conditio humana* as such (Norberg, Bergsten, and Lundmann 2001). Several forms of consolation can be distinguished that affect the emotional state either directly or indirectly. Consolation can be given by sharing a person's sorrow by just being around, keeping the person company, letting them express their grief, and touching or hugging them, thus showing care and compassion. In addition to physical attention, consolation can also be given verbally: one can express shared regret about the loss, but also try to infuse hope for a brighter, better future in the sad person. Or one can try to alter their evaluation of the past event that caused their distress by demonstrating that it is either not as bad or as important as was assumed, or that it is even actually positive. Finally, distraction and comforting can also be a form of consolation, whereby one tries to create some pleasurable and positive experiences for the grieving person in order to ease their pain, at least for a while, and to make them think of things other than the one that caused their distress. All of these forms are far from being mutually exclusive. They aim, however, not at altering the situation itself – which is often not possible, as in the case of death or loss – but at helping people endure it emotionally and mentally and finally even overcome their distress.[12] The people providing consolation either may or may not be affected by the sorrowful event or situation. Particularly in the first case, forms of joint and shared grieving and mutual support are common and effective, despite their appearance as mere expressions of grief.

In addition to direct and personal forms of consolation, mediated forms of consolation have also emerged and been practiced in European cultural history. Most importantly, there is a long tradition of consolation literature in Europe, with a first peak in Boethius' late antique *Consolatio philosophiae* (e.g., Kassel 1958; von Moos 1971, 1972; Bolton 1977; Chadwick 1981; Smith 1995). Here, the relevant writings serve as substitutes for direct personal intercourse in the absence of an empathic other. Their use, however, also requires a certain activity on the part of the person in need of consolation: they have to actively turn to a writing that might be of help in their present state. Thus, consolation media become means of self-consolation practices, and get very close to what psychology describes as individual and self-directed coping strategies, such as cognitive reappraisal of the situation, emotional response modulation, or attention deployment (Shifriss, Bodner, and Palgi 2015, 795).

Next to philosophical and religious literature, music can be seen as another consolation medium with a long history of related practices, discourses, and repertoires (Farias 1992; Atkinson and Atkinson 1993; Roch 1998). That Orpheus played his lyre and sang to it in the wilderness of Thrace after he had lost Eurydice for ever can be seen not only as an expression of his grief, but also as a form of self-soothing (Ovid, *Metamophoses* 12.10). The same is true for funeral music worldwide, which combines the expression of sorrow with the attempt to console (Robertson 1967; Agawu 1988; Johnston 2000; Gamliel 2006; Hodges 2009; Kremer 2015). Music's use scenarios are even broader than those of literature, since both those who offer and those who seek consolation can turn to it. Music can be used as an intensified form of personal attention when it is played and sung in the presence of a person, or a group of persons in grief; consolation compositions can pick up on themes from the consolation literature, but add emotional and aesthetic layers to their arguments to render them more persuasive and effective; and music can be used by people in need of consolation via playing or singing pieces that they think will help them, or since the invention of recording and playback techniques, by listening to a recording they choose. This type of listening to music with the aim of regulating one's own state has been called a 'technology of the self' by Tia DeNora in her book on everyday music listening in the age of music recording (2000, 46–74).

More recently, seeking consolation via music has also been researched from a psychological angle. Besides the few studies dedicated explicitly to this topic (e.g., Hanser et al. 2016; ter Bogt et al. 2017), other relevant empirical studies mostly come from research concerning the functions and effects of music (listening), and in particular, musically mediated emotion regulation and mood management strategies (e.g., Saarikallio and Erkkilä 2007; Saarikallio 2010; Shifriss, Bodner, and Palgi 2015; Van den Tol and Edwards 2015; Eerola et al. 2018; Schäfer, Saarikallio, and Eerola 2020). Saarikallio and Erkkilä identified 'solace' as one of seven to eight main musical mood regulation strategies (2007, 96).[13] People resort to it when feeling sad and troubled, in particular as a consequence of 'losses, grief over loved ones, problems with relationships, ill health of oneself or a significant other, periods of sadness and general stress' (Hanser et al. 2016, 131). They seek music that mirrors their mood and situation in terms of expressivity and lyrical content, to feel understood and comforted. Very much like what has been described as parasocial interactions in mediated encounters with artists (Horton and Wohl 1956), they perceive a piece of music or its persona as if it were an empathic human other sharing their distress and providing comfort. Other studies support the idea of music being used as a social surrogate that is experienced as expressing empathy and understanding, and makes people feel less lonely (Van den Tol and Edwards 2015; ter Bogt et al. 2017). The aspect of sharing grief and showing compassion can also be linked to another branch of music psychological studies that have corroborated the effect of joint music making on building and strengthening group cohesion and feelings of belonging (e.g., Kang, Scholp, and Jiang 2017; Bullack et al. 2018).

However, Saarikallio and Erkkilä's two other mood regulation strategies related to sadness, depression, and similar negative states, 'diversion' and 'discharge' (2007, 96), can also be understood as forms of consolation, albeit via different routes. While diversion involves happy and pleasant music, and aims at forgetting about the current negative mood and its causes, and lifting one's spirits up,[14] discharge is associated with mood-congruent music and is supposed to express one's own negative state. The important role of mood-congruent, that is, sad and melancholic music for musical consolation was also corroborated by Hanser et al. (2016). In addition, that study found that the feeling of being comforted by music was not simply associated with becoming happier, but with experiencing a complex mix of emotions, part of which were being moved, but also sadness, the feelings of beauty, tenderness, admiration, and even empowerment. Consequently, solace or consolation have been discussed as a partial explanation of the apparent paradox of deriving pleasure from listening to sad music (Van den Tol and Edwards 2015; Eerola et al. 2018). In their review of existing theories and empirical evidence, Eerola et al. underline the idea that, on a psychosocial level, musical consolation through sad music works primarily via a parasocial relationship, that is, experiencing music as 'an imaginary friend who provides support and empathy after the experience of a social loss' (2018, 109). Furthermore, their discussion of the links between the pleasure of feeling moved and the aesthetic appreciation of sad music can also shed light on why people may feel comforted by (sad) music.

Results from music psychological studies are less differentiated when it comes to identifying relevant features of consoling music. It is only clear that both lyrics (if there are any) and musical features play an essential role, and that they interact strongly with individual factors: pieces work particularly well for consolation purposes if listeners hear their lyrics as describing their own situations and feelings, or find them in any other way personally meaningful (ter Bogt et al. 2017). Also, people need to have a strong personal relationship

202   *Melanie Wald-Fuhrmann*

with a piece, either via personal memories attached to it, a high degree of preference, or an identification with the artists and/or their fans (Van den Tol and Edwards 2015; Hanser et al. 2016).

Musical Consolation during COVID-19

All of the above discussed elements of interpersonal and mediated consolation can also be found when analysing coronamusicking practices and repertoires, as well as people's responses to it. The relationship between (corona)musicking and consolation is already clear from the results of the aforementioned study by Fink et al. (2021). During the first lockdown, we asked people via an extensive online questionnaire whether they had changed their music-related behaviours and attitudes, and whether they experienced music as something that helped them cope emotionally with the crisis and/or served as a social surrogate. Both were the case for the majority of participants.[15] Of the reasons why people chose to listen to or make music, those aimed at reducing negative states and inducing positive ones increased the most in importance. In addition to 'provides comfort and support,' other items also directly related to consolation such as 'puts me in a good mood,' 'reduces my stress and anxiety,' 'supports me in a bad mood,' 'distracts me from problems and worries,' and 'makes me reminiscent of more positive times' ranked among the top ten (of well over 30) more important functions overall. This became even clearer when we compared the quartile of participants who reported to experience negative emotions much more often during the lockdown than before with the opposite quartile, that is, those who did not experience more negative emotions, or even less of them. The group comprised of those people who needed consolation the most did not only give higher ratings to the questions related to socio-emotional coping via music but differed particularly regarding the degree of change in importance of consolation-relevant musical functions. For them in particular, the potential of music to provide comfort and support, to distract, to reduce loneliness and stress, or to support in a bad mood had become much more important than before the pandemic. But what is more, the item most likely to predict how much people felt that music helped them cope during the lockdown was interest in how other people used music during the pandemic, which is to say interest in coronamusicking.

To get a deeper and more differentiated understanding of how people felt comforted and consoled via coronamusicking, one can analyse the individual, mostly verbalised responses of performers and listeners to individual pieces and performances. Here, a number of relevant common topics can be identified – topics that were also present in the public discussion of these phenomena. In the following, examples will be given from media reports on balcony singing, as well as YouTube user comments on corona contrafacta and virtual choir and ensemble performances of existing pieces. Joint musicking from balconies and windows sparked enormous media interest in spring 2020.[16] Soon, shared narratives evolved that sought to explain why people did this, and how it helped them in the present crisis. The most important elements of those narratives were the emotional effect of singing for and with others, as well as the social aspect of this practice. In two *Guardian* articles from 14 March 2020, Italian balcony singing was interpreted as a means 'to lift spirits' (Kearney 2020), 'to boost morale' (Thorpe 2020), and to express mutual solidarity during the just begun nationwide lockdown. On the same day, an article in *Elle* also referred to balcony singing in Italy as a way 'to keep each other in good spirits,' but also to make people feel less lonely, and provide mutual consolation (Weaver 2020). The German *Frankfurter Allgemeine Zeitung* used the metaphor of singing as a non-physical form of hugging that can provide consolation and infuse courage, helping people overcome loneliness and feel connected (Krueger

2020). Nagina Bains, who reviewed the function of music during the COVID-19 pandemic in the *Indian Express*, summed all of this up by calling music a 'therapeutic companion' and a 'constant source of comfort and companionship' (2020).

Overall, these early reports all support the interpretation given above of balcony singing as Umgangsmusik that can provide consolation. In particular, they point to the close connection of the active and participatory form of such musicking with its effects on emotional and mental stabilisation through the experience of empathic social exchange and support. More concretely, balcony singing is a musical form of interpersonal mutual consolation, where everyone gives and receives support at the same time because everyone is in basically the same situation and equally in need of consolation. Singing with and for each other therefore combines social, emotional, and cognitive elements of consolation in the form of personal attention, being there for each other, and acknowledging the source of emotional distress, while at the same time expressing hope and confidence. In addition, joint musicking is generally well known to have the potential to induce strong positive experiences in participants and can make people feel a sense of agency, doing something meaningful, and being part of something important and beautiful, all of which provide further sources for feeling comforted.

The way corona contrafacta afforded consolation worked rather differently, as one can infer from YouTube user comments:[17] particularly often, people referred to the topical content of the new lyrics and the videos together with related aspects of the COVID-19 pandemic. In this way, they showed that they found their own experiences expressed in the song and that they identified with its message – that is, those pieces made them feel understood and that they were not the only ones going through hard times, which also is a core component of consolation. At the same time, the unvarnished, even exaggerated, descriptions of lockdown situations provided an opportunity for discharging the accompanying negative emotions, which is another consolation mechanism. Second, people commented on the quality of the pieces and their performance, and voiced their aesthetic appreciation of them, as in the case of the singer Kira Haas' (also known as Dovelybell) parody of 'Maria' from *The Sound of Music* (see Figure 15.1).[18] These often very strong aesthetic responses such as admiration, wonder, awe, and fascination must be seen as a main beneficial effect of corona parody songs on people's emotional and mental states, insofar as these responses are positively valenced and highly appreciated subjective states. Here, the change of negative into (more) positive emotions in the process of consolation results at least in part from intense aesthetic experiences.

Another frequently mentioned emotional response was smiling and laughing. Such reactions can be seen as genre-specific, since parody songs are regarded as a humorous art form in contemporary popular culture. The top comments on Kathy Mak's parody of Natalie Imbruglia's 'Torn,' for example, all refer to laughing responses, either in words or via emojis (see Figure 15.2). At the same time, some of them also reflect on the idea that laughing has an important function in the context of the pandemic. This topic can also be found in other comments that see laughter as something that makes the pandemic and the lockdown 'more endurable,'[19] as well as less dire and frightening; reduces the related stress;[20] and helps people to stay mentally healthy.[21] Related to this are comments in which people say that they feel cheered up or that their spirits have been uplifted. A user who self-identified as a nurse even attests to the health value of laughing. The proverb that laughter is the best medicine is quoted by this user, and alluded to in many other comments to corona contrafacta, along with other medical metaphors. In sum, laughing can be seen as providing yet another route to enhance the emotional and mental well-being of a person. At the same time, being able to laugh about an actually rather dire and challenging situation also falls

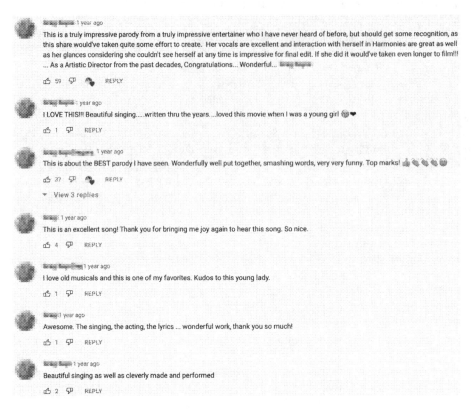

Figure 15.1 User comments expressing very positive aesthetic evaluations of a *Sound of Music* parody by the singer Dovelybell on YouTube.

under the category of cognitive reappraisal and the ability to shield oneself somewhat from the stressor.

Another frequent response the comments address is being grateful to the musicians for having created and shared their song. Gratefulness is motivated by the commenter's own feeling of joy, being cheered up, supported, and entertained, or directed to the content of the songs if people agree with it and find it important in terms of the information provided or the admonition communicated by it. Both variants can be found, for example, in comments on Chris Mann's 'Covid/Back,' a parody of Justin Timberlake's 'SexyBack' released in August 2021, that urges people to get vaccinated.[22] Users thanked Mann because his songs helped them cope emotionally with the situation (e.g., 'I have this stuck in my head, but it was absolutely needed today especially. Thank you!' A user who introduces themselves as being a COVID-19 survivor: 'Chris, your parodies are the best! It made me laugh even when I was at my lowest point and every breath was a struggle. Thank you so much for keeping it real with humour. '). In addition, another user relates their gratitude for the song's message: 'I am 70 years old but this got me up and dancing because of the awesome beat! I was vaccinated early on. Thanks, Chris for telling it like it is! GET VACCINATED, ya'll!!!'

Many of the musicians who produced corona contrafacta reflected explicitly about why they did this and what they hoped to achieve with their songs. Their ideas are very much in line with what their audiences actually experienced and also revolve around the ideas of

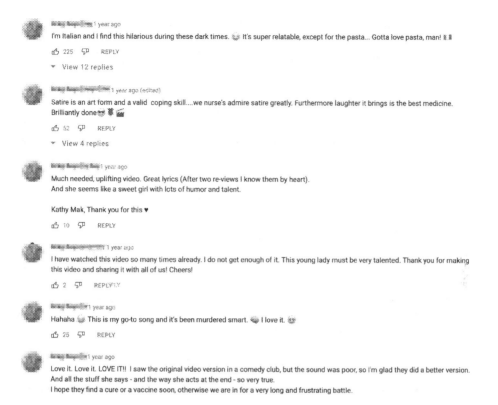

Figure 15.2 Top user comments to Kathy Mak's parody of Natalie Imbruglia's 'Torn' on YouTube.

sharing experiences, showing empathy, bringing some joy, providing consolation, and the relief of humour. Shirley Șerban, a singer from New Zealand who created a series of such songs explains her motivation in the About section of her YouTube channel as follows:

> Hi, I'm Shirley from New Zealand. Creating songs to make myself and others smile has been one of my favourite pastimes since I was a child. With the onset of COVID-19, I have been creating and sharing 'relyriced' classic songs, as well as a few originals about current events to help keep myself happy – and I've found that others enjoy them too. The purpose – to educate and bring joy in these dark times. Yes, the pandemic is serious, but yes, we can all do with a bit more joy. So welcome to my channel. Make a hot drink, get yourself cosy and enjoy … sing along and smile. We will get through this – thank you for allowing me to help entertain you along the way.
>
> (Șerban, n.d.)

Similarly, Kathy Mak commented on why she created her 'Torn' parody: 'I thought that if I'm feeling this way, many other people would be too,' she said, explaining that she hopes the song will offer some relief amid a stressful situation: 'In a way, it's been kind of a coping mechanism.'[23] Neil Diamond, in a parody of his own 'Sweet Caroline,' addresses his audience directly before he starts singing: 'Hi, everybody. This is Neil Diamond, and I know we're going through a rough time right now. But I love ya, and I think maybe if we sing together, well, we'd just feel a little bit better. Give it a try, okay?'[24]

206 *Melanie Wald-Fuhrmann*

User comments on corona-motivated splitscreen performances of earlier songs on the internet touched on similar aspects, in particular strong aesthetic appreciation and grate-fulness, together with the experience of social support and empathy. However, as a result of a rather different expressive focus from the songs described earlier, further emotional responses fall much less often into the categories of laughing, but belong instead to the equally consolation-relevant realms of feeling moved, breaking into tears, comfort and relief, and feelings of amazement and awe. A virtual choir and orchestra performance of 'You'll Never Walk Alone' from the musical *Carousel*, under the direction of the American choir conductor Harrison Sheckler that has more than 1,675,000 views can serve as an example. Sheckler described his motivation to do this performance in the credits below the video:

> During early lockdown, I found comfort in playing the beautiful song from the musical *Carousel*, 'You'll Never Walk Alone.' I embarked on collaborating online like so many others are doing. What started to fill the void of music collaboration has evolved to new meaning for me with the lengthened quarantine. Hopefully, the words, 'you'll never walk alone,' along with the visual of 300 people joining together offers the audience some comfort and peace during this time. Stay safe and healthy my friends!
>
> (Sheckler 2020)

He also asks his audience to share the video 'to help spread a little hope during this time!' (Sheckler 2020).

Of the top ten comments on this video, four refer to tears, but as a genuinely positive response: 'This brought tears to my eyes and peace to my heart. The musicality and videog-raphy, especially matching the voices of the first three singers, is captivating. I will listen to this each night before I go to bed. Thank you, thank you, thank you'; 'If this doesn't bring a tear to your eye and a smile on your face, you have no soul. Well done. Absolutely Beautiful. Thank you for sharing'; 'This made me cry, and feel less alone. Thank you. Beautiful';

> I was having one of those days and had been crying quite often today because of the state of our world. I cried tears of joy at this uplifting rendition. My late father-in-law used to sing this and he had such a beautiful rich voice. Thank you so much for sharing. We are NOT Alone. xo.[25]

Interestingly, all users related their being moved, feeling less alone and reassured to their appreciation of the aesthetic qualities of the performance, not (just) its topical content.

Conclusions

Analysing the evidence on the positive emotional effects of coronamusicking, in the light of research on social and individual forms of mood management, allows us to precisely under-stand the emergence and effects of these musical practices. As a form of Umgangsmusik explicitly targeting the pandemic and its accompanying hardships, coronamusicking can substitute almost all elements of human consolation during times of spatial distancing. It also harnesses most known musical mood regulation potentials by working simul-taneously as participatory action and collectively meaningful communication. In those coronamusicking forms that imply separate roles for performers and audiences, it is not just the audience that receives consolation, but also the performers, because they can main-tain their self-image as musicians, experience agency, and can expect to receive positive and supportive feedback from their audience.

In particular, consolation through coronamusicking works via participatory practices (serving as a provider of mutual attention within a grief-stricken community); emotional expressivity (a musical way to show compassion); its potential to evoke emotional responses (a musical way to affect the emotions of a person in need of consolation); topical lyrics that describe how it feels to deal with the pandemic and the lockdown (a way to allow people to feel understood and that others share their distress); and also via its perception as an intentional communication addressed at oneself, in particular regarding vocal music (social surrogacy, but also a substitute of verbal forms of consolation).

In addition, a number of genuinely aesthetic routes to consolation were found: first of all, aesthetic appreciation as an art-specific way of affording intensely positive emotions. Other important responses that are also bound to the aesthetic form of the repertoires and practices in question include being moved, humour, and ironic self-distancing, which are unevenly distributed across sub-types of coronamusicking. Finally, one can expect people to gain hope and encouragement from the example of others who have found ways of staying creative and at the same time helping others, despite restrictive circumstances. In this sense, coronamusicking appears as a set of musical practices of great societal importance. It can help people individually, and on the group level, to maintain or regain well-being and mental health. It provides people with a means to share their experiences, thoughts, and feelings about the pandemic, and its many accompanying aspects. It offers a forum for communities to arrive at a shared understanding of this unprecedented event, and an attitude towards it. This social aspect captured by Umgangsmusik also distinguishes it from purely individual musical coping strategies, which might have relatively similar effects on the personal level, but lack the situation-specific dimension of collective debate and processing that is of fundamental significance not only in an event such as a pandemic, but potentially also for other major societal challenges. Perhaps we will find that *Umgangsmusik*, having regained such a significant role in the context of COVID-19, will retain this prominence also beyond the pandemic.

To conclude, this chapter aimed at providing a societally relevant interpretation of coronamusicking as Umgangsmusik and mediated consolation practices, but also novel contributions to existing research in the fields of consolation, musical mood management, and solace. These contributions consist mainly in illuminating the important role of the aesthetic routes and the character as Umgangsmusik, as well as the genre-specificity of afforded consolation strategies. And whereas, so far, studies on musical mood management have almost exclusively focussed on music listening, the focus on coronamusicking also helps understand the relevance of active and, particularly, participatory forms of music making. In return, understanding the use of music and other media as forms of mediatised self-consolation could add an important theoretical differentiation to the discussion about the history, forms, and impact of consolation.

Acknowledgements

The author would like to gratefully acknowledge the help and support of the following persons: Jan Eggert, Hannah Fiehn, Emily Gernandt, and Nikita Kudakov, student assistants at the Max Planck Institute for Empirical Aesthetics, for their help in collecting and annotating a large number of coronamusicking examples; Professor Dr Tobias Bulang and PD Dr Carsten Wergin from Heidelberg university, who invited me to contribute to their interdisciplinary conference on consolation in Heidelberg in September 2021, that became a rich source of inspiration and conceptual clarification; Dr Guy Schwegler, University of Lucerne, and his colleagues from the Research Committee Sociology of Arts and Culture whose

208　*Melanie Wald-Fuhrmann*

invitation to a panel on 'Culture in Times of Crisis: The Music Sector Facing COVID-19' in the context of the annual congress of the Swiss Sociological Association in June 2021 gave me an opportunity to discuss the interpretation of coronamusicking as Umgangsmusik; and finally, my colleagues Dr Lara Pearson and PD Dr Christian Grüny, as well as my husband Professor Dr Wolfgang Fuhrmann (Leipzig University) for their welcome comments to an earlier version of this chapter.

Notes

1　In what follows, those types of musicking that are less prominently covered in other chapters of this volume will be described in more detail than the others, particularly forms of repertoire creation.

2　Research on the work situation and well-being of musicians during lockdown can be found in Hansen, Davidson and Wald-Fuhrmann 2021; see also Spiro et al. 2021; Cohen and Ginsborg 2021; and Teixeira et al. 2021.

3　See www.globalcitizen.org/en/media/togetherathome/ and www.onelove.asia/, respectively.

4　An example of a jointly created playlist initiated by a radio station is *Emotional Rescue Songs* on RBB's radio eins in Germany. Two calls were issued before Good Friday in 2020 and in early 2021.

5　Creating playlists on streaming platforms can be seen as the digital complement to mixtapes in the age of the cassette recorder. See Glennon 2019.

6　Two particularly extensive playlists on Spotify provide the basis for these statements: *Corona Hits 2020*, curated by user Laura Nürnberger, with 217 songs and more than 10,000 likes at the time of this writing, https://open.spotify.com/playlist/2xaPVlnPIPfuX82qf9whPK?si=1daecbc2bfeb4 a78; and *Coronavirus Awesome Mix*, curated by user Chuck Stoltz, with 339 songs and 20,000 likes, https://open.spotify.com/playlist/4Jpc4oj6UlWF7BzEAD9cHI?si=c3f9882930664fed (last accessed 6 December 2021).

7　What is described today in popular culture as a parody song, is called a contrafactum in musicological terminology, while the term parody is typically reserved to more substantial reworkings of pre-existing musical material (Falck and Picker 2001; Tilmouth and Sherr 2001). Adapting existing pieces to present needs is a relatively old phenomenon, regularly used by composers between the fifteenth and eighteenth centuries. Only in nineteenth-century popular culture did contrafacta and other musical parodies develop a mainly humorous overtone. In both serious and comical contexts, however, a contrafactum or parody can unfold its full meaning only if the audience knows the original and can relate its lyrics, meaning, and context to the new ones.

8　www.youtube.com/watch?v=BtulL3oArQw. 101,141,108 views on 2 November 2021.

9　https://en.wikipedia.org/wiki/Ghen_Cô_Vy.

10　The song 'The Twelve Days of Christmas' was also adapted into a corona contrafactum by the Marsh family in December 2020. www.youtube.com/watch?v=YAZ7D8NhWbU.

11　Another corona adaptation of this song by Maestro Ziikos is also a critique of president Donald Trump's irresponsible corona politics. www.youtube.com/watch?v=giDttMU-k00.

12　There is only a very limited number of empirical and (social) psychological research on consolation, some of which is summarized in ter Bogt et al. (2017, 156s).

13　See also Saarikallio 2008 and Saarikallio 2010.

14　The use of sad versus happy music to counter bad moods was also examined by Shifriss, Bogner, and Palgi 2015.

15　Similar results were also found in other empirical studies on positive effects of music use during COVID-19, most of them published in Hansen, Davidson, and Wald-Fuhrmann (2021).

16　In the following, examples are mainly taken from leading newspapers and (online) magazines of different, mostly English and German speaking countries.

17　The basis for the following descriptions are the top ten comments on the 50 most frequently watched corona contrafacta. The ranking of comments is generated by a YouTube algorithm that apparently involves the numbers of likes and replies a comment gets from other users.

18　www.youtube.com/watch?v=M4jR_9-YPK8.

19 A user about Chris Mann's parody of Madonna's 'Vogue': 'Laughing at Corona makes it more endurable!' www.youtube.com/watch?v=3xt58OVnmXU.
20 Another user about the same song: 'Chris, you are wildly talented and hilarious in a biting way that really shores me up! Thank you for the videos that shut down the stress!' www.youtube.com/watch?v=3xt58OVnmXU.
21 A user in a comment on the 'Broadway Coronavirus Medley' by Zach Timson: 'I love this. In a time where everyone is concerned; fears about the future etc. it is good to have something to laugh at – not discounting the serious impacts this has had but to keep our sanity – as they say "you have to laugh or you would cry." God Bless.' www.youtube.com/watch?v=n1OCZRann8w.
22 www.youtube.com/watch?v=Xed9liS-sOQ.
23 www.youtube.com/watch?v=6IOS7-4n12c.
24 www.youtube.com/watch?v=qxnETrhOIAE.
25 'You'll Never Walk Alone' Virtual Choir/Orchestra 15 Countries: 300 People, www.youtube.com/watch?v=6gpoJNv5dlQ.

References

Agawu, Victor Kofi. 1988. 'Music in the Funeral Traditions of the Akpafu.' *Ethnomusicology* 32 (1) (Winter): 75–105. https://doi.org/10.2307/852226.
Ahrens, Kira F., Rebecca J. Neumann, Bianca Kollmann, Michael M. Plichta, Klaus Lieb, Oliver Tüscher, and Andreas Reif. 2021. 'Differential Impact of COVID-Related Lockdown on Mental Health in Germany.' *World Psychiatry* 20 (1): 140–141. https://doi.org/10.1002/wps.20830.
Atkinson, Jo B. and Colin B. Atkinson. 1993. 'Changing Attitudes to Death: Nineteenth-Century Parlour Songs as Consolation Literature.' *Canadian Review of American Studies* 23 (2) (September): 79–100. https://doi.org/10.3138/CRAS-023-02-03.
Bains, Nagina. 2020. 'How Music Became a Therapeutic Companion in the Times of COVID-19.' *Indian Express*, 22 May 2020. https://indianexpress.com/article/lifestyle/art-and-culture/music-constant-therapeutic-companion-covid-19-lockdown-6422049/.
BBC. 2020. 'Coronavirus: Wuhan Residents Shout from Balconies in Solidarity.' *BBC News*, 28 January 2020. www.bbc.com/news/av/world-asia-china-51278756.
Besseler, Heinrich. (1925) 1978. 'Grundfragen des musikalischen Hörens.' In *Aufsätze zur Musikästhetik und Musikgeschichte*, edited by Peter Gülke, 29–53. Leipzig: Reclam.
Besseler, Heinrich. (1959) 1978. 'Das musikalische Hören der Neuzeit.' In *Aufsätze zur Musikästhetik und Musikgeschichte*, edited by Peter Gülke, 104–173. Leipzig: Reclam.
Bolton, Diane K. 1977. 'The Study of the Consolation of Philosophy in Anglo-Saxon England.' *Archieves d'historie doctrinale et littéraire du Moyen-Âge* 44 : 33–78.
Bullack, Antje, Carolin Gass, Urs M. Nater, and Gunter Kreutz. 2018. 'Psychobiological Effects of Choral Singing on Affective State, Social Connectedness, and Stress: Influences of Singing Activity and Time Course.' *Frontiers in Behavioral Neuroscience* 12. https://doi.org/10.3389/fnbeh.2018.00223.
Buß, Christian. 2020. 'Krisenhits 2020: Das sind die besten Corona-Songs des Jahres. Hammerhead: "'Ich sauf allein" (Bonustrack).' *Der Spiegel*, 23 December 2020. www.spiegel.de/kultur/musik/krisenhits-2020-die-besten-corona-songs-des-jahres-a-64c8cd76-afa7-444b-a3e2-684d4c0773d3.
Chadwick, Henry. 1981. *Boethius: The Consolations of Music, Logic, Theology and Philosophy*. Oxford: Oxford University Press.
Cohen, Susanna and Jane Ginsborg. 2021. 'The Experiences of Mid-Career and Seasoned Orchestral Musicians in the UK during the First COVID-19 Lockdown.' *Frontiers in Psychology* 12. https://doi.org/10.3389/fpsyg.2021.645967.
Curcio, Michela. 2020. 'Canzoni anti-pandemia da Bono a Mahmood: il coronavirus si combatte con la musica.' *inChiostronline*, 30 March 2020. www.unisob.na.it/inchiostro/index.htm?idrt=9441.
Daffern, Helena, Kelly Balmer, and Jude Brereton. 2021. 'Singing Together, Yet Apart: The Experience of UK Choir Members and Facilitators during the Covid-19 Pandemic.' *Frontiers in Psychology* 12. https://doi.org/10.3389/fpsyg.2021.624474.
DeNora, Tia. 2000. *Music in Everyday Life*. Cambridge: Cambridge University Press.

210 *Melanie Wald-Fuhrmann*

Eerola, Tuomas, Jonna K. Vuoskoski, Henna-Riikka Peltola, Vesa Putkinen, and Katharina Schäfer. 2018. 'An Integrative Review of the Enjoyment of Sadness Associated with Music.' *Physics of Life Reviews* 25 (August): 100–121. https://doi.org/10.1016/j.plrev.2017.11.016.

Falck, Robert and Martin Picker. 2001. 'Contrafactum.' In *Grove Music Online*. https://doi.org/10.1093/gmo/9781561592630.article.06361.

Farias, Pazlo Fernando de Moraes. 1992. 'History and Consolation: Royal Yorùbá Bards Comment on Their Craft.' *History of Africa* 19: 263–297. https://doi.org/10.2307/3172001.

Fekadu, Mesfin. 2020. '40 Songs about the Coronavirus Pandemic: Listen Here.' *Chicago Tribune*, 4 May 2020. www.chicagotribune.com/entertainment/music/ct-ent-coronavirus-songs-20200504-r4j dtacc2jakpoecfwqah3hzzu-story.html.

Fink, Lauren K., Lindsay A. Warrenburg, Claire Howlin, William M. Randall, Niels Chr. Hansen, and Melanie Wald-Fuhrmann. 2021. 'Viral Tunes: Changes in Musical Behaviours and Interest in Coronamusic Predict Socio-Emotional Coping during COVID-19 Lockdown.' *Humanities and Social Sciences Communications* 8. https://doi.org/10.1057/s41599-021-00858-y.

Gamliel, Tova. 2006. 'Women's Wailing in a Men's Circle: A Case Study of a Yemenite-Jewish Tradition in Israel.' *Journal of Ritual Studies* 20 (1): 51–72.

Glennon, Mike. 2019. 'Consumer, Producer, Creator: The Mixtape as Creative Form.' *Organised Sound* 24 (2): 164–173. https://doi.org/10.1017/S1355771819000207.

Hansen, Niels Chr., John M. Treider, Dana Swarbrick, Joshua S. Bamford, Johanna N. Wilson, and Jonna K. Vuoskoski. 2020. Coronamusic Database. https://osf.io/y7z28/.

Hansen, Niels Chr., Jane Davidson, and Melanie Wald-Fuhrmann, eds. 2021. *Social Convergence in Times of Spatial Distancing: The Role of Music during the COVID-19 Pandemic.* Frontiers Research Topic. www.frontiersin.org/research-topics/14089/social-convergence-in-times-of-spatial-distanc ing-the-role-of-music-during-the-covid-19-pandemic.

Hansen, Niels Chr., John M. Treider, Dana Swarbrick, Joshua S. Bamford, Johanna N. Wilson, and Jonna K. Vuoskoski. 2021. 'A Crowd-Sourced Database of Coronamusic: Documenting Online Making and Sharing of Music during the COVID-19 Pandemic.' *Frontiers in Psychology* 12. https://doi.org/10.3389/fpsyg.2021.684083.

Hanser, Waldie E., Tom F.M. ter Bogt, Annemieke J.M. Van den Tol, Ruth E. Mark, and Ad J.J.M. Vingerhoets. 2016. 'Consolation through Music: A Survey Study.' *Musicae Scientiae* 20 (1): 122–137. https://doi.org/10.1177/1029864915620264.

Hodges Jr, William Robert. 2009. 'Ganti Andung, Gabe Ende (Replacing Laments, Becoming Hymns): The Changing Voice of Grief in the Pre-Funeral Wakes of Protestant Toba Batak (Northern Sumatra, Indonesia).' PhD diss., University of California Santa Barbara.

Horton, Donald and Richard Wohl. 1956. 'Mass Communication and Para-Social Interaction: Observation on Intimacy at a Distance.' *Psychiatry* 19 (3): 215–229. https://doi.org/10.1080/00332 747.1956.11023049.

Johnston, Gregory. 2000. 'Lamentation to Consolation: Aspects of Music and Rhetoric in Funerary Compositions of the German Baroque.' In *Künste und Natur in Diskursen der Frühen Neuzeit*, edited by Hartmut Laufhütte, Barbara Beker-Cantarino, Martin Burcher, Ferdinand van Ingen, Sabine Solf, and Carsten P. Warncke, 2: 913–935. Wiesbaden: Harrassowitz.

Kang, Jing, Austin Scholp, and Jack J. Jiang. 2017. 'A Review of the Physiological Effects and Mechanisms of Singing.' *Journal of Voice* 32 (4) (July): 390–395. https://doi.org/10.1016/j.jvo ice.2017.07.008.

Kassel, Rudolf. 1958. *Untersuchungen zur griechischen und römischen Konsolationsliteratur*. München: Beck.

Kearney, Christina. 2020. 'Italians Sing Patriotic Songs from their Balconies during Coronavirus Lockdown.' *Guardian*, 14 March 2020. www.theguardian.com/world/2020/mar/14/italians-sing-patriotic-songs-from-their-balconies-during-coronavirus-lockdown.

Kremer, Joachim. 2015. 'Trauer, Erinnerung und Trost. Musikalische Memoria in der Frühen Neuzeit.' In *Grab und Memoria im frühen Landschaftsgarten*, edited by Annette Dorgerloh, 61–87. Paderborn: Fink.

Krueger, Karen. 2020. 'Der Gesang der italienischen Seele.' *Frankfurter Allgemeine Zeitung*, 16 March 2020. www.faz.net/aktuell/feuilleton/quarantaene-reaktionen-auf-corona-in-italien-16680202.html.

Moos, Peter von. 1971–1972. *Consolatio. Studien zur mittellateinischen Trostliteratur über den Tod und zum Problem der christlichen Trauer*. München: Fink.

Norberg, Astrid, Monica Bergsten, and Berit Lundmann. 2001. 'A Model of Consolation.' *Nursing Ethics* 8 (6) (November): 544–553. https://doi.org/10.1177/096973300100800608.

Onderdijk, Kelsey, Freya Acar, and Edith Van Dyck. 2021. 'Impact of Lockdown Measures on Joint Music Making: Playing Online and Physically Together.' *Frontiers in Psychology* 12. https://doi.org/10.3389/fpsyg.2021.642713.

Pieh, Chrisoph, Sanja Budimir, Jaime Delgadillo, Michael Barkham, Johnny R. Fontaine, and Thomas Probst. 2021. 'Mental Health during COVID-19 Lockdown in the United Kingdom.' *Psychosomatic Medicine* 83 (4): 328–337. https://doi.org/1 0.1097/PSY.0000000000000871.

Reuter, Annie. 2020. '12 Country Songs Written in Quarantine about the Coronavirus Pandemic.' *Forbes*, 12 June 2020. www.forbes.com/sites/anniereuter/2020/06/12/12-country-songs-written-in-quarantine-about-the-coronavirus-pandemic/.

Robertson, Alec. 1967. *Requiem: Music of Mourning and Consolation*. London: Cassell.

Roch, Eckhard. 1998. 'Trost der Musik: Der ästhetische Paradigmenwechsel zwischen Aufklärung und Romantik.' In *Das gebrochene Glücksversprechen: Zur Dialektik des Harmonischen in der Musik*, edited by Otto Kolleritsch, 38–55. Wien: Universal Edition.

Rossi, Rodolfo, Valentina Socci, Dalila Talevi, Sonia Mensi, Cinzia Niolu, Francesca Pacitti, Antinisca Di Marco, Alessandro Rossi, Alberto Siracusano, and Giorgio Di Lorenzo. 2020. 'COVID-19 Pandemic and Lockdown Measures Impact on Mental Health Among the General Population in Italy.' *Frontiers in Psychiatry* 11. https://doi.org/10.3389/fpsyt.2020.00790.

Saarikallio, Suvi. 2008. 'Music in Mood Regulation: Initial Scale Development.' *Musicae Scientiae* 12 (2): 291–309. https://doi.org/10.1177/102986490801200206.

Saarikallio, Suvi. 2010. 'Music as Emotional Self-Regulation throughout Adulthood.' *Psychology of Music* 39 (3): 307–327. https://doi.org/10.1177/0305735610374894.

Saarikallio, Suvi and Jaakko Erkkilä. 2007. "The Role of Music in Adolescents' Mood Regulation.' *Psychology of Music* 35 (1): 88–109. https://doi.org/10.1177/0305735607068889.

Schäfer, Katharina, Suvi Saarikallio, and Tuomas Eerola. 2020. 'Music May Reduce Loneliness and Act as Social Surrogate for a Friend: Evidence from an Experimental Listening Study.' *Music & Science* 3: 1–16. https:/doi.org/10.1177/2059204320935709.

Şerban, Shirley. n.d. 'About.' YouTube. www.youtube.com/c/ShirleyŞerban/about.

Small, Christopher. 1998. *Musicking: The Meanings of Performing and Listening*. Middletown: Wesleyan University Press.

Smith, Abraham. 1995. *Comfort One Another: Reconstructing the Rhetoric and Audience of 1 Thessalonians*. Louisville: Westminster John Knox Press.

Sheckler, Harrison. 2020. ' "You'll Never Walk Alone" Virtual Choir/Orchestra 15 Countries: 300 People.' YouTube. www.youtube.com/watch?v=6gpoJNv5dlQ.

Shifriss, Roni, Ehud Bodner, and Yuval Palgi. 2015. 'When You're Down and Troubled: Views on the Regulatory Power of Music.' *Psychology of Music* 43 (6): 793–807. https://doi.org/ 10.1177/0305735614540360.

Spiro, Neta, Rosie Perkins, Sasha Kaye, Urszula Tymoszuk, Adele Mason-Bertrand, Isabelle Cossette, Solange Glasser, and Aaron Williamon. 2021. 'The Effects of COVID-19 Lockdown 1.0 on Working Patterns, Income, and Wellbeing among Performing Arts Professionals in the United Kingdom (April–June 2020) 2021.' *Frontiers in Psychology* 11. https://doi.org/10.3389/fpsyg.2020.594086.

Teixeira, Nisio, Graziela Mello Vianna, Ricardo Lima, Carlos Jáuregui, Lucianna Furtado, Thiago Pereira Alberto, and Rafael Medeiros. 2021. 'Covid-19 Impact on the Music Sector in Belo Horizonte (Minas Gerais, Brazil).' *Frontiers in Sociology* 6. https://doi.org/10.3389/fsoc.2021.643344.

ter Bogt, Tom F.M., Alessio Vieno, Suzan M. Doornwaard, Massimiliano Pastore, and Regina J.J.M. van den Eijnden. 2017. ' "You're Not Alone": Music as a Source of Consolation among Adolescents and Young Adults.' *Psychology of Music* 45 (2): 155–171. https://doi.org/10.1177/0305735616650029.

212 *Melanie Wald-Fuhrmann*

Thorpe, Vanessa. 2020. 'Balcony Singing in Solidarity Spreads across Italy during Lockdown.' *Guardian*, 14 March 2020. www.theguardian.com/world/2020/mar/14/solidarity-balcony-singing-spreads-across-italy-during-lockdown.

Tilmouth, Michael and Richard Sherr. 2001. 'Parody (i).' *Grove Music Online*. www.oxfordmusiconline.com/grovemusic/view/10.1093/gmo/9781561592630.001.0001/omo-9781561592630-e-0000020937.

Turino, Thomas. 2008. *Music as Social Life: The Politics of Participation*. Chicago: University of Chicago Press.

Van den Tol, Annemieke J. M. and Jane Edwards. 2015. 'Listening to Sad Music in Adverse Situations: How Music Selection Strategies Relate to Self-Regulatory Goals, Listening Effects, and Mood Enhancement.' *Psychology of Music* 43 (4): 473–494. https://doi.org/10.1177/0305735613517410.

Volkmann, Linus. 2020. '"Sing this Corona to me": Die Coronavirus-Songs im Überblick.' *Musikexpress*, 29 October 2020. www.musikexpress.de/sing-this-corona-to-me-die-songs-zum-virus-im-ueberblick-1545623/.

Weaver, Hilary. 2020. 'Something to Smile About: Quarantined Italians Fight Social Isolation by Singing from Their Homes.' *Elle*, 14 March 2020. www.elle.com/culture/career-politics/a31562041/quarantined-italians-fight-social-isolation-singing-homes/.

Wikipedia. n.d. 'Parody in Popular Music.' https://en.wikipedia.org/wiki/Parody_in_popular_music#1980–present.

16 The Pandemic as a Catalyst for Remotivity in Music

Mark Thorley

Introduction

The year 2020 was completely unlike what most people were expecting. A prediction in 2019 that much of the world would be working from home, that hospitals would be overflowing with critically ill patients, and that planes would be grounded would have been met with disbelief. Most people were more concerned about threats that could be considered to be associated with modern life. The most recent global shock grounded in modern life was that of the 2008 financial crash. This occurred because of the complex bundling of financial instruments, the workings of which are well beyond the average person's comprehension. Banks and companies were declared bankrupt, others were nationalised, and the personal finances of many individuals suffered, with many of them never recovering. Similarly, the manner in which our personal data is stored and used is well beyond most people's understanding. For this reason, whilst many people are concerned with the risk of their data being compromised or their identities stolen, they still engage with online services that increase such a risk. It is ironic then, that the coronavirus pandemic was not a phenomenon of modern times. Instead, it came about through virus mutation, which has always taken place, with the influenza outbreak of 1918 being the most recent example on a similar scale. Although being unable to breathe is easily imagined, a focus on man-made risks and a degree of complacency meant that much of the world was unprepared for a health-related challenge to their lives.

The complexity and interconnectedness of the world in 2020 did serve to make the outcome completely of the time. Extensive global travel meant that the virus spread quickly through countries and continents. Whilst the technology, if you will, of transmission was new and frighteningly effective, governments struggled to respond to the threat with similar speed and effectiveness. Those with an authoritarian style, and prior experience of outbreaks such as SARS, were best able to respond. The more liberal responded too slowly, however, in a bid to minimise damage to their economies and disruption to normal life.

Within this time of global upheaval, professions were affected in different ways and had to respond accordingly. Health workers grappled with the immediate challenges of limited resources and treating increasing numbers of patients with an unpredictable disease. As their populations came under increasing threat from COVID-19, politicians struggled to respond to the threat to the health and economic well-being of their countries' populations. Often, protecting health came at the expense of economic functioning, and vice versa. Musicianship was one occupation which was affected in a particularly brutal way. Creating, recording, and performing takes place with groups, both in terms of those involved in the production, and the audience. Performing music was therefore stopped virtually overnight in many countries. Similarly halted was the production of recorded music that would ordinarily take place with groups of people. Although some governments offered financial support to businesses, musicians often fell through this safety net due to their being

DOI: 10.4324/9781003200369-20

214   *Mark Thorley*

commonly self-employed. As a consequence of this scenario, many were left with time on their hands and often responded creatively. The COVID-19 pandemic of 2020 was far more disruptive than the 1918 influenza pandemic. As William Robin notes,

> a columnist in Musical America back then estimated that the financial damage to music from the influenza outbreak amounted to around $5 million nationwide, the equivalent of approximately $85.5 million today. In 2020, the Met alone stands to lose that much, or more, if the coronavirus outbreak keeps it closed into the fall.
>
> (2020)

Adaption and Response in Challenging Times

Musicians' work has often reflected the times in which they lived, and this has been particularly marked in times of unrest or upheaval. Responding to the influenza pandemic after the First World War, Robert B. Smith and Malvin F. Franklin wrote 'Influenza Blues.' Interestingly, the song seems to equate being outside in the rain with catching influenza. Similarly, the Vietnam War caused Marvin Gaye to move from his mainstream soulful style towards a grittier offering with his album *What's Going On?* released in 1971. It is therefore no surprise that the upheaval of the coronavirus pandemic resulted in musicians responding creatively. There are key differences on this occasion though, and these centre around the technological fragility of working lives, the abrupt change, the manner of their isolation, and the new tools with which they could respond. Although musicians were isolated and lost income, they had new ways of connecting with others around the globe.

The History of Remote and Home Working in Music

Although remote working seems very current, the term has been used in recording parlance for decades, most often in the phrase remote recording. This refers to taking a recording set-up out of the recording studio to the site of a live performance. This could be a live performance venue, or alternatively a remote building (such as a country house) that provides unique acoustics or ambiance for the performance. Making the recording facility remote allows the best of two physical environments to be brought together in ways not otherwise possible. This approach was pioneered in the 1960s by Wally Heider with his mobile recording truck in California. The work he undertook recording remotely received consistent praise, both within the industry and in reviews (Goodhope 2012; Wennekes 2013).

Any consideration of remote working in a home studio has to acknowledge the work of Joe Meek. Meek was a true pioneer of recording with his innovations in overdubbing, reverb, using the studio as an instrument, and the recognition of producer identity – all from a flat in Islington, London (Cleveland 2001). Despite his critical success, Meek suffered from anxiety and depression. This sadly resulted in him shooting his landlady (who apparently had complained about the noise) before committing suicide at the age of 37. Like Meek, many people working at home now experience feelings of isolation, anxiety, and depression. This suggests the need to reflect upon the fact that whilst technology empowers practitioners, it can also isolate them. Another example of early studio home working is that of Rudy Van Gelder. At the age of 30, he persuaded his parents to let him put a window in the lounge wall to make their house in Hackensack, New Jersey into a recording studio. The compromised acoustic environment together with Van Gelder's approach was less than perfect though. Charles Mingus refused to record there and Steve Hoffman's (2020) criticism of the approach has been widely shared. Nevertheless, this home studio became the location for many iconic jazz records, particularly those recorded for the Blue Note label.

This Time It's Different

Although global shocks such as the pandemic of 1918, or the financial crisis of 2008 have happened before, the coronavirus pandemic of 2020 has some key characteristics for music. These extend from the situation in which consumers found themselves, through to the particular challenges and opportunities that music practitioners faced.

Production and Consumption

As coronavirus spread throughout the world, more countries went into lockdown, travel stopped and a large part of economic activity ceased. Essential businesses' activity (food, construction, health) was often allowed, but the creative industries, including music making, stopped suddenly. The music industry postponed new releases because their production was often halted. Live performances and festivals were also cancelled. As pubs and restaurants closed for business, performing rights royalties dropped. The lives of music consumers also changed. Many worked from home, some were furloughed from work, and others lost their jobs completely. Although the production of music was therefore halted, music consumption continued, despite the fact that music consumers' lives had been disrupted in previously unimaginable ways.

Changes caused by the rapid spread of coronavirus, and the resulting halt of economic activity throughout the world, occurred quickly. Goldman Sachs' May 2020 *Music in the Air* report highlights the changes (Yang et al. 2020), some of which are surprising. For example, although it would be expected that music streaming would increase as more people had time on their hands, the opposite actually happened. Streaming went down 11 per cent, attributable to a decrease in time spent commuting or in offices (40 per cent of music consumption takes place in cars or at work). Music streaming on TV and games consoles actually doubled though, as more consumers were confined to their homes. For the same reasons, music video streaming increased by 8.1 per cent in the week ending 9 April 2020, compared to average consumption in the eight weeks prior to 19 March, whilst audio streaming was down 6.5 per cent when compared with the pre-COVID-19 average in the same week (Yang et al. 2020, 14–15). Changes in music consumers' everyday lives therefore had a significant effect on their listening (and viewing) habits.

The Response of Music Practitioners

Musicians have always responded creatively to the events they find around them. The recorded music industry also has a history of remote working, whether it be from a remote recording truck or in a home studio. There are, however, a number of unusual factors in the 2020 pandemic that influenced the response of music practitioners.

The Disruption of a Technologically Enabled Global Network

Music making is now intrinsically linked with the latest technology, which lies within a global ecosystem. A composer is likely to use software designed in Silicon Valley to write and record their compositions. A performing musician depends upon electronic instruments, live sound systems, and lighting, possibly built in Asia, to provide the audience with the experience to which they have become accustomed. A plethora of secondary technologies also supports music practitioners' activities. These include social media, online ticketing systems, digital and traditional media outlets, as well as hospitality and travel technologies (flying, hotels,

216 *Mark Thorley*

etc.). Compared with the 1918 pandemic, for example, the music practitioner's working life is now tied up in a complex web of connections facilitated by technology rather than face-to-face meetings or performances.

The pandemic of 2020 disrupted much of the activity that these technologies had enabled. Air travel was quickly suspended, hotels could not be booked, and live venues were not able to host events due to the risk of contagion. The disruption to the lives of musicians was therefore much greater in its breadth and depth than any they had experienced before. Many of their activities were halted, and the vast majority of channels and territories were affected in exactly the same way.

Income Stream Disruption

Intertwined with technological disruption, musicians' income streams were also disrupted. Live performance was halted and so all musicians, ranging from those employed with salaries in high-profile international orchestras through to singer songwriters playing in bars, faced an uncertain financial future. Making a living from music has always been a precarious affair and there are high levels of self-employment (Brown 2007). This disruption also applied to related professions such as sound engineering, sound design, and production. Similarly, recording projects which involved travel and working in the same recording studio were halted. Music practitioners often have 'portfolio careers' where they combine their 'core' practice (such as performance) with, for example, private teaching (Latukefu and Ginsborg 2019). Even in this case, however, the impact was still significant as often all of their income streams were affected at the same time.

In some countries, and under certain conditions, furlough funding and self-employment business grants were available. In these cases, music practitioners had time on their hands without the usual avenues for their work. Where these were not available, however, music practitioners needed to concentrate more on work which could create income either immediately or later on.

Living and Working in Isolation

Lockdown measures were imposed in much of the world to halt the spread of coronavirus through respiratory transmission. This seems like a draconian and somewhat crude response in an age of sophisticated medicine, though it was deemed as the most effective response in order to protect health services from being overwhelmed, and to allow vaccines and other treatments to be developed. Compared with the influenza of 1918, which was relatively short-lived, lockdowns had long durations with often uncertain outcomes. This meant that music practitioners were isolated at home, unable to meet other people, and certainly not able work with others professionally.

New Instruments of Communication, Production, and Collaboration

In the past, musicians' tools were limited to the musical instruments that they had mastered. Their response to a disruptive scenario, whether local or global, was fundamentally musical – writing a song or piece of music, and sharing this with an audience directly. In 2020, this situation has changed significantly because of the tools at their disposal. Most of these new tools (which are arguably also instruments of expression) come out of developments in computer processing and storage improvements that have taken place since the early 1990s. These have offered new opportunities to musicians in terms of production, communication, and collaboration.

In terms of production, the advent of MIDI (musical instrument digital interface) in the 1980s, and the subsequent development of early tape-based digital recording media, heralded a move to a lower cost of recording. Previously, recording costs were exorbitant and professional recordings could only be made with the help of a financial backer such as a record company or manager. Technological developments presented musicians with new instruments that they could use as part of their means of expression. Chadabe (2004) and O'Grady (2020) have referred to this as the democratisation of music production. Taking this principle forward means that the musicians' response to the pandemic of 2020 goes beyond the purely musical (composing and performing), but also into the realm of production, particularly in home (or remote) studios.

In addition to the tools of production, access to social media tools at low or no cost enables musicians to better communicate with their fans and audience (Frick, Teskouras, and Li 2014; Waldron 2018). Prior to social media, communication from musicians came in two forms. If a musician was in a contractual relationship with, say, a record company, then the record company would manage the communication, often by using public relations staff. The message was out of the musician's control. Alternatively, musicians who were not in a contractual relationship with another party could manage their own communications, promoting concerts or recordings themselves. The costs of doing this, and getting through media gatekeepers, often made this a challenge. Social media has enabled musicians to connect directly with fans (and potential fans) and consumers of their music. Importantly, the musicians themselves are in control of the message.

Along with social media have come a number of tools that, though not necessarily intended to, now enable collaboration in music. These include generic communication tools such as videoconferencing platforms (Microsoft Teams, Zoom, and Google Hangouts, for example). As well as allowing musicians to communicate with each other and their fans, they also help the practice of co-production or peer-production. In *Wikinomics*, Tapscott and Williams (2008) outline how technology can achieve better results compared to more traditional working methods. Collaborators isolated by geography can achieve improvements in quality, innovation, and also lower costs. Prior to the pandemic, music practitioners may have used videoconferencing to discuss projects with clients or collaborators. However, most video conferencing platforms include audio processing such as filtering and compression designed to optimise speech transmission in noisy office environments. During the pandemic, as music practitioners experimented with videoconferencing in order to collaborate, they found that the standard audio settings degraded the quality of music transmission significantly. For this reason, Zoom introduced a 'high fidelity music mode' in September 2020. This feature enabled music practitioners to share music with others at a higher quality by bypassing the standard mode's audio processing. Such a move shows how the changes in music practitioners' practices during the pandemic encouraged generic platforms to improve their functionality to facilitate new ways of working.

Cloud storage offerings such Dropbox and WeTransfer can also play a part in this as they facilitate collaboration in projects. Creative practitioners, including musicians, can upload digital assets (audio, video, or other) to the cloud for their collaborators to access. In the past, if musicians were collaborating internationally, the need to exchange a physical media storage device (such as tape or hard disk) added cost and inconvenience to the creative process. Lastly, new tools which are specifically designed for music collaboration are now also available. Some of these are part of digital audio workstation software. Cubase VST Connect and VST Transit bring collaborators together as if they were sitting in front of the same workstation, via cloud storage. Others include Audiomovers, a separate piece of software that can be used as a plug-in with other digital audio workstation software such as ProTools or Logic.

218   *Mark Thorley*

These factors bring a new focus on the site (or location) where collaboration takes place. Traditionally, collaboration would take place in a shared physical location, such as a rehearsal room or a recording studio. New technology has however moved the site of collaboration from a physically shared space into a virtual world, such that the site is now quite vague. Rehearsal rooms and recording studios are also technological constructions, based on the physical building and the bespoke equipment choices inside. The adoption of technology to music collaboration is not therefore totally new, though the type of technology is.

New approaches and tools build upon the already collaborative aspect of music production. Negus (1992) explains that much of the work that recording industry personnel undertake is collaborative or collective activity, co-ordinated around shared objectives and commercial aims. Emerging technology tools, though new, can be thought of as a further development of this concept. Emerging technology also develops the nature of collaboration, which Kealey (1979) originally described for music mixing. Kealey identified three modes: 'the craft union mode, the entrepreneurial mode, and the art mode' which involved increased collaboration between the sound mixer and musicians as recording technology became more complex (1979, 7). The 'craft-union mode' was a formal, distant relationship whilst the 'art-mode' became highly collaborative (ibid., 19–22). The approaches discussed here take this even further, insofar as collaboration can take place not only between technical and creative processes in the same room, but also across the world.

New Modes of Expression through Remotivity

Musicians have used their skills of expression in particular ways during the pandemic. Although they have often worked remotely, the uniqueness of this situation has challenged them differently, whilst at the same time presented new opportunities emanating from the behaviour of music consumers. This demands a new way of working, and for this reason, *remotivity* is proposed here as a word to reflect the set of novel capabilities and activities associated with working remotely. Music is produced through a number of stages – composition, performance, recording, mixing, and mastering, followed by marketing and distribution. Each stage is overseen by a specialist practitioner, and as digital technology has developed, different stages have been affected at different times. Digital technology developments originally impacted performance, recording, and mixing. For example, MIDI provided new instruments of expression in the 1980s, with digital tape and hard disk recording following later, giving rise to home (or project) studios. In terms of marketing and distribution, the advent of MP3, digital distribution, and most recently streaming, have all been facilitated by global networks with greater speed and capacity.

The opportunities for music practitioners to engage with others around the world grew as global networks developed and greater computing power affected the marketing and distribution aspects of the production process. Whether working with collaborators or engaging with audiences, the opportunities are now much greater. The isolation and disruption of the pandemic presented music practitioners with a sudden need to make use of this opportunity, and seek new ways of doing work if they were to continue in their practice. In doing so, a number of approaches or models emerged, including relocating studios back to home environments, live streaming, new performance platforms, collage, and performances.

Studios Relocating to Home

Although equipment is now much smaller, there is still considerable emphasis on having a professional environment as an indicator of expertise, despite the additional cost. For this reason, many practitioners still have premises to do their work, often in music centres such as London, Los Angeles, or Nashville. Even clients who are looking for low-cost solutions can be impressed by the commitment and prestige of professional premises. One of the more obvious outcomes of the pandemic was practitioners in any stage of the production process packing up their equipment and relocating it to their homes. Based on this phenomenon, the website recordproduction.com started a *Self-Isolation Home Studio Setup* series during the first UK lockdown. There are six or seven videos in the self-isolation series including Dan Armstrong, Mike Exeter, Russel Cottier, and Owen Davis and they follow a similar format to other videos on the site.

The degree of success in such a move depends on how much of the practitioner's value comes from the environment, and how much from their expertise. At one end, an audio technician who owns and hires out rehearsal and recording rooms would find such a move challenging as the facility itself is more of the value than their expertise. Even amongst seemingly similar practitioners, the situation can also be different. Taking the example of remote mixing, although a high-end remote mix engineer does the same work as a lower-priced one, their clients and the value they offer to them can be quite different. Although both integrate their role into that of producing and tracklaying, at the higher end, the remote mix engineer mitigates against risk and adds another sonic dimension to the recording through expertise, technology, and environment (Thorley 2019). The success of any pandemic relocation therefore depends upon the nature of that value offering to the client. In the pandemic, it also depends upon how badly their clients' activities have been hit by lockdowns, income stream disruptions, and so on.

Live Streaming from Home

Working on music remotely is now not limited to the recording process. Whereas faster networks tended to cause changes earlier in the production chain, faster broadband is now providing performers with an opportunity to connect directly with music consumers. New modes that facilitate this are therefore emerging, and were given a boost by the pandemic. For music performers, concerts and festivals shut down very suddenly due to the high risk of coronavirus spread between concert goers in close proximity and large numbers. This had a greater impact because in recent years, the relationship between live performance and recording has shifted. Whereas loss-making tours were once used to drive profitable sales, the opposite is now true, and touring plays a greater role in financial success (Klein, Meier, and Powers 2017; Williamson and Cloonan 2013). This made the impact of the concert and festival cancellations all the more severe. Such a situation was not limited to musicians with recording contracts, however, as even small-scale singer songwriters performing in small bars, pubs, and restaurants were affected.

As a response to this, many musicians turned to live streaming to allow their fans to watch their performances from home. The simplest way of doing this that emerged was using Facebook Live or YouTube. At this most basic level, performers could change settings in Facebook Live or YouTube to stream sound and images from their laptops. Those with more technical proficiency used broadcasting software, which facilitates multiple cameras and the addition of text and effects to improve the experience for viewers. Free versions of broadcast software such as OBS exist, and whilst the integration with Facebook Live or YouTube is more challenging, the experience for the audience can be made much better.

Many of these performances had a low budget feel to them however, due to the speed with which they had been put together. Nevertheless, they suited performers looking for a quick and effective way to reach out to audiences. The increase in music video consumption

220 *Mark Thorley*

outlined earlier also probably meant that audiences were more forgiving over the quality of output, and in many cases were receptive to new online experiences. Many of the live streams had a low-fidelity, almost homey feel to them, portraying an unusual intimacy between performer and audience. Well-established global artists such as Elton John and Coldplay were also streaming their performances in the same way, particularly at the start of the lockdowns. The highly-esteemed *Tiny Desk Concert* series produced by National Public Radio similarly moved from in-person performances on 11 March 2020, and became the *Tiny Desk (Home) Concert* series. These shifts led to a narrowing of experience for consumers, between a global artist's work and that of a minor performer. Elton John's performance, for example, was in his garden, with many performance imperfections, including incomprehensible lyrics. Rather than a polished production then, performers reached the audience with a rougher, more intimate experience. This takes the concept of the democratisation of music recording and production even further. However, rather than music practitioners being able to produce music due to a lowering of the costs of production, here democratisation occurred because all artists had the same limits on their activities. In many ways, this continues and reflects the concept of lo-fi production, an aesthetic emanating from the 'Do-It-Yourself' approach and resulting aesthetic (Borlagdan 2010).

The increase in activity at all levels was reflected on the online concert discovery and tour promotion platform, bandsintown.com. Between 22 and 28 April 2020, there was a 21.5 per cent sequential growth in weekly livestream events posted on the platform, following an 11.75 per cent increase the week before (Yang et al. 2020, 64). As the market became more saturated, live streamed concerts later moved to a more refined form, with larger budgets and greater planning. Dua Lipa's *Studio 2054* was perhaps the highest profile and most successful live streamed concert. It cost $1.5 million, took five months to put together, and sold 284,000 tickets. It was streamed via the Livenow platform, with regional contracts made for different territories. For example, in India, it was streamed via Gaana to view for free. In China, where there are more complex regulations regarding streaming and ticketing, it streamed on Tencent for free. Both of these companies paid to stream the content. India had 950,00 viewers whilst China had nearly two million. Wendy Ong, President of Dua Lipa's management company TaP, stated:

> There's been a lot of live shows we've seen online during lockdown that weren't very exciting. This was an attempt to be multidimensional, set build, have special guests, and have it shot in a way where it's more like TV or a movie.
>
> (Millman 2020)

Although this event with such a sizeable audience had a positive effect on sales for Dua Lipa's next live tour, the question of how exactly to monetise live performance streaming remains unanswered (Yang et al. 2020, 65).

New Performance Platforms

The live stream concert is an immediately understandable concept as it repurposes the performance genre to digital delivery. However, the pandemic also saw more activity which facilitated performance on new platforms. Live performances took place in games, with one of the most significant being rapper Travis Scott's concert in the game *Fortnite*. Despite only lasting 15 minutes, the performance on 23 April 2020 had over 12.3 million attendees, not including YouTube and Twitch viewers. The previous record for a concert on *Fortnite* was 10.7 million for Marshmello's virtual performance in February 2019. Scott was also added to Epic Games' *Icon* series, so that players could buy clothes and accessories. According to

Variety Magazine, 'the prodigious turnout, no doubt, was helped by stay-at-home quarantine conditions in many parts of the world, as well as the promise of the premiere of a new track by Scott' (Spangler 2020). These in-game concerts were so successful that Epic Games built a sound studio in Los Angeles later in the year that was specially designed with the sophistication necessary to deal with the specific challenges of these productions. This was used for the first time on the 12 September for a Dominic Fike concert. Nate Nanzer, *Fortnite*'s head of global partnerships explained the development: 'This is a tour stop. If you're on tour, you want to stop on the *Fortnite* stage. It's a unique way to get in front of an audience that maybe you're not reaching through other means' (Webster 2020).

Collage Performances

A further way in which musicians who were primarily concerned with performance responded was through the use of what could be termed collaged performances. Such performances consisted of geographically isolated performers using technology to seemingly perform together. An interesting example is that of David Sanborn's *Sanborn Sessions*. These sessions originally took the form of guest musicians (Kandice Springs, Bob Jones, Michael McDonald, and others) performing from Sanborn's home studio. However, during the pandemic and the resulting lockdown, these were facilitated remotely and featured artists such as Sting and Marcus Miller. The sessions took the form of a guest musician discussing how they will arrange and play the song with Sanborn and his band. The actual performance follows, but the manner in which it was edited gave the impression that the players were playing together. This was not the case however – the recordings and filming actually took place separately. The featured artist recorded their performance first (usually a vocal with an instrument), to which the other musicians would then add their parts. This adds considerable complexity to the technical elements and to the performance itself. The guest musician needs to be able to record their vocal and instrument in sufficiently high quality, and be aware of tempo irregularities. Similarly, the musicians put their parts down one by one, and so cannot play off of each other. Nevertheless, to the viewer on YouTube, it creates the impression that they are playing together. The preproduction discussion which is shown before the performance provides an interesting background, particularly where there is some discussion of lockdown.

The Practice of Remotivity

Although the technology to do much of this has existed for some time, and some musicians have a background of working remotely, the pandemic and resulting lockdowns cut off the usual options, which meant that music practitioners had to be creative in their responses. Whilst there was little time to strategise, the risk of any experimentation or new behaviour also changed completely. This was due to the lack of alternatives, the imminent danger of completely losing livelihoods, and the shift in behaviour of music consumers. This section therefore explores those capabilities that were necessary to practice remotivity in music.

A Creative and Innovative Mindset

All of the new models or approaches highlighted here required a creative and innovative mindset. This was necessary in order to solve the problem of connecting with audiences and clients despite the new challenges. Music practitioners who were involved in any of the approaches taken could have chosen to do nothing and wait for the pandemic disruption to pass, perhaps even changing their profession in the short or long term. The problem with such an approach is that music consumption not only continued, but actually went up in cases such as music video streaming. By not engaging with the new reality, a music practitioner would not only lose out on potential income, they would also be in danger of losing their existing audience or clients.

Instead, what was needed was an approach that could focus on a clear objective, and then to find creative ways to achieve that objective. This was as much the case for high-profile acts such as Dua Lipa as a low-key singer songwriter. For the high-profile musician, a whole team of people contributed to this creative mindset solution. This included the more obvious collaborators, such as musicians and arrangers focused on developing an attractive set, through to business and legal affairs personnel who look to maximise income and exposure through investment decisions. For the more low-key singer songwriter instead, whilst the mindset still needs to be the same, it is more likely their own ideas and implementation that make any solutions effective.

Technical Expertise to Repurpose Tools

The second necessary part is the expertise to repurpose a collection of technical tools in order to facilitate the remote working situation. The same principle applies whether at the highest end performance or the lowest-key – a technological solution needs to be decided upon and managed in order to communicate with the audience or clients. Taking the example of low-key musicians who begin to stream their live performances: this can take the relatively simple form of streaming via Facebook Live or YouTube. However, if such a performer is familiar with the technicalities of live performance, they are likely to want to use a more sophisticated set-up. This may mean using several microphones, a mixer, and effects, the output from which is routed to a computer for conversion to a digital signal so that it can be streamed. Use of a higher quality camera (or cameras) also presents the need for some kind of broadcast software such as OBS. In the case of Facebook Live, the user needs to go into the publishing menu of Facebook Live and change the settings to live stream. They then need to go into OBS and set the stream to go into Facebook Live, as well as setting the scenes and sources before starting streaming. Given that a scene is a set of sources, and a source could be any number of cameras or microphones, set-up and use starts to become somewhat challenging, particularly if the person doing so then has to perform. If such an artist is to retain the interest of their existing audience base (or gain new fans), however, they are likely to need to retain the level of production quality that those fans expect. Although the early pandemic successfully presented some fairly crude performances online, the more that music practitioners started to do this, the more the production standards had to rise.

The higher profile the musical artist, the more sophisticated and challenging the technical approach is likely to be. Travis Scott's concert in *Fortnite*, for example, involved not only elements of musical production, but also game design and programming. Scott's performance was presented by an avatar, which is a representation of his stage persona, but different from reality in the proportions of particular elements. This is a complex technical solution, but the result is a novel experience which is reflected in the number of people who engaged with the concert that took place in the game.

Maintaining a Stable Client or Audience Base

The third element of remotivity is having an established client, fan base, or audience. The nature of that following will differ between practitioners. For example, a music producer is likely to have clients in the form of record companies, publishing companies, or management companies. These relationships may well have been nurtured through face-to-face interactions taking place over a number of years. A self-financed singer songwriter trying to establish themself would be in a different position, however. In that instance, their following comes in the form of an audience interested in listening to their work.

During the pandemic of 2020, the amount of information and online communication was such that it would have been difficult to build any sort of following from scratch. As the online world becomes more and more crowded, this situation is likely to continue. Whilst the technology of production and communication is democratising in many ways, the challenge of reaching clients or audiences becomes greater, because that technology is available to so many other people. It then becomes more a case of offering new experiences or services to remote audiences or clients, rather than music practitioners trying to launch themselves as somehow new. This approach goes a long way towards maintaining any existing audience or client base.

A Differentiated Expertise

The last aspect of the remote turn is being able to demonstrate a unique selling point, or point of differentiation, in what the music practitioner is offering. This is likely to have come from work undertaken and expertise developed before working remotely. It is important to note that offering a remote service is not a sufficient point of differentiation in itself. This is because whilst working remotely opens the music practitioner up to a worldwide market, it is also a crowded stage, and the practitioner has to have something to offer that is differentiated from other practitioners in the field.

In such a global market, the practitioner's expertise as a function of their location, culture, or background often becomes their point of differentiation. For example, Chris Lord Alge is probably the world's most successful remote mix engineer: his expertise is based upon his background (Thorley 2019). His path to mix specialist started with using his mother's tape recorder, then becoming tea boy, assistant engineer, house engineer, freelance engineer and producer, then specialist (ibid.). So although he ostensibly just does mixing, he has an in-depth understanding of the technical, logistical, creative, and psychological issues involved in making records. He is therefore not afraid to make changes, which can seem like overstepping the limits of his responsibility to mix (ibid.). If his expertise were just limited to mixing, he would have neither the ability nor the credibility to do this. However, his extensive and varied experience with the other processes of record production, often in the centres of record production in New York and Los Angeles, allow him to integrate the expertise he developed through his earlier experiences.

Differentiating by virtue of location or background also raises the concept of the branding associated with a music practitioner. Using an expert who is recognised as a brand is one way of mitigating against risk, particularly where they are very specialised. Marshall (2013) has explored the role of stars in the recording industry stating that the use of stars is a way to make the unpredictability of music markets more predictable. In this case, the star (or brand) recording engineer, alongside the star (or brand) remote producer heralded by Joe Meek, contributes to risk mitigation in the same way as a star lead singer. They do need capability though, and this is based upon the development of their expertise, which underpins their differentiated offering. Without such demonstrable capability, it would be an anonymised remote service, which would introduce greater rather than less risk.

Summary

As I have shown, the speed and depth of disruption which came from the coronavirus pandemic led to unprecedented changes in the lives of music practitioners. Live music activity

224　*Mark Thorley*

came to a standstill and production activities such as recording, which relied upon people being together, were also halted. These disruptions were all the more significant compared with previous pandemics due to the particular situational characteristics of the lives of music practitioners. These included being part of a technologically enabled global network, the immediate disruption to income streams, living and working in isolation, and the availability of new instruments of expression.

At the same time, music consumers continued to listen to music, albeit in different ways. This did not just present a need for music practitioners to keep going in order to not lose profile or future income streams. In addition, it also presented an opportunity for presenting music or services in different ways. The unprecedented nature of the pandemic allowed the presentation of riskier and less polished interactions to audiences and clients. Much of the technology to do this was already available, and there is a rich history of remote working in the practice of music. A number of new approaches or models therefore emerged. These included returning to home to work, live streaming from home, launching new performance platforms, and producing collage performances. I have argued here that this way of working can be defined as remotivity, and that there are a number of elements which contribute to its effectiveness in music practice. These include having a creative and innovative mindset, acquiring the technical expertise to use new tools, having an existing following or client base, and having a differentiated offering based on prior experience.

As the world comes slowly out of lockdown, it is worth briefly considering the sustainability of these changes to working. Addressing this question, a Goldman Sachs report notes that the increased shift from offline to online consumption in the pandemic led to an increase in the importance of social media for discovery and promotion, and closed the gap between artists and fans (Yang et al. 2020, 4–5). Furthermore, there have been significant acquisitions which suggest that the new ways will remain. In terms of live streaming, Noonchorus was bought by Mandolin (Needham 2021) and Boiler Room was bought by Dice (Malt 2021). In the production space, Audio Movers was bought by Abbey Road Studios (Gumble 2021). These developments suggest that in the future, the need for capabilities associated with remotivity are likely to increase for music practitioners if they are to survive and be successful in a global marketplace.

References

Borlagdan, Joseph. 2010. 'The Paradox of 'Do-It-Yourself' in Unpopular Music.' In *Philosophical and Cultural Theories of Music*, edited by Eduardo de la Fuente and Peter Murphy, 175–199. Leiden: Brill.

Brown, Ralph. 2007. 'Enhancing Student Employability? Current Practice and Student Experience in HE Performing Arts.' *Arts and Humanities in Higher Education* 6 (1): 28–49. https://doi.org/10.1177/1474022207072198.

Chadabe, Joel. 2004. 'Electronic Music and Life.' *Organised Sound* 9 (1): 3–6. https://doi.org/10.1017/S1355771804000020.

Cleveland, Barry. 2001. *Creative Music Production: Joe Meek's Bold Techniques*. Boston: Mix Books.

Frick, Thomas, Dimitrios Tsekouras, and Ting Li. 2014. 'The Times They Are A-Changin: Examining the Impact of Social Media on Music Album Performance.' *Academy of Management Proceedings* 2014 (1). https://doi.org/10.5465/ambpp.2014.16984abstract.

Goodhope, Pat. 2012. 'Santa Monica 1960.' *International Association of Jazz Record Collectors Journal* 45 (4): 72.

Gumble, Daniel. 2021. 'We Want to Be Part of This World: Abbey Road Acquires Remote Production Start-up Audiomovers.' *Audio Media International*, 16 March 2021. https://audiomediainternational.com/we-want-to-be-part-of-this-new-world-abbey-road-acquires-remote-production-start-up-audiomovers/.

Hoffman, Steve. 2020. 'What Rudy Van Gelder Did "Wrong." Steve Hoffman (forum), 26 July 2020. https://forums.stevehoffman.tv/threads/what-rudy-van-gelder-did-wrong-article.980394/.

Kealey, Edward. 1979. 'From Craft to Art: The Case of Sound Mixers and Popular Music.' *Sociology of Work and Occupations* 6 (1): 3–29. https://doi.org/10.1177/009392857961001.

Klein, Bethany, Lesley M. Meier, and Devon Powers. 2017. 'Selling Out: Musicians, Autonomy, and Compromise in the Digital Age.' *Popular Music and Society* 40 (2): 222–238. https://doi.org/10.1080/03007766.2015.1120101.

Latukefu, Lotte and Jane Ginsborg. 2019. 'Understanding What We Mean by Portfolio Training in Music.' *British Journal of Music Education* 36 (1) (March): 87–102. https://doi.org/10.1017/S0265051718000207.

Malt, Andy. 2021. 'Dice Acquires Boiler Room.' *Complete Music Update*, 4 October 2021. https://completemusicupdate.com/article/dice-acquires-boiler-room/.

Marshall, Lee. 2013. 'The Structural Functions of Stardom in the Recording Industry.' *Popular Music and Society* 36 (5): 578–596. https://doi.org/10.1080/03007766.2012.718509.

Millman, Ethan. 2020. 'Dua Lipa's Very Expensive Concert Is the Future of Livestreaming.' *Rolling Stone*, 1 December 2020. www.rollingstone.com/pro/news/dua-lipa-livestream-cost-viewership-1096950/.

Needham, Jack. 2021. 'Livestream Firm Mandolin Buys Virtual Concert Platform Noonchorus.' *Music Business Worldwide*, 22 September 2021. www.musicbusinessworldwide.com/livestream-firm-mandolin-buys-virtual-concert-platform-noonchorus/.

Negus, Keith. 1992. *Producing Pop: Culture and Conflict in the Popular Music Industry*. London: Arnold.

O'Grady, Pat. 2020. 'Sound City and Music from the Outskirts: The De-Democratisation of Pop Music Production.' *Creative Industries Journal* 14 (3): 211–225. https://doi.org/10.1080/17510694.2020.1839281.

Recordproduction. n.d. www.recordproduction.com.

Robin, William. 2020. 'The 1918 Pandemic's Impact on Music? Surprisingly Little.' *New York Times*, 6 May 2020. www.nytimes.com/2020/05/06/arts/music/1918-flu-pandemic-coronavirus-classical-music.html.

Spangler, Todd. 2020. 'Travis Scott Destroys "Fortnite" All-Time Record with 12.3 Million Live Viewers.' *Variety*, 24 April 2020. https://variety.com/2020/digital/news/travis-scott-fortnite-record-viewers-live-1234589033/.

Tapscott, Don and Anthony D Williams. 2008. *Wikinomics: How Mass Collaboration Changes Everything*. London: Atlantic books.

Thorley, Mark. 2019. 'The Rise of the Remote Mix Engineer: Technology, Expertise, Star.' *Creative Industries Journal* 12 (3): 301–313. https://doi.org/10.1080/17510694.2019.1621596.

Waldron, Janice L. 2018. 'Online Music Communities and Social Media.' In *The Oxford Handbook of Community Music*, edited by Brydie-Leigh Bartleet and Lee Higgins, 109–130. New York: Oxford University Press.

Webster, Andrew. 2020. 'Fortnite Is Launching a Concert Series It Hopes Will Become a "Tour Stop" for Artists.' *The Verge*, 8 September 2020. www.theverge.com/2020/9/8/21423004/fortnite-party-royale-concert-series-dominic-fike.

Wennekes, Emile. 2013. 'Let Your Bullets Fly, My Friend: Jimi Hendrix at Berkeley.' In *The Music Documentary: Acid Rock to Electropop*, edited by Benjamin Halligan, Robert Edgar, and Kirsty Fairclough-Isaacs, 87–99. New York: Routledge.

Williamson, John and Martin Cloonan. 2013. 'Contextualising the Contemporary Recording Industry.' In *The International Recording Industries*, edited by Lee Marshall, 11–29. London: Routledge.

Yang, Lisa, Heath P. Terry, Piyush Mubayi, and Heather Bellini. 2020. *The Show Must Go On*. London: Goldman Sachs. www.goldmansachs.com/insights/pages/infographics/music-in-the-air-2020/report.pdf.

17 Music in Lockdown

On Sonic Spaces during the COVID-19 Pandemic, March–June 2020

Esteban Buch

Boccaccio's Lesson

During the first COVID-19 lockdown between March and June 2020, Giovanni Boccaccio's *Decameron* was often evoked in the media, together with Albert Camus' *La Peste*, and a few other classics. In March, the case was made in the *Newstatesman* that this book, completed in 1353, 'shows us how to survive coronavirus' (Spicer 2020). As part of 'an epidemic of advice,' storytelling and musicking were recommended as therapeutic resources in the face of the pandemic, echoing the 'innovation of narrative prophylaxis,' as Boccaccio's contribution to early Renaissance medical treatises was described (Marafioti 2005, 69). The same article was highlighted by a Boccaccio scholar in *JSTOR's Daily*, a section whose motto is 'where news meets its scholarly match' (Wills 2020). In June, the Library of Congress presented a Boccaccio Project, with videos of ten musicians playing works composed ad hoc, such as Miya Masaoka's *Intuit (a way to stay in this world)*, performed by cellist Kathryn Bates. The composer explained:

> This piece is a positive and optimistic hope for a way to be present in the outside world of lockdown, and a wish to find a healthy balance between our own interiority and the outside world, which, under such extraordinary circumstances, we seek relief.
>
> (LOC 2020)

It is not hard to see why. Boccaccio's book is both a collection of a hundred short stories mostly dealing with love and sex in a libertine and anticlerical mood, and a first-hand account of the so-called black death, the horrible epidemic of bubonic plague that in 1348 killed more than half of Florence's population, and in a few years decisively altered Europe's demographic and social map. In 2020, the interest had less to do with the tales than with the *cornice* or frame tale, the narrative about a group of people – a *brigata* of seven young women and three young men – voluntary isolating in a beautiful country villa, and devoting themselves to tell stories, sing ballads, and dance, alongside other pleasures such as drinking and eating, *siesta* sleeping, and exploring the surrounding nature. They also experiment with a proto-democratic, women-led political organisation that makes each one the sovereign for one day, ruling over the collective life and the storytelling ritual. (Servants are not included, though.)

The first day of *The Decameron* starts with a breathtaking account of the plague in Florence, a topic to which the author never returns, consistent with his characters' idea of avoiding bad news as part of their survival strategy. This ominous picture ranges from the repellent symptomatology of the sick to the desperate and selfish attitudes of the sane, in the midst of a total collapse of laws human and divine, and of institutions such as the family, the church, and the state. As part of the people's techniques for surviving – or dying – music

DOI: 10.4324/9781003200369-21

features pre-eminently in both extremes of their spectre of behaviours, namely total self-isolation and total recklessness (Beck 1993). Concerning the first, Boccaccio writes:

> Among whom there were those who thought that to live temperately and avoid all excess would count for much as a preservative against seizures of this kind. Wherefore they banded together, and, dissociating themselves from all others, formed communities in houses where there were no sick, and lived a separate and secluded life, which they regulated with the utmost care, avoiding every kind of luxury, but eating and drinking very moderately of the most delicate viands and the finest wines, holding converse with none but one another, lest tidings of sickness or death should reach them, and diverting their minds with music [*con suoni*] and such other delights as they could devise.
>
> (Boccaccio [1353] 1903, 7)

Listening to *suoni* is the only example of the 'delights' that, alongside drinking, eating, and talking, occupy the people who isolate themselves in their homes. This implied playing the musical instruments themselves, a skill that was part of the education of the upper classes, even if in normal times it was their privilege to have professionals doing it for them. Yet Boccaccio clearly stresses the passive side of the experience, the listening rather than the playing, and makes no mention of singing. His description is that of a static musical atmosphere (Böhme 2017) that turns a habitation under lockdown into a more pleasant, hence a safer space to be – at least according to medical theories of the time.

The opposite stance is the active singing that the reckless combine with heavy doses of alcohol, a cultivation of humour, and a predatory attitude towards spaces public and private:

> Others, the bias of whose minds was in the opposite direction, maintained, that to drink freely, frequent places of public resort, and take their pleasure with song [*l'andar cantando attorno*] and revel, sparing to satisfy no appetite, and to laugh and mock at no event, was the sovereign remedy for so great an evil: and that which they affirmed they also put in practice, so far as they were able, resorting day and night, now to this tavern, now to that, drinking with an entire disregard of rule or measure, and by preference making the houses of others, as it were, their inns, if they but saw in them aught that was particularly to their taste or liking.
>
> (Boccaccio [1353] 1903, 7)

Contrary to the static *suoni* of the self-quarantined communities, *l'andar cantando attorno* knows no physical boundaries, and emanates not from self-chosen company but from an opportunistic alliance of wandering individuals. Its sonic impact is out of control, as is the plague itself. On the other hand, while these people embrace a fatalistic attitude towards death, they also make self-expression a thrilling, if desperate, attempt to boost the intensity of life in a shrinking liveable space and time.

Boccaccio stops short of moral condemnations, but it is easy to sense that the prudent and self-preserving attitude of the first group is better than the suicidal and egoistic attitude of the second. Yet neither group has any guarantee of surviving. In fact, Boccaccio suggests, the best thing to do is to flee the city, as the *brigata* does. Now, his recipe for successful retirement in the countryside retains something of both musical practices charted in the city. Even if music is secondary to storytelling in *The Decameron*, the singing of *ballate*, accompanied by a lute and/or combined with dance, allows exalted emotions to circulate as shared sonic objects, without trespassing the boundaries and norms that the community has adopted for itself. Indeed, the ballads sung at the end of each day 'constitute the secular liturgy and

228   *Esteban Buch*

ritual ceremony of musical commentary, without which the recreation of the social and moral order which the *brigata* brings about could not succeed' (Ciabattoni 2019, 150).

This secular liturgy allows voicing extraordinary concerns about life and the self in times of ordeal. Nora Beck underlines the 'important role' of women throughout this process, quoting from *Decameron* passages such as:

> They then had supper, in the course of which there was much laughter and merriment, and when they had risen from table, at the queen's request Emilia began to dance whilst Pampinea sang the following song, the others joining in the chorus.
>
> (Beck 1993, 14)

In another passage, at the end of the first day, Emilia sings, accompanied on the lute by Dioneo while the other eight people listen, a sublime praise of narcissism:

> When in the mirror I my face behold,
> That see I there which doth my mind content,
> Nor any present hap or memory old
> May me deprive of such sweet ravishment.
>   (Boccaccio [1353] 1903, 66)

Mood Music in Times of COVID-19

Even if mention of *The Decameron* in the media was only ever episodic, during the first COVID-19 wave many people seem to have implicitly followed Boccaccio's advice on the therapeutic virtues of music and stories during lockdown. The emerging scholarly literature on the psychological impact of the confinement supports the notion that listening to music at home helped many to endure the situation, at least according to data and video materials essentially coming from Europe and the Americas, as a consequence of social and technical inequalities in both production and research resources. Anyway, a study of 5,000 people from 'three continents' (France, Germany, Italy, the UK, the USA, and India) concluded that

> More than half of the respondents reported engaging with music to cope [with the situation]. People experiencing increased negative emotions used music for solitary emotional regulation, whereas people experiencing increased positive emotions used music as a proxy for social interaction.
>
> (Fink et al. 2021)

In another study done in Spain, 3–18 April 2020, a majority of the 1,868 respondents 'reported using music to cope with the lockdown, finding that it helped them to relax, escape, raise their mood or keep them company' (Cabedo-Mas, Arriaga-Sanz, and Moliner-Miravet 2021, 1).

Moreover, a study based on 1,031 respondents from Spain, Italy, and the USA ranked music as the first 'most helpful' activity to cope with lockdown, followed by 'talking with friends, family and colleagues' and 'exercising outdoors'; it must be said that the actual figure for music was less than 15 per cent, suggesting a broad dispersion of individual strategies (Mas-Herrero et al. 2020). An Australian survey on a much smaller population (127 respondents) observed that 'at the within-person level, life satisfaction was positively associated with music listening and negatively associated with watching TV/videos/

movies' (Krause et al. 2021). Still another survey, done in the UK with 233 respondents, gave similar results concerning the use of music 'to cope' with the situation, while pointing to a strong correlation between music listening and 'substance use,' that is, 'the turning to drugs or alcohol in order to suppress the process of giving attention to, and dealing with, the stressor' – a finding unique to that study, which the authors see as proof that 'music can also be used in unhealthy ways' (Henry, Kayser, and Egermann 2021).

Put together, these results are consistent, if also somehow unsurprising. Music did help many individuals to cope with the situation. Indeed, the suggestion was made that 'the aesthetic emotions prompted by music might improve and intensify communication, thereby allowing the emergence of collective strategies to reduce [cognitive] dissonance' in the face of the pandemic (Sarasso et al. 2021). That might well be true, and partly account for the capacity to face a situation no one had ever experienced before. On the other hand, all of these studies elaborate on pre-COVID-19 literature, stressing the ordinary role of music in mood regulation. This being a well-established use of music before the pandemic, especially since streaming platforms foregrounded dedicated hubs such as 'Relax,' 'Night,' or 'Wellbeing' (Knobloch and Zillmann 2002; Anderson 2015; Chodos 2019), it was unlikely to suddenly disappear in the face of the COVID-19 crisis.

The fact that a majority of people reported using music for coping with lockdown must also be squared with a reduction in global streaming activity reported by Spotify and other platforms (Sim et al. 2020). This has been explained by the reduction of commuting time spent in cars and on public transportation. Yet 'individuals who continued to go to work onsite also declared listening to music less often than before,' according to a survey of 2,963 respondents in France: 70 per cent of people declaring to do so, compared to 92 per cent in 2018 (Jonchery and Lombardo 2020, 17). As we see, the overall picture of listening practices is far from being completely clear, especially if one also takes into account potential biases, like the fact that 69.3 per cent of the people who responded to one of the Spanish questionnaires were women, and 74.8 per cent had university education (Cabedo-Mas, Arriaga-Sanz, and Moliner-Miravet 2021, 3).

Also, throughout this literature few things are said about what listening to music actually meant in terms of musical genres, technical and spatial configurations, and so on, to say nothing of the kinds of personal meanings attached to the music, which might give some indication of *how* exactly the music helped individuals to cope. The lower activity on streaming platforms was counterpoised with a reported increase in YouTube consumption, suggesting that purely sonic experiences might have receded in favour of audiovisual ones (Sim et al. 2020); yet this interesting remark disregards the frequent use of YouTube as just another streaming platform for music only. Additionally, Timothy Yu-Cheong Yeung (2020) looked at the proportion of older tracks in Spotify consumption to suggest that the COVID-19 pandemic could 'trigger nostalgia,' but his implicit definition of nostalgia as the feeling aroused by any record older than three years is debatable. On the other hand, authors not especially interested in music did detect a surge of 'nostalgia-inspired leisure,' as part of a 'significant interest in the past' (Gammon and Ramshaw 2021).

Another puzzling question is what people worldwide did, if anything, with their personal collections of music (vinyl LPs, CDs, mp3s, etc.), a traditional way of gathering resources in view of private, repetitive listening experiences (Buch 2015). In March 2020, a *New Yorker* journalist asked some musicians what they were listening to and received answers like Maggie Rogers' 'I've been dancing in my kitchen to records I loved in college,' or Lenny Kaye's 'yes, gotta dance, so get out the disco twelve-inches' (Petrusich 2020). Bernard Loutte, a French journalist, wrote a book made of short texts inspired each day by a different 45 rpm record from his collection (Kaganski 2021). But the evidence on such

230 *Esteban Buch*

matters is hardly more than anecdotal. In short, for all the merits of early research by scholars and journalists, the inescapable conclusion is that little precise is known about the actual uses of music under lockdown.

That being said, the mere existence of these studies, carried out in different countries by scholars and respondents who themselves had to cope with a difficult and unprecedented situation, as well as the presence of many positive comments on music in general in the media and social networks, to say nothing of laments about the suppression of concerts and other live musical activities, suggest that the pandemic activated a broad, inorganic *belief* in music's social and psychological usefulness. And in the media at least, praise for music as a resource for intimate wellbeing was perhaps less preeminent than praise for music as a therapy for the social bond. Some even claimed that music could be a substitute for the social bond itself, perceived as utterly damaged by the radical limitation of social interactions, which was the very purpose of lockdown. More specifically, music was called to remedy a damaged social bond by providing alternative *spaces*, as a compensation for the cancelling of most physical spaces for social interaction, including playing and listening to music together. By the same token, even if most music events during lockdown were avowedly unpolitical, the insistent mention of music in public discourse could arguably sound like a response to the partial suspension of public space itself.

The Musical Politics of Balconies

On 13 March 2020, *La Repubblica* published the article, 'Coronavirus, Italy at the Balconies: Songs against Fear,' describing a 'musical flashmob' launched on Facebook, which invited people to sing or play music from their balconies and windows the same day at a given time (Scorza Barcellona 2020).

> And the result is a melody, which is different in every neighbourhood and city, which runs through the regions from North and South, and which now follows the notes of the *Inno di Mameli* [the Italian national anthem], now hums 'Napule è' by Pino Daniele.
> (ibid.)

Videos of the event show Italians producing all kinds of music, from solo trumpets, to opera tenors, to kids battering cans, to DJs moving the inhabitants of an entire building to dance. *Il Messaggero* noted that 'the paradox is that the quarantine produces contact rather than isolation' and claimed that 'the bodies are at home, but the feelings go out' (Ajello 2020).

Italian music balconies featured in international media, and some videos became viral. A few days later, in Germany, a flashmob with the 'Ode to Joy' 'against the virus' was organised along similar lines, reenacting an old tradition of using Beethoven in the social and political arena (Buch 2020). 'Beethoven again brings the people together, even if the famous "embrace of millions" is not to be physically enacted for the moment,' the *Deutsche Welle* commented (Deutsche Welle 2020). 'All can participate, as singing unites and makes us feel good,' claimed the Evangelical Church in Germany, calling to sing and play the children's song 'Der Mond ist aufgegangen' (The moon has risen), rather than Beethoven's tune (DPA 2020). Also, in other places like France, some musicians performed from their balconies and posted the videos on the internet. On 17 March, Tomás Gubitsch, an Argentine-born guitarist, and his son Noé, a violinist, started a 'balcony tour' consisting of one song performed every day from their home in the Paris *banlieue*, and streamed live on a Facebook account; playing an omnivorous repertoire of pieces from very different genres and origins eventually amounted to a 53-day, static 'world tour' (Olivares Palma 2020).

During the first lockdown, popular attention and media coverage focused less on what ordinary people were doing every day with music inside their homes, and more on extraordinary events that took place at the borders of these quarantined spaces, namely in the balconies and windows where private homes connect with open and/or collective space. These intermediary zones allowed for different kinds of interaction among the people, from passive spectatorship to active participation, depending on the musical genres, acoustic settings, and technical resources, and also, on local sociability and idiosyncrasies. In Italy, a sense of national pride was often fuelled through these performances, not least by singing the national anthem; in Germany, a few associated Beethoven's melody to a European spirit that, at a political and practical level, was then eclipsed by the nation-states, starting with lockdown itself. Overall, music events suddenly emerged as sensory manifestations of emotional tuning and collective solidarity. Indeed, these sonic vignettes of a resilient humanity might well remain in many memories as true icons of the period.

Yet, the actual practice of singing at the balcony would hardly be more than an anecdote, had it not inspired media images such as this single melody uniting a whole suffering country, or these feelings navigating freely among bodies stuck at home. In a context of temporary limitations to individual liberties, insidious alterations of the perception of time (Irons 2020) and even of the content of dreams (Jarvis 2021; Pesonen et al. 2020), music at the borders of private space became an available symbol of collective and personal freedom (Dhami, Weiss-Cohen, and Ayton 2020). Of course, such sonic fictions gathered social momentum through the representations of balconies and windows on the internet, which in turn people accessed through the screens of their cell phones, tablets, and computers. And what is a screen if not a kind of balcony or window, that is, an audiovisual space connecting the interior and the exterior of private physical spaces?

Thus, music in balconies appeared as both a metaphor and a metonym of ways of bringing people together, as if following Boccaccio's advice to not only listen to music, but also to sing and play it. And the *locus amoenus* for this kind of encounter was less a series of preexisting physical spaces like real balconies, than virtual spaces created ad hoc. Hence the dissemination of videos showing split screens with singers, players, and even dancers confined in their homes, yet performing together, sometimes as a duo or a trio, sometimes as an orchestra, a choir, or a ballet. The audiovisual setting of these objects corresponded to the idea of their production, as a unified sound seemed to emerge from a multitude of separated rectangles, each being like a small window on a private space. More often than not, these images put homes under their best light and sound, situating their inhabitants in the best possible position in a real-world scale where small rooms, 'an unusable balcony,' and a 'poor-quality view from the apartment' were correlated with severe depressive symptoms (Amerio et al. 2020). 'The bodies are at home, but the feelings go out,' as the *Messaggero* said (Ajello 2020), might also be the motto of these collective virtual performances, whose immediate accessibility through the internet contrasted with the fact that they were actually located nowhere – not in any single physical place at least.

In that sense, music in lockdown arguably challenged ordinary ontologies of music itself, such as the notion of sonic objects produced in stable locations, and whose atmospheric qualities are transposed to the receptor's own space and time. Granted, awareness of recording studio operations had already often challenged the naturalistic myth of phonography giving access to the performers' actual acoustic environment. Also, the production of 'network music,' simultaneously produced by people located in distant places, emerged quite early in the history of the internet as an example of its utopian, democratic potential (Haworth 2020). Yet, during the pandemic the viral spreading of split-screen music videos arguably implied a qualitative leap in the perception of music production as a kind of assemblage,

232   *Esteban Buch*

not only of performers and technical devices such as instruments, but also of contingent fragments of space and time. It might also have contributed to the ongoing destabilisation of classical ontologies of classical musical works, as once discussed by Lydia Goehr (1992) and others (de Assis 2018), by arguably enhancing the phenomenological proximity with fictions (Kania 2013; Buch 2013) and quasi-things (Griffero 2017).

On the other hand, videos featuring classical masterpieces like Beethoven's *Ninth Symphony* or Strauss's *Also sprach Zarathustra* were like statements reaffirming the heroic role of cultural patrimony, now viewed as a resource in the face of the pandemic. Sometimes, musicians and music institutions made these masterpieces a way of plainly showing that they still existed. Indeed, computer screens were for many 'a kind of *portal*' allowing for the performance of 'a professional identity' (Gourlay 2020, 808). Particularly successful was the French National Orchestra's rendition of Ravel's *Boléro*, uploaded on 30 March 2020, which after the verbal self-presentation of the players reduced the nearly 17 minutes piece to less than five. 'The mixing engineer is the real MVP [most valuable player] here,' dryly stated an anonymous viewer on YouTube, among mostly enthusiastic comments (France Musique 2020).

In fact, in the first months of 2020 the available technologies did not allow for real-time sonic interaction. Instead, each musician had to record their part without listening to the others until the video was completed. And when the aim was not to share a video with the public, but just to entertain the practice for itself, performing and listening never came together at all. An early critic of virtual choir encounters discarded them as pure simulation, pointing among other things at 'the lack of a shared acoustic' (Datta 2020, 250). A broader study detected mixed feelings in the participants: on one side, 'solo singing resulted in significantly lower reported wellbeing, but with choir singing creating a more coherent "meaningful" social group than the sport activity'; on the other, 'participants reported missing the social aspect more than the other six components that were assessed, which were related to aesthetic experience, flow, and physical aspects of singing.' Particularly damaging for morale, and a reason for nearly half of the respondents diminishing their participation with time, was the impossibility to 'share the same room acoustic' (Daffern, Balmer, and Brereton 2021). Still, the practice continued, producing music that nobody could actually hear, and yet was somehow real, at least in the participants' imagination.

Self-expression and 'Coronamusic'

During lockdown, streaming music performances of classical music from empty and silent concert halls were not only a practical way to maintain some scheduled events, but also a powerful, if paradoxical, display of the disruption of art rituals and their ordinary emotional dynamics. Soloists performing works of long-dead composers for people they could not see or hear, and for whom the rewards of applause and smiling faces were replaced by a string of emojis, or a written comment on a screen, or nothing at all for that matter, showed what was lost with the pandemic, as much as the will to carry on living and experiencing the pleasures of art no matter what. A case in point was Daniel Barenboim's performance of Beethoven's *Diabelli Variations* at the Berliner Philharmonie on 10 April 2020, as a surviving bit of the grandiose, and marred, commemoration of the 250th anniversary of the composer's birth.

Indeed, balcony singing, streaming concerts, split-screen videos, and online choir practices all represented the continuity of music life, hence of life itself. They addressed the pandemic in a kind of negative way, as a dramatic new context against which every effort was made to keep on doing things as before. If they arguably expressed 'a way to be present in the outside world of lockdown,' as Masaoka said of her cello piece (LOC 2020), they did so mostly by trying to keep bad news at bay, as if following Boccaccio's lesson. This resonates with the respondents who in the above-mentioned studies contrasted the enhancement of wellbeing through music listening with the stress of watching pandemic-dominated news.

Now, even if both things are conflated in some studies, watching news is quite different from watching fiction. According to one survey, some people under lockdown appreciated dystopic movies, such as Steven Soderbergh's 2011 *Contagion*, as a way of coping with fear through catharsis (Testoni et al. 2021). However, more than the 15 respondents of this study would be needed to compare the catharsis strategy with the massive consumption of audiovisual fiction unrelated to the pandemic, mostly through platforms like Netflix, whose streaming volume increased notoriously. It is safe to say that music activities overall were *not* aimed at catharsis, nor at picturing negative emotions like mourning, fear, pain, or sadness, that were raised by the pandemic. Quite the opposite.

A team led by Niels Chr. Hansen gathered, through crowdsourcing, an important corpus of what they called 'coronamusic,' broadly defined as the result of 'the diverse practices of listening to, playing, dancing to, composing, rehearsing, improvising, discussing, exploring, and innovating musical products during lockdown with explicit or implicit reference to the novel coronavirus' (Hansen et al. 2021). Most of them expressed positive emotions such as togetherness, happiness, gratitude, and so on. Significantly, nearly 40 per cent did it by way of humour, often through parodies of existing songs. The Knack's 1979 hit 'My Sharona' was a privileged case, thanks to the rhyme with *corona*. According to an Italian study, independently of music, humour was indeed a frequent reaction to the global situation, with the caveat that 'participants who lived at a greater objective geographical distance from the Italian epicentre of the pandemic showed higher enjoyment' of it (Bischetti, Canal, and Bambini 2020, 12).

In addition to many examples of 'coronamusic' that are not part of Hansen's corpus, the general picture is a set of exclusively uplifting 'mood management' sonic devices. This is all the more striking if one remembers that, according to the World Health Organization, by 1 June 2020 the death toll worldwide was already over 370,000 people (WHO 2022). Moreover, according to one study 'the participants who were more severely impacted by the COVID-19 pandemic reported higher levels of subjective engagement in music-related activities' (Mas-Herrero et al. 2020). Yet, mourning songs like New York-based rapper Lil Tjay's 'Ice Cold,' released on 1 May 2020, are an exception, almost an anomaly.

Apart from a few videos associated with patriotism and a couple of anti-Chinese messages, politics in general and the authorities' handling of the crisis in particular are absent from musicians' reactions at this early stage of the pandemic. While applauding health workers at the balcony became in March a daily sonic ritual in many cities around the world, few attempts were made to express this solidarity through music. Exceptions worth noting are a UN-promoted version of Queen's 'We Are the Champions,' or more confidential gestures like Emily's contribution from Stuttgart to the 'Ode to Joy' flashmob. On another level, sung instructions to children and young people for washing hands and avoiding social contact were sponsored by state health departments in places like New Zealand and Vietnam.

Even if the media presented music as the agent of continued social interactions, in actual practice the public sphere was rather downsized to the domestic spaces pictured in the videos. And this included showing individuals facing solitude, as much as showing them mitigating it through jokes and virtual contacts. If one counts the many concerts live streaming from home that were not uploaded as videos, the number of split-screen representations of collaborative ventures seems outweighed by that of solitary artists,

234   *Esteban Buch*

performing from their homes in front of a camera, perhaps reminding one Boccaccio's Emilia singing 'when in the mirror I my face behold.' And few of them suggested that being alone could be a problem. Even a song by st. Pedro that addressed the disruption of a couple's sex life gave the solution in its very title: 'Phone Sex.'

Yet, it would be unfair to draw a picture opposing calculated efforts at showing resilience through solidarity to actual narcissism and solitude; or to stigmatise the musicians' apparent disinterest in other people's death and suffering as somehow characteristic of their group. Rather, by remaining on the internet over time, their reactions during these fateful weeks now offer to the researcher a window on how the 2020 pandemic temporarily redefined for all, at least in parts of Europe and America, the borders between private and public spaces; between the scene and the out-scene; between professional life and life *tout court*; and between being alone and being with others. Contrary to the image of music as the cohesive element of a faltering social bond, the situation arguably implied a partial destabilisation of prevalent ontologies of sound in normative spatial configurations, such as concert halls and recording studios. Moreover, together with favouring a temporary redistribution of the places in which music happens, the pandemic inflected the ways in which sound itself bears witness to life. It is no wonder that many salient innovations about sound in lockdown were not any musician's feat, but the rediscovery of silence and natural sounds in urban areas where they were no longer part of the landscape (Muñoz et al. 2020; Manzano Pastor et al. 2021).

References

Ajello, Mario. 2020. 'Coronavirus, i balconi di casa nuovi social dell'Italia in trincea.' *Il Messaggero*, 15 March 2020. www.ilmessaggero.it/italia/coronavirus_news_balcone_musica_inno_d_italia_ter razzo_video-5111690.html.

Amerio, Andrea, Andrea Brambilla, Alessandro Morganti, Andrea Aguglia, Davide Bianchi, Francesca Santi, Luigi Costantini, Anna Odone, Alessandra Costanza, Carlo Signorelli, Gianluca Serafini, Mario Amore, and Stefano Capolongo. 2020. 'COVID-19 Lockdown: Housing Built Environment's Effects on Mental Health.' *International Journal of Environmental Research and Public Health* 17: 5973. https://doi.org/10.3390/ijerph17165973.

Anderson, Paul Allen. 2015. 'Neo-Muzak and the Business of Mood.' *Critical Inquiry* 41 (Summer): 811–840.

Beck, Nora M. 'Singing in the Garden: An Examination of Music in Trecento Painting and Boccaccio's "Decameron."' PhD diss., Columbia University, 1993. ProQuest (304066764).

Bischetti, Luca, Paolo Canal, and Valentina Bambini. 2020. 'Funny but Aversive: A Large-Scale Survey of the Emotional Response to Covid-19 Humor in the Italian Population during the Lockdown.' *Lingua* 249. https://doi.org/10.1016/j.lingua.2020.102963.

Boccaccio, Giovanni. (1353) 1903. *The Decameron*. Translated by James Rigg. London: Private Publisher.

Böhme, Gernot. 2017. *The Aesthetics of Atmospheres*. London: Routledge.

Buch, Esteban. 2013. 'Relire Ingarden: l'ontologie des œuvres musicales, entre fictions et montagnes.' In *Roman Ingarden: ontologie, esthétique, fiction*, edited by Christophe Potocki and Jean-Marie Schaeffer, 177–194. Paris: Editions des Archives Contemporaines.

Buch, Esteban. 2015. 'On the Evolution of Private Record Collections. A Short Story.' In *Musical Listening in the Age of Technological Reproduction*, edited by Gianmario Borio, 41–52. London: Routledge.

Buch, Esteban. 2020. 'Beethoven at 200 + 50: The Changing Meaning of Commemorations.' In *Ignition: Beethoven. Reception Documents from the Paul Sacher Foundation*, edited by Felix Meyer and Simon Obert, 143–156. Woodbridge: The Boydell Press.

Cabedo-Mas, Alberto, Cristina Arriaga-Sanz, and Lidon Moliner-Miravet. 2021. 'Uses and Perceptions of Music in Times of COVID-19: A Spanish Population Survey.' *Frontiers in Psychology* 11. https://doi.org/10.3389/fpsyg.2020.606180.

Chodos, Asher Tobin. 2019. 'What Does Music Mean to Spotify? An Essay on Musical Significance in the Era of Digital Curation.' *INSAM Journal of Contemporary Music, Art and Technology* 1 (2) (July): 36–64.

Ciabattoni, Francesco. 2019. 'Music in Trecento Italy and the Soundtrack of Boccaccio's Decameron.' *MLN* 134 Supplement (September): S-138–S-151.

Daffern, Helena, Kelly Balmer, and Jude Brereton. 2021. 'Singing Together, Yet Apart: The Experience of UK Choir Members and Facilitators during the Covid-19 Pandemic.' *Frontiers in Psychology* 12. https://doi.org/10.3389/fpsyg.2021.624474.

Datta, Anita. 2020. '"Virtual Choirs" and the Simulation of Live Performance Under Lockdown.' *Social Anthropology* 28 (2): 249–250. https://doi.org/10.1111/1469-8676.12862.

de Assis, Paulo, ed. 2018. *Virtual Works – Actual Things: Essays in Music Ontology.* Leuven: Leuven University Press.

Deutsche Welle. 2020. 'Mit Beethovens "Ode an die Freude" gegen die Coronakrise.' 23 March 2020. www.dw.com/de/mit-beethovens-ode-an-die-freude-gegen-die-coronakrise/a-52886153.

Dhami, Mandeep K., Leonardo Weiss-Cohen, and Peter Ayton. 2020. 'Are People Experiencing the 'Pains of Imprisonment' during the COVID-19 Lockdown?' *Frontiers in Psychology* 11. https://doi.org/10.3389/fpsyg.2020.578430.

DPA. 2020. 'Musik als Zeichen in der Krise. Vom Balkon aus: Beethovens "Ode an die Freude."' *Zeit*, 22 March 2020. www.zeit.de/news/2020-03/22/vom-balkon-aus-beethovens-ode-an-die-freude.

Fink, Lauren, Lindsay A. Warrenburg, Claire Howlin, Williamn Randall, Niels Chr. Hansen, and Melanie Wald-Fuhrmann. 2021. 'Viral Tunes: Changes in Musical Behaviours and Interest in Coronamusic Predict Socio-Emotional Coping during COVID-19 Lockdown.' *Humanities and Social Sciences Communications* 8. https://doi.org/10.1057/s41599-021-00858-y.

France Musique, 2020. 'Le Boléro de Ravel par l'Orchestre national de France.' 30 March 2020. www.youtube.com/watch?v=Sj4pE_bgRQI&t=2s.

Gammon, Sean and Gregory Ramshaw. 2021. 'Distancing from the Present: Nostalgia and Leisure in Lockdown.' *Leisure Sciences* 43 (1–2): 131–137 https://doi.org/10.1080/01490400.2020.1773993.

Goehr, Lydia. 1992. *The Imaginary Museum of Musical Works: An Essay in the Philosophy of Music.* Oxford: Clarendon Press.

Gourlay, Lesley. 2020. 'Quarantined, Sequestered, Closed: Theorising Academic Bodies under Covid-19 Lockdown.' *Postdigital Science and Education* 2: 791–811. https://doi.org/10.1007/s42438-020-00193-6.

Griffero, Tonino. 2017. *Quasi-Things. The Paradigm of Atmospheres.* Albany: SUNY Press.

Hansen, Niels Chr., John Melvin Treider, Dana Swarbrick, Joshua Bamford, Johanna Wilson, and Jonna Katariina Vuoskoski. 2021. 'A Crowd-Sourced Database of Coronamusic: Documenting Online Making and Sharing of Music during the COVID-19 Pandemic.' *Frontiers in Psychology* 12. https://doi.org/10.3389/fpsyg.2021.684083.

Haworth, Christopher. 2020. 'Network Music and Digital Utopianism: The Rise and Fall of the Res Rocket Surfer Project, 1994–2003.' In *Finding Democracy in Music*, edited by Robert Adlington and Esteban Buch, 144–163. London: Routledge.

Henry, Noah, Diana Kayser, and Hauke Egermann. 2021. 'Music in Mood Regulation and Coping Orientations in Response to COVID-19 Lockdown Measures within the United Kingdom.' *Frontiers in Psychology* 12. https://doi.org/10.3389/fpsyg.2021.647879.

Irons, Rebecca. 2020. 'Quarantime: Lockdown and the Global Disruption of Intimacies with Routine, Clock Time, and the Intensification of Time-Space Compression.' *Anthropology in Action* 27 (3): 87–92. https://doi.org/10.3167/aia.2020.270318.

Jarvis, Brooke. 2021. 'Did Covid Change How We Dream?' *New York Times*, 3 November 2021. www.nytimes.com/2021/11/03/magazine/pandemic-dreams.html.

Jonchery, Anne and Philippe Lombardo. 2020. 'Pratiques culturelles en temps de confinement.' *Culture études* 6 (6): 1–44. https://doi.org/10.3917/cule.206.0001.

Kaganski, Serge. 2021. 'Voyage autour de sa chambre – sur *45 tours de confinement* de Bertrand Loutte.' *AOC*, 10 February 2021. https://aoc.media/critique/2021/02/09/voyage-autour-de-sa-chambre-sur-45-tours-de-confinement-de-bertrand-loutte/.

Kania, Andrew. 2013. 'Platonism vs. Nominalism in Contemporary Musical Ontology.' In *Art and Abstract Objects*, edited by Christy Mag Uidhir, 197–212. Oxford: Oxford University Press.

236 *Esteban Buch*

Knobloch, Sylvia and Dolf Zillmann. 2002. 'Mood Management via the Digital Jukebox.' *Journal of Communication* 52: 351–366. https://doi.org/10.1111/j.1460-2466.2002.tb02549.x.

Krause, Amanda, James Dimmock, Amanda Rebar, and Ben Jackson. 2021. 'Music Listening Predicted Improved Life Satisfaction in University Students during Early Stages of the COVID-19 Pandemic.' *Frontiers in Psychology* 11. https://doi.org/10.3389/fpsyg.2020.631033.

LOC (Library of Congress). 2020. 'Boccaccio Project Participants Miya Masaoka and Kathryn Bates.' *The Boccaccio Project.* www.loc.gov/concerts/boccaccio-project/masaoka-bates.html.

Manzano Pastor, Jerónimo, José Antonio Almagro Vida, Rafael García Quesada, Francesco Aletta, Tin Oberman, Andrew Mitchell, and Jian Kang. 2021. 'The "Sound of Silence" in Granada during the COVID-19 Lockdown.' *Noise Mapping* 8 (1): 16–31. https://doi.org/10.1515/noise-2021-0002.

Marafioti, Martin. 2005. 'Post-Decameron Plague Treatises and the Boccaccian Innovation of Narrative Prophylaxis.' *Annali d'Italianistica* 23, *Literature & Science*: 69–87.

Mas-Herrero, Ernest M., Neomi Singer, Laura Ferreri, Michael McPhee, Robert Zatorre, and Pablo Ripolles. 2020. 'Rock 'n' Roll but Not Sex or Drugs: Music Is Negatively Correlated to Depressive Symptoms during the COVID-19 Pandemic via Reward-Related Mechanisms.' *PsyArXiv.* https://doi.org/10.31234/osf.io/x5upn.

Muñoz, Patricio, Bruno Vincent, Céline Domergue, Vincent Gissinger, Sébastien Guillot, Yann Halbwachs, and Valérie Janillon. 2020. 'Lockdown during COVID-19 Pandemic: Impact on Road Traffic Noise and on the Perception of Sound Environment in France.' *Noise Mapping* 7 (1): 287–302. https://doi.org/10.1515/noise-2020-0024.

Olivares Palma, Eduardo. 2020. 'Música: la increíble "gira estática" de Tomás Gubitsch.' *La Francolatina*, 4 June 2020. www.lafrancolatina.com/spip.php?article3366.

Pesonen, Anu-Katriina, Jari Lipsanen, Risto Halonen, Marko Elovainio, Nils Sandman, Juha-Matti Mäkelä, Minea Antila, Deni Béchard, Hanna M. Ollila, and Liisa Kuula. 2020. 'Pandemic Dreams: Network Analysis of Dream Content during the COVID-19 Lockdown.' *Frontiers in Psychology* 11. https://doi.org/10.3389/fpsyg.2020.573961.

Petrusich, Amanda. 2020. 'Music to Endure the Coronavirus Quarantine.' *New Yorker*, 20 March 2020. www.newyorker.com/culture/culture-desk/music-to-endure-the-coronavirus-quarantine.

Sarasso, Pietro, Irene Ronga, Marco Neppi-Modona, and Katiuscia Sacco. 2021. 'The Role of Musical Aesthetic Emotions in Social Adaptation to the Covid-19 Pandemic.' *Frontiers in Psychology* 12. https//doi.org/10.3389/fpsyg.2021.611639.

Scorza Barcellona, Gaia. 2020. 'Coronavirus, l'Italia sul balcone: canzoni contro la paura.' *La Repubblica*, March 13, 2020. www.repubblica.it/cronaca/2020/03/13/news/coronavirus_italia_al_balcone_canzoni_contro_la_paura-251221289/.

Sim, Jaeung, Daegon Cho, Youngdeok Hwang, and Rahul Telang. 2020. 'Virus Shook the Streaming Star: Estimating the COVID-19 Impact on Music Consumption.' *SSRN.* http://dx.doi.org/10.2139/ssrn.3649085.

Spicer, André. 2020. 'The Decameron – The 14th-Century Italian Book That Shows Us How to Survive Coronavirus.' *Newstatesman*, 8 March 2020. www.newstatesman.com/2020/03/coronavirus-survive-italy-wellbeing-stories-decameron.

Testoni, Ines, Emil Rossi, Sara Pompele, Iliaria Malaguti, and Hod Orkibi. 2021. 'Catharsis through Cinema: An Italian Qualitative Study on Watching Tragedies to Mitigate the Fear of COVID-19.' *Frontiers in Psychiatry* 12. https://doi.org/10.3389/fpsyt.2021.622174.

WHO (World Health Organization). 2022. *WHO Coronavirus (COVID-19) Dashboard.* https://covid19.who.int.

Wills, Matthew. 2020. 'Boccaccio's Medicine.' *JSTOR Daily*, 18 March 2020. https://daily.jstor.org/boccaccios-medicine/.

Yu-Cheong Yeung, Timothy. 2020. 'Did the COVID-19 Pandemic Trigger Nostalgia? Evidence of Music Consumption on Spotify.' *Centre for Economic Policy Research Press* 44. https://cepr.org/sites/default/files/news/MusicConsumption.pdf.

18 What a Blackbird Has Told Me

Latent Acoustic Learning in the Times of COVID-19

Theodoros Lotis

In an era of extensive urban noise pollution overloaded with meaningless saturated acoustic information, the sudden and dramatic drop of long-standing anthropogenic noises during the COVID-19 lockdown allowed for soft sounds which were previously masked, such as the songs of birds, to emerge on the acoustic surface. This chapter argues that the ubiquitous lockdown not only changed the perceived sonic environment by revealing hidden soundscapes, but also redefined our qualitative hearing; it unveiled a forgotten latent acoustic memory, inactive for a long period due to the prevalence of lo-fi soundscapes and the salami sound effect, or *flat line* effect (Schafer 1994, 78), in urban areas. The reminiscences of transparent soundscapes emerged once again, refreshing our memory, unfolding the necessity of acoustic transparency in our environment, and sensitising our ears and brain to vulnerable sounds. Even after decades of living in noisy environments, the latent memory of transparent soundscapes persists and can be revealed by the single song of a bird. The chapter also suggests that constant noise and the salami sound effect may interfere with our latent learning processes, as well as with our ability to create auditory cognitive maps of the environment, and with the motivation to enjoy and explore the full range of soundscapes.

How Silent Urban Soundscapes Can Become

Over the course of the lockdown due to the COVID-19 pandemic, we have all experienced minor and major changes in our sonic environments. In many urban areas, noise levels have dropped significantly due to the reduction of anthropogenic noises. Although further research is needed, some early data have already revealed that noise levels in urban areas were dramatically lower during the lockdown. An indicative example from research on Granada's soundscapes shows that 'during the COVID-19 pandemic the situation completely changed because of the absence of human activity, giving place to a great decrease in noise levels … The considerable drop in SPLs during the lockdown in 2020 evidences natural non-anthropic sounds, mainly birds' (Manzano et al. 2021, 27–28); moreover, 'in terms of recorded sound pressure levels, the lack of human activities between the 2019 (pre-lockdown) and 2020 (during lockdown) measurements campaigns at four typical touristic destinations in Granada accounted for reductions ranging between 13.3 and 30.5 dB(A)' (ibid., 30). Additionally, early studies in several European cities indicated a notable decrease in urban noise due mainly to the reduction of commuting and traffic noise (Aletta et al. 2020; Asensi, Pavòn, and de Arcas 2020; Basu et al. 2021; Carra 2020; Parker and Spennemann 2020; Rumpler, Venkataraman, and Göransson 2020; Smith et al. 2020; Spennemann and Parker 2020).

Another article published in *Science* argued that the noise levels across the San Francisco Bay area in the United States during lockdown were similar to those of the 1950s. As a

DOI: 10.4324/9781003200369-22

238 *Theodoros Lotis*

consequence, people became 'newly aware of more conspicuous animal sounds, such as bird songs, particularly in normally noisy areas': they could 'hear birds at twice the previous distance, or effectively four times more birds than usual,' since 'restricted human movement reduced use of motorized vehicles, effectively unmasking bird songs otherwise obscured by associated noise pollution' (Derryberry et al. 2020, 575–576). As a result, socially distanced people confined to their homes, balconies, and courtyards in previously noisy areas were able to hear the bird songs and other susceptible sounds in their close surroundings. Many realised that the environment was quieter than it used to be and that they could hear more subtle sounds. Birds and a variety of insects and smaller species reclaimed their acoustic space of favoured frequencies,[1] which was previously masked by overlapping noise.

Blackbirds on My Roof

During the first lockdown in Greece (March–May 2020) I realised that a family of blackbirds had built their nest on my rooftop. I could clearly hear their chirping, the sounds of their wings, and later, the calls of the nestlings in the nest. During the same period, I decided to record the sonic environment from my balcony for one minute and one second every evening.[2] Throughout these recording sessions I realised that I was perceiving a hi-fi version of what had become by then a lo-fi sonic environment. However, listening to the chirping of a bird does not imply the perception of its whole sonic activity (small noises in the nest, subtle noises of its movements, wing flapping, communication with other birds at a distance, etc.). The quietness imposed by the restrictions on movement manifested the fact that our prior-to-lockdown daily acoustic experience included only a part of the whole soundscape. A variety of isolated, fragmented sounds (known as sound signals) were often cut off from their context due to the omnipresent anthropogenic noises. Sound signals are not, however, the only collateral damages in a lo-fi sonic environment. The motivations, the stimuli, and the associations that emerge from all sounds and nourish our learning processes are also buried beneath the noise floor. In front of our ears, the lockdown unfolded the establishment of our daily global sonic regime.

Basic Terminology

Before proceeding further, I shall identify some key terms as described by Raymond Murray Schafer in his book *The Soundscape: Our Sonic Environment and the Tuning of the World.* Schafer identifies soundscape with the sonic environment: 'technically, any portion of the sonic environment regarded as a field for study. The term may refer to actual environments, or to abstract constructions such as musical compositions and tape montages, particularly when considered as an environment' (1994, 274–275). Soundscapes are dynamic environments. They constantly change according to variations in seasonal weather conditions, human and animal behaviour, social, cultural and economic activities, traffic congestion, and any alteration in the behaviour or the state of the biophony, geophony and anthropophony (Brooks et al. 2014; see also Truax 1999).

Depending on their spectral characteristics and the density of sonic events, soundscapes can be characterised as hi-fi or lo-fi. Schafer defines a hi-fi sonic environment as 'one in which sounds may be heard clearly without crowding or masking' (1994, 272). On the opposite side, lo-fi environments are identified by unfavourable signal-to-noise ratio: 'Applied to soundscape studies a lo-fi environment is one in which signals are overcrowded, resulting in masking or lack of clarity' (ibid.). In sonic environments, *keynote* sounds constitute a relatively constant base line from which other sounds emerge. According to Schafer, 'keynote sounds are those which are heard by a particular society continuously or frequently enough to form a background against which other sounds are perceived. ... Often keynote sounds are not consciously perceived, but they act as conditioning agents in the perception of other sound signals' (ibid., 272). Keynote sounds 'become listening habits in spite of themselves. ... they help to outline the character of men living among them' (ibid., 9). *Sound signals* are pronounced sounds usually associated with specific connotations. Schafer describes a sound signal as 'any sound to which the attention is particularly directed. In soundscape studies sound signals are contrasted by keynote sounds, in much the same way as figure and ground are contrasted in visual perception' (ibid., 275).

Noise Pollution

Today, a large number of keynote sounds have changed significantly compared to those of past centuries. Urban environments are overpopulated by mechanical and techno-logical sounds to such an extent that 'it is no longer possible to know what, if anything, is to be listened to' (Schafer 1994, 71). A large part of the world's sonic environments have moved into an all-time lo-fi state with ubiquitous and 'imperialistic' properties (ibid., 77). Technologically and mechanically generated noise is continuously rising while the perception of subtle sounds and signals is generally declining. While producing low noise-emission machinery should mitigate this increase in noise levels, the extent of this mitigation seems to be far from reassuring, at least in Europe (Suter 1991, 11; see also Clark and Paunovic 2018; Dreger et al. 2019; EEA 2020; WHO 2018).

Continuous Noise and the Salami Sound

Noise pollution occurs as a consequence of society's urge for increased mobility, product-ivity, and product preservation and storage. Backup and portable generators, industrial and domestic refrigerators, air conditioners and fans, motors, and compressors 'create low-information, high-redundancy sounds' (Schafer 1994, 78), adding not only to the opacity of urban sonic spectra – mostly at their lower and middle parts – but also to the general noise levels within the soundscape. Constant engine noise, long-lasting drones with little or no variation in time, and continuous noises of equivalent energy intrude into the soundscape as the new 24/7 keynote sounds in urban areas. I already referred to these keynote sounds as *salami sounds*. The continuous, prolonged, and unchanging sound energy is their founda-tional and 'predominating feature' (ibid.). Such constant and broadband spectra reduce the acoustic horizon of the environment, the perception of distant sounds, and consequently, the overall sense of spatial distribution (Truax 2001, 26). As an example, an 'average backup generator measures between 60 and 70 decibels' while a 'portable generator has a noise level between 70 and 100 decibels when heard from the industry standard of seven meters.'[3] If we consider that most roads in cities accommodate several shops with power generators, air conditioners, refrigerators, and fans, and that whenever two broadband noise waveforms of the same intensity are added together they result in an increase of approximately three decibels (Hansen 2001, 35), we can safely assume that tackling the phenomenon of the salami sound is both demanding and challenging.

Although there is sufficient scientific evidence that noise exposure can impair public health and the environmental quality of life (Basner et al. 2014; Kerns et al. 2018; Münzel et al. 2020; Passchier-Vermeer and Passchier 2000), central policy makers are still reluc-tant to adopt effective action plans for the reduction of noise in public areas. This becomes clear when one delves into the labyrinth of conflicting European directives and regulations.

240   *Theodoros Lotis*

Despite the stated aim of the EU's Environmental Noise Directive 2002/49/EC (consolidated in 2021) 'to define a common approach intended to avoid, prevent or reduce on a prioritised basis the harmful effects, including annoyance, due to exposure to environmental noise' (EU 2002, Art. 1.1), parts of the legislation are vague or poorly defined. For example, Directive 2002/49/EC

> carefully avoided the setting of any binding noise limit value or non-binding target value, the exceedance of which would require the elaboration of an action plan. It defined the term 'limit value' but hastened to add that this value was to be determined by Member States: 'the exceeding of which causes competent authorities to consider or enforce mitigation measures.'
>
> (ERA n.d.)

The sound power level requirements of air conditioners and power generators outdoors are established respectively in Regulation (EU) 206/2012 and Directive 2000/14/EC. In Regulation 206/2012 the requirements for maximum sound power levels of air conditioners and comfort fans are set to 65–70 dB(A) (EC 2012).[4] Manufacturers of air conditioners and comfort fans are urged to label the noise levels of their products without, however, being subject to efficient limitations. Moreover, a review of Regulation 626/2011 concerning the air conditioners and comfort fans published in 2018 indicates another unspecified issue in the EU's legislation regarding adequate noise prevention and mitigation strategies: 'The prescription about noise is not that clear: "The noise level of fans and regulators at all speeds shall be within reasonable limits"' (Huang et al. 2018). The prescription of the 'reasonable' limits was left unspecified.[5]

The salami sound introduces a new soundscape context within which humans, animals, and birds are called to adapt their listening strategies. Low-information background sounds create homogeneous sonic environments with poor acoustic definition. However, listening is the key issue in sonic perception because 'it is the primary interface between the individual and the environment. It is a path of information exchange, not just the auditory reaction to stimuli' (Truax 2001, xviii). Low-information high-redundancy soundscapes obstruct the interactivity between the individual and the environment because 'such environments do not encourage more active types of listening, and their prevalence may prevent listeners from experiencing any alternative' (ibid., 27). The perception of acoustic information by the listener is highly dependent on the soundscape's context, which is 'essential for understanding the meaning of any message' (ibid., 11). Whenever the salami sound effect is prevalent in the soundscape, the listener's perception slips away on a doubtful and non-contextual experience.

Let us examine the following example from the one-minute-and-one-second recordings during the lockdown. The spectrogram in Figure 18.1 illustrates the spectral characteristics of the captured sounds. The microphones were placed at a distance of about 60 metres from the sources.

The salami sound, which was emitted by the power generator of a nearby corner shop, occupies the low and middle areas of the spectrum (part of it constitutes the recording noise). Its spectral energy is continuously distributed over a fairly large range of frequencies up to 1,700 Hz. On the upper parts of the spectrum we can hear chirping birds far above the salami sound. The chirping is perceived clearly by the listener. At about 850 Hz, within the salami sound, resides an almost imperceptible sound, which is possibly the chirping of another bird. This sound is completely masked by the noise of the power generator, which obscures the auditory image and 'lessens the clarity or definition of the acoustic information' (Truax 2001, 96). As a consequence, the acoustic definition is blurred and the listener

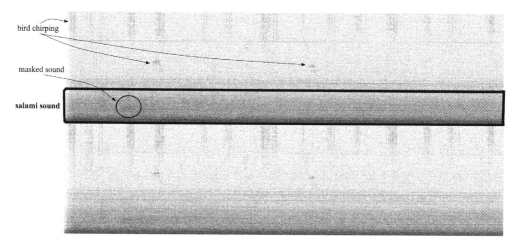

Figure 18.1 Spectrogram of birds chirping and the salami sound.

is no longer able to hear all the sounds clearly and to perceive their distinctive and varied acoustic features. The constant noise interferes both with the communication of the birds (Brumm and Zollinger 2013; Slabbekoorn and Boer-Visser 2006) and the perception of the listener. This fact leads to the following question: what is the relationship between humans and the sounds of the environment, and what happens when those sounds are obstructed, masked, or interrupted?

Latent Learning and Cognitive Maps

Latent learning – from Greek λήθη (lēthē) and Latin *latentem* (concealed, secret, unknown) – is the subconscious observation of the environment with no particular motivation to obtain any information from it (Tavris and Wade 2000; Hothersall 2004). Early experiments with rats (Tolman and Honzik 1930; Tolman, Ritchie, and Kalish 1946; Tolman 1948, 1949) demonstrated that latent learning is a form of learning that stores environmental information on a daily basis. This information emerges and affects later behaviours when the right incentives are presented. Baldwin Hergenhahn and Matthew Olson define this phenomenon as 'learning that is not translated into performance. In other words, it is possible for learning to remain dormant for a considerable length of time before it is manifested in behaviour' (2016, 298). Although reinforcement, motivation, and associations with the stimuli are usually absent or irrelevant, latent learning constantly provides us with invaluable knowledge of our surrounding environment, explaining – at least to some extent – how we perceive and store environmental information. Moreover, both humans and animals seem to share the same ability to exploit this knowledge at a later date when adequate motivation appears. As scientists suggest (Nelson 1978; Medina, Garcia, and Mauk 2001; MacLeod 1988), even after a memory is apparently forgotten, its latent residue can still be present because 'a latent (residual) memory persists and can be revealed by facilitated acquisition in a subsequent learning task' (Philips et al. 2006, 224).

In his paper *Cognitive Maps in Rats and Men*, Edward Tolman introduced the concept of cognitive map as a 'tentative map, indicating routes and paths and environmental relationships, which finally determines what responses, if any, the animal will finally release' (1948, 192). Although cognitive maps are usually related to the representation of a space in

242 *Theodoros Lotis*

the brain, a cognitive map can refer to 'any visual representation of a person's (or a group's) mental model for a given space, process or concept' (Gibbons 2019). In general, cognitive maps are mapping/memory systems responsible for context-dependent memories across the hippocampal area of the brain. Cognitive maps contain information about the spatial environment. For example, if we want to exit from a cinema we remember where the exit door is in relation to our position. The knowledge of the position of the door is acquired prior to our exit without the presence of any motivation. In terms of operational processes, 'for a cognitive map to be useful, the organism must have a mechanism for connecting map coordinates to fixed aspects of the environment that can be identified by perceptual systems' (Epstein et al. 2017, 1508). Cognitive maps provide us with a framework for identifying our position in relation to other objects or stimuli in our surroundings.

Auditory Cognitive Maps: A Hypothesis

Every day, various acoustic stimuli knock on the doors of our perception carrying information about the spatial sonic environment. As with objects in a landscape, the acoustic stimuli of a soundscape enrich our latent learning process by creating what I might call *auditory cognitive maps*. I shall compare the process of latent learning and the auditory cognitive maps to Truax's idea of *background listening*. According to Truax, background listening is a common experience during which

> the sound usually remains in the background of our attention. It occurs when we are not listening for a particular sound, and when its occurrence has no special or immediate significance to us. However, we are still aware of the sound, in the sense that if asked whether we had heard it, we could probably respond affirmatively.
>
> (2001, 24)

We can assume that background listening is an action with no motivation or reinforcement, an activity that just happens as a part of our daily latent learning process.

During the latent learning processes, one hears a variety of sounds with no purpose or any specific goal in mind. For example, one hears the bird songs without any incentives unless one is an ornithologist. However, there might be a neutral satisfaction in this activity; a sense of exploration and curiosity through which auditory cognitive maps are developed as representatives of the sonic environment. A great number of subtle sounds, such as the wing flapping of birds, the buzz of insects or the distant song of a bird are either completely masked or obstructed by the salami sound effect. As a result, background listening is partially or completely blocked by the presence of constant noise. Latent learning thus becomes the first unwilling victim under the dominance of noise.

My hypothesis is that the obstruction of the latent learning process by the salami sound effect may interfere with our ability to create auditory cognitive maps of the environment. As a consequence, the capacity to develop an accurate sense of location, direction, and distance within our environment, as well as the motivation to enjoy and explore the full range of a soundscape, may also be affected. Constant noise is a condition which favours narrow-strip auditory cognitive maps where the representation of sonic events in spatial-temporal context becomes blurred and deficient in clarity and in coherence.[6] Furthermore, constant noise distracts our sense of exploration and curiosity since it impedes the latent learning processes by blocking out the acoustic stimuli. As a result, it interferes with our long-term skills of holistic soundscape evaluation and global environmental understanding.

Due to the decrease of anthropogenic sounds during the COVID-19 lockdown, many of the previously masked sounds emerged once again from obscurity. Traffic noise and mechanical sounds shrank away so that more bird songs and insect sounds could be heard. Soundscape analyses from urban areas in Europe showed that 'eventfulness, acoustic complexity, and acoustic richness increased significantly over the time period, while the amount of technological sounds decreased' and that 'an increased presence of birdsong emerge as a novel characteristic element of the local urban soundscape' (Lenzi, Sádaba, and Lindborg 2021). Moreover, long forgotten motivations and associations related to acoustic stimuli emerged once again during the quiet periods of the lockdown.

However, even during the most silent hours of the lockdown, power generators, compressors, and air conditioners operated in the open air on a 24/7 basis, if only to remind us that electrical energy and cooling and refrigeration comforts should be available at any time and at any cost. While hearing and recording the evening soundscapes from my balcony in the midst of the first lockdown period in 2020, it became obvious to me that the salami sound emitted by the power generator of the corner shop obstructed a large part of the soundscape. Sounds in the low and middle parts of the spectrum were completely or partially masked (see Figure 18.1). Consequently, my understanding of the environment was impaired and biased because of the constant noise and the salami sound effect. Noise pollution and the regular exposure to elevated sound levels and constant noise can have deleterious effects on human-environmental interactions. Along with the climate crisis and environmental destruction, constant noise and the salami sound effect present us with a new kind of extinction: the disappearance of the perception of what exists but cannot be perceived.

What the Blackbird Has Told Me. Or, How Am I Related to the Blackbird?

Living with constant noise is a process of adaptation. The concealment of acoustic information establishes a new status quo between humans and the sonic environment. Adaptive people are continually revising their own auditory cognitive maps to match the loss of information that they encounter in their everyday listening experiences. Such a deprived acoustic environment could conceivably change their view of the world and the way they interact with it and feel about it (Epting and Paris 2006). Constant noise and the salami sound effect re-evaluate the relationship between societies and the sonic environment. Although noises of the technosphere were often invasively present, the silent periods of the lockdown changed the perceived sonic environment, redefined our qualitative hearing, and unveiled a forgotten latent acoustic memory that had been inactive for a long time due to the prevalence of lo-fi soundscapes in urban areas: the memory of transparent soundscapes within which the operating process of latent learning remained unhindered.

Once again, people were able to explore 'the sound in an environment in its complexity, ambivalence, meaning, and context' (Brooks et al. 2014, 30). The memory of transparent soundscapes has emerged refreshing our latent memory, unfolding the necessity of acoustic transparency and sensitising our ears and brain to vulnerable sounds. The subtle sounds of the blackbird family, the song of distant birds, the sound of wind and leaves and all the subtle sounds that suffer under the regime of constant noise emerged afresh into more transparent soundscapes. Silence and immobility during lockdown allowed us to reaffirm our relation with the sonic environment through 'the function of memory and knowledge (mnemonic-cognitive), which concerns the ability to investigate the significance of mnemonic phenomena and constant learning' and through 'the physiology of memory (mnemonic-physiology), which concerns the acquisition, stabilisation, and recall of memories, as well

244   *Theodoros Lotis*

as the identification and correlation of the essential components of memory' (Triarchou 2015, 127).

How Do I Reply to the Blackbird? Conclusion

Both the sonic environment and our capacity to interact with it are greatly affected by our economic and social practices, and by our impassioned devotion to machines and the processes of producing, consuming, preserving. Noise emission (an offspring of this devotion) is 'the ignored pollutant' (King and Murphy 2016, 214) for it represents a socially affordable price for the well-being of our societies. As Schafer puts it:

> When a society fumbles with sound, when it does not comprehend the principles of decorum and balance in soundmaking, when it does not understand that there is a time to produce and a time to shut up, the soundscape slips from hi-fi to lo-fi condition and ultimately consumes itself in cacophony.
>
> (1994, 237)

However, the COVID-19 pandemic and its restrictions on movement revealed that human interaction with nature is also an essential element for human and social well-being. For such interaction to exist, latent learning operations should continue to run smoothly and unaffected by the intrusion of constant noise and the salami sound effect. Indeed, as a WHO working group declared back in 1971: 'Noise must be recognized as a major threat to human well-being' (Suter 1991, 1; see also Suess 1973).

Acoustic sustainability, or 'our ability as a culture to live within a positively functioning soundscape that has long-term viability' (Truax 2016, 3), is of vital importance in maintaining human-environment interaction through latent learning processes. The hypothesis that latent learning may be affected by the constant noise and the salami sound effect needs further development and validation by an interdisciplinary approach that involves ecologists and psychologists, architects and sound designers, businesspeople and policy makers, and the general public. According to Suter, 'this approach could very well provide the key to understanding a great deal more about the general impact of noise on society' (1991, 36). It is also important to define areas of future research that will build a platform for better understanding the interconnections between constant noise, latent learning, and latent memory.

At the moment, little research has been done on the subject. Proof of the direct cause-and-effect relationship between the constant noise and latent learning are insufficient, if not absent. Meanwhile, palliative measures such as the EU's Environmental Noise Directive and the Member States' legislation can only scratch the surface of the problem. Although the official European and national policies on noise mitigation remain at best only partially successful, several action plans can be adopted by individuals and communities, including the reconsideration of daily social practices in order to re-establish the hereditary right of all species to listen to and to interact with their sonic environment, 'to create "space" for the sounds that are potentially present to emerge and be heard' (Truax 1999, 5), and to allow latent learning to create auditory cognitive maps that relate us with these sounds. However, these actions require a complete redefinition of the notion of environmental normality. As Slavoj Zizek stressed in an article for *RT.com* published on December 2020,

> instead of dreaming about a 'return to (old) normality' we should engage in a difficult and painful process of constructing a new normality. This construction is not a medical

or economic problem, it is a profoundly political one: we are compelled to invent a new form of our entire social life.

And that includes noise as well.

Acknowledgments

I thank Konstantina Diamantidou, PhD for useful discussions and comments on latent learning.

Notes

1 Birds adjust the signal frequency of their songs when they live in a habitat with a persistent and consistent noise source: 'Another way to improve the signal-to-noise ratio of a signal is to shift the frequency content of the signal away from frequency bands containing the noise (or at least from the band with the highest intensity noise)' (Brumm and Zollinger 2013, 218). During experiments with playback of white noise and city-like noise different birds have been reported to 'sing with higher minimum frequency in response to temporary playback of traffic-like noise, returning to lower minimum frequency songs after cessation of the noise playback' (ibid.). See also, Berger-Tal et al. 2019; Roca et al. 2016.
2 These recordings are available at: www.theodoroslotis.com/quarantine-soundscapes.
3 The values are indicative. This information is published by AlltimePower Company on their website.
4 Sound power level means the A-Weighted sound power level – dB(A) – indoors and/or outdoors measured at standard rating conditions for cooling or heating. A-Weighting is the most common weighting that is used in noise measurement. However, A-Weighted sound level discriminates against low frequencies by cutting off the lower parts of the spectrum.
5 For a summary of the EU's environmental legislation, see King and Murphy 2016.
6 For a detailed analysis of cognitive maps and the representation of items and events in spatial-temporal context see O'Keefe and Nadel 1978 and Eichenbaum 2015.

References

Aletta, Francesco, Tin Oberman, Andrew Mitchell, Huan Tong, and Jian Kang. 2020. 'Assessing the Changing Urban Sound Environment during the COVID-19 Lockdown Period Using Short-term Acoustic Measurements.' *Noise Mapping* 7 (1): 123–134. https://doi.org/10.1515/noise-2020-0011.

Asensio, Cesar, Ignacio Pavón, and Guillermo de Arcas. 2020. 'Changes in Noise Levels in the City of Madrid during COVID-19 Lockdown in 2020.' *The Journal of the Acoustical Society of America* 148: 1748–1755. https://doi.org/10.1121/10.0002008.

Basner, Mathias, Wolfgang Babisch, Adrian Davis, Mark Brink, Charlotte Clark, Sabine Janssen, and Stephen Stansfeld. 2014. 'Auditory and Non-Auditory Effects of Noise on Health.' *The Lancet* 383 (9925): 1325–1332. https://doi.org/10.1016/S0140-6736(13)61613-X.

Basu, Bidroha, Enda Murphy, Anna Molter, Arunima S. Basu, Srikanta Sannigrahi, Miguel Belmonte, and Francesco Pilla. 2021. 'Investigating Changes in Noise Pollution Due to the COVID-19 Lockdown: The Case of Dublin, Ireland.' *Sustainable Cities and Society* 65 (February). https://doi.org/10.1016/j.scs.2020.102597.

Berger-Tal, Oded, Bob B. M. Wong, Ulrika Candolin, and Jesse Barber. 2019. 'What Evidence Exists on the Effects of Anthropogenic Noise on Acoustic Communication in Animals? A Systematic Map Protocol.' *Environmental Evidence* 8 (suppl. 1). https://doi.org/10.1186/s13750-019-0165-3.

Brooks, Bennett M., Brigitte Schulte-Fortkamp, Kay S. Voigt, and Alex U. Case. 2014. 'Exploring Our Sonic Environment through Soundscape Research and Theory.' *Acoustics Today* 10 (1) (Winter): 30–40.

Brumm, Henrik and Sue Anne Zollinger. 2013. 'Avian Vocal Production in Noise.' In *Animal Communication and Noise*, edited by Henrik Brumm, 187–227. Animal Signals and Communication 2. Berlin: Springer. https://doi.org/10.1007/978-3-642-41494-7_7.

246 *Theodoros Lotis*

Carra, Sébastien, Vincent Gissinger, Sébastien Guillot, Yann Halbwachs, Valérie Janillon, and Bruno Vincent. 2020. 'Confinement COVID-19: impact sur l'environnement sonore.' *Acoucité. Observatoire de l'environnement sonore*, 11 May 2020. www.acoucite.org/confinement-covid-19-imp act-sur-lenvironnement-sonore/.

Clark, Charlotte and Katarina Paunovic. 2018. 'WHO Environmental Noise Guidelines for the European Region: A Systematic Review on Environmental Noise and Cognition.' *International Journal of Environmental Research and Public Health* 15 (2). https://doi.org/10.3390/ijerph1 5020285.

Derryberry, Elizabeth P., Jennifer N. Phillips, Graham E. Derryberry, Michael J. Blum, and David Luther. 2020. 'Singing in a Silent Spring: Birds Respond to a Half-Century Soundscape Reversion during the COVID-19 Shutdown.' *Science* 370, iss. 6516 (October): 575–579. https://doi.org/ 10.1126/science.abd5777.

Dreger, Stefanie, Steffen Andreas Schüle, Lisa Karla Hilz, and Gabriele Bolte. 2019. 'Social Inequalities in Environmental Noise Exposure: A Review of Evidence in the WHO European Region.' *International Journal of Environmental Research and Public Health* 16 (6). https://doi.org/ 10.3390/ijerph16061011.

EEA (European Environment Agency). 2020. 'Number of Europeans Exposed to Harmful Noise Pollution Expected to Increase.' 5 March 2020. www.eea.europa.eu/highlights/number-of-europe ans-exposed-to.

Eichenbaum, Howard. 2015. 'The Hippocampus as a Cognitive Map ... of Social Space.' *Neuron* 87 (1) (July): 9–11. https://doi.org/10.1016/j.neuron.2015.06.013.

Epstein, Russell, A., Eva Zita Patai, Joshua B. Julian, and Hugo J. Spiers. 2017. 'The Cognitive Map in Humans: Spatial Navigation and Beyond.' *Nature Neuroscience* 20 (11): 1504–1513. https://doi. org/10.1038/nn.4656.

Epting, Franz and Mark E. Paris. 2006. 'A Constructive Understanding of the Person: George Kelly and Humanistic Psychology.' *The Humanistic Psychologist* 34 (1): 21–37. https://doi.org/10.1207/ s15473333thp3401_4.

EC (European Commission). 2012. Commission Regulation No. 206/2012 of 6 March 2012 Implementing Directive 2009/125/EC of the European Parliament and of the Council with Regard to Ecodesign Requirements for Air Conditioners and Comfort Fans (OJ L 72, 10.3.2012, pp. 7– 27). https://eur-lex.europa.eu/legal-content/EN/TXT/PDF/?uri=CELEX:02012R0206-0170109& from=EN.

ERA (Academy of European Law). n.d. 'Air Quality and Noise Legislation Module 3: Environmental Noise Directive (END) Directive 2002/49/EC.' https://ec.europa.eu/environment/legal/law/5/e_l earning/module_3_6.htm.

EU (European Parliament and Council). 2002. Directive 2002/49/EC of 25 June 2002 Relating to the Assessment and Management of Environmental Noise (OJ L 189, 18.7.2002). https://eur-lex.eur opa.eu/eli/dir/2002/49/oj.

Gibbons, Sarah. 2019. 'Cognitive Maps, Mind Maps, and Concept Maps: Definitions.' *Nielsen Norman Group*, 14 July 2019. www.nngroup.com/articles/cognitive-mind-concept/.

Hansen, Colin H. 2001. 'Fundamentals of Acoustics.' In *Occupational Exposure to Noise: Evaluation, Prevention and Control*, edited by Berenice Goelzer, Colin H. Hansen, and Gustav A. Sehrndt, 23–52. Geneva: World Health Organization.

Hergenhahn, Baldwin R. and Matthew Olson. 2016. *An Introduction to Theories of Learning*, ninth edition. New York: Routledge.

Hothersall, David. 2004. *History of Psychology*, fourth edition. New York: McGraw-Hill Education.

Huang, Baijia, Jan Viegand, Peter Martin Skov Hansen, Philippe Riviere, and Florian Dittmann. 2018. 'Review of Regulation 206/2012 and 626/2011. Air Conditioners and Comfort Fans.' www. applia-europe.eu/images/Library/Review_Study_on_Airco_05-2018.pdf.

Kerns, Ellen, Elizabeth A. Masterson, Christa L. Themann, and Geoffrey M. Calvert. 2018. 'Cardiovascular Conditions, Hearing Difficulty, and Occupational Noise Exposure Within US Industries and Occupations.' *American Journal of Industrial Medicine* 61 (6): 477–491. https://doi. org/10.1002/ajim.22833.

King, Eoin and Enda Murphy. 2016. 'Environmental Noise – "Forgotten" or "Ignored" Pollutant?' *Applied Acoustics* 112 (November): 211–215. https://doi.org/10.1016/j.apacoust.2016.05.023.

Lenzi, Sara, Juan Sádaba, and PerMagnus Lindborg. 2021. 'Soundscape in Times of Change: Case Study of a City Neighbourhood during the COVID-19 Lockdown.' *Frontiers in Psychology* 24. https://doi.org/10.3389/fpsyg.2021.570741.

MacLeod, Colin. 1988. 'Forgotten but Not Gone: Savings for Pictures and Words in Long-term Memory.' *Journal of Experimental Psychology: Learning, Memory, and Cognition* 14 (2): 195–212. https://doi.org/10.1037//0278-7393.14.2.195.

Manzano, Jeronimo V., Jose A. A. Pastor, Rafael G. Quesada, Francesco Aletta, Tin Oberman, Andrew Mitchell, and Jian Kang. 2021. 'The "Sound of Silence" in Granada during the COVID-19 Lockdown.' *Noise Mapping* 8: 16–31. https://doi.org/10.1515/noise-2021-0002.

Medina, Javier F., Keith S. Garcia, and Michael D. Mauk. 2001. 'A Mechanism for Savings in the Cerebellum.' *Journal of Neuroscience* 21(11): 4081–4089. https://doi.org/10.1523/JNEUROSCI.21-11-04081.2001.

Münzel, Thomas, Swenza Kröller-Schön, Matthias Oelze, Tommaso Gori, Frank P. Schmidt, Sebastian Steven, Omar Hahad, Matin Röösli, Jean-Marc Wunderli, Andreas Daiber, and Mette Sørensen. 2020. 'Adverse Cardiovascular Effects of Traffic Noise with a Focus on Nighttime Noise and the New WHO Noise Guidelines.' *Annual Review of Public Health* 41: 309–328. https://doi.org/10.1146/annurev-publhealth-081519-062400.

Nelson, Thomas O. 1978. 'Detecting Small Amounts of Information in Memory: Savings for Nonrecognized Items.' *Journal of Experimental Psychology: Human Learning and Memory* 4 (5): 453–468. https://doi.org/10.1037/0278-7393.4.5.453.

O'Keefe, John and Lynn Nadel. 1978. *The Hippocampus as a Cognitive Map*. Oxford: Oxford University Press.

Parker, Murray and Dirk H. Spennemann. 2020. 'Anthropause on Audio: The Effects of the COVID-19 Pandemic on Church Bell Ringing and Associated Soundscapes in New South Wales (Australia).' *The Journal of the Acoustical Society of America* 148 (5). https://doi.org/10.1121/10.0002451.

Passchier-Vermeer, Willy and Wim F. Passchier. 2000. 'Noise Exposure and Public Health.' *Environmental Health Perspectives* 108 (suppl. 1): 123–131. https://doi.org/10.1289/ehp.00108s1123.

Philips, Gary T., Ekaterina I. Tzvetkova, Stephane Marinesco, and Thomas J. Carew. 2006. 'Latent Memory for Sensitization in Aplysia.' *Learning & Memory* 13 (2) (March–April): 224–229. https://doi.org/10.1101/lm.111506.

Roca, Irene T., Louis Desrochers, Matteo Giacomazzo, Andrea Bertolo, Patricia Bolduc, Raphaël Deschesnes, Charles A. Martin, Vincent Rainville, Guillaume Rheault, and Raphaël Proulx. 2016. 'Shifting Song Frequencies in Response to Anthropogenic Noise: A Meta-analysis on Birds and Anurans.' *Behavioral Ecology* 27 (5) (September–October): 1269–1274. https://doi.org/10.1093/beheco/arw060.

Rumpler, Romain, Siddharth Venkataraman, and Peter Göransson. 2020. 'An Observation of the Impact of CoViD-19 Recommendation Measures Monitored through Urban Noise Levels in Central Stockholm, Sweden.' *Sustainable Cities and Society* 63 (December). https://doi.org/10.1016/j.scs.2020.102469.

Schafer, R. Murray. 1994. *The Soundscape: Our Sonic Environment and the Tuning of the World*. Rochester: Destiny Books.

Slabbekoorn, Hans and Ardie den Boer-Visser. 2006. 'Cities Change the Songs of Birds.' *Current Biology* 16 (23): 2326–2331. https://doi.org/10.1016/j.cub.2006.10.008.

Smith, Lauren M., Linyan Wang, Kuba Mazur, Michael Carchia, Glen DePalma, Reza Azimi, Samantha Mravca, and Richard L. Neitze. 2020. 'Impacts of COVID-19-related Social Distancing Measures on Personal Environmental Sound Exposures.' *Environmental Research Letters* 15 (10). https://iopscience.iop.org/article/10.1088/1748-9326/abb494/pdf.

Spennemann, Dirk H. R. and Murray Parker. 2020. 'Hitting the "Pause" Button: What Does COVID-19 Tell Us about the Future of Heritage Sounds?' *Noise Mapping* 7: 265–275. https://doi.org/10.1515/noise-2020-0022.

Suess, Michael. 1973. *The Long-term Planning of a Noise Control Program*. In *Proceedings of the International Congress on Noise as a Public Health Problem*, edited by W. Dixon Ward, 73–75. Washington, DC: U.S. Environmental Protection Agency.

Suter, Alice H. 1991. 'Noise and Its Effects.' Report prepared for the Administrative Conference of the United States. www.nonoise.org/library/suter/suter.htm.

Tavris, Carol and Carole Wade. 2000. *Psychology in Perspective*. London: Pearson.

Tolman, Edward C. 1949. 'There Is More than One Kind of Learning.' *Psychological Review* 56 (3): 144–155. https://doi.org/10.1037/h0055304.

Tolman, Edward C. 1948. 'Cognitive Maps in Rats and Men.' *Psychological Review* 55 (4): 189–208. https://doi.org/10.1037/h0061626.

Tolman, Edward C., Benbow F. Ritchie, and Donald Kalish. 1946. 'Studies in Spatial Learning. I. Orientation and the Short-cut.' *Journal of Experimental Psychology* 36 (1): 13–24. https://doi.org/10.1037/h0053944.

Tolman, Edward C. and Charles H. Honzik. 1930. 'Introduction and Removal of Reward, and Maze Performance in Rats.' *University of California Publications in Psychology* 4: 257–275.

Triarchou, Lazaros. 2015. Νευροβιολογικές Βάσεις στην Εκπαίδευση. Athens: Hellenic Academic Libraries Link. http://hdl.handle.net/11419/5170.

Truax, Barry. 1999. *Handbook for Acoustic Ecology*. Vancouver: Cambridge Street Publishing.

Truax, Barry. 2001. *Acoustic Communication*. Westport: Greenwood Publishing Group.

Truax, Barry. 2016. 'Music, Soundscape and Acoustic Sustainability.' www.sfu.ca/~truax/Sustainability.pdf.

WHO (World Health Organization). 2018. *Environmental Noise Guidelines for the European Region*. www.euro.who.int/en/publications/abstracts/environmental-noise-guidelines-for-the-european-region-2018.

Zizek, Slavoj. 2020. 'There Will Be No Return to Normality after Covid. We Are Entering a Post-Human Era & Will Have to Invent a New Way of Life.' *RT.com*, 8 December 2020. www.rt.com/op-ed/508940-normality-covid-pandemic-return/.

19 The Sounds and Silence of COVID-19 Quarantine

Media Representation, Debility, and Neoliberal Biopolitics[1]

James Deaville

Introduction

Sound and music are vital components in audiovisual newscasting: while sound serves within the diegesis of the individual stories (regardless of the platform used to disseminate the news item), news music functions extradiegetically, primarily as an ident for the particular newscast but also as bumpers that lead into or out of commercial breaks (Graakjær 2015).[2] Despite their significance for establishing location and narrative context, the sounds from the sites of reportage remain largely overlooked features of screen news. As Hadar Levy and Amit Pinchevski note, 'the soundscape … may offer new insights into the news genre and its practices, and work to counteract the visual bias dominating the academic discourse of political communication and journalism studies' (2017, 3355). Moreover, the soundscape of the newscast and its individual segments and reports can help to orient the public towards the item and acoustically insert 'liveness and drama' into the stories and their framing (ibid. 3359). The two scholars significantly draw a distinction between the 'intended sounds' and 'unintended sounds' of televised news, between the narration of reporters (and anchors) and the sounds they choose to highlight, and incidental background sounds.[3]

By extending Levy and Pichevski's findings to North American and European audiovisual news coverage of the COVID-19 pandemic from the earliest global sites of infection, it is possible to demonstrate how both foregrounded sounds/music and ambient background soundscapes can convey meanings and interpretations to news consumers (Deaville 2006, 2007, 2012). Whether through conscious decision or serendipitous occurrence, the sounds featured in (and excluded from) the frame of audiovisual reportage in television and the internet can influence the audience through blatant or latent expressions of bias (Deaville 2006). The news media's representations of the Wuhanese public through silence, of the Iranian regime through the coughing of its health minister, and of Italian culture through balcony music can all be regarded as examples of such aural stereotyping and – ultimately – sonic colonisation, exacerbated by the COVID-19 pandemic (Deaville and Lemire 2021).

This chapter will analyse sight and sound in Western audiovisual news coverage from Wuhan and Italy during the weeks following their respective lockdowns on 23 January and 9 March 2020. After an introduction to the role of sound in disaster news reporting, the chapter considers the situation for the sounds of news during the earlier pandemic of H1N1 in 2009. The discussion proceeds to study the forces at work behind audiovisual media representations of the crisis in the Chinese and Italian epicentres through the theoretical lenses of Naomi Klein's 'disaster capitalism' (2007) and Jasbir K. Puar's concept of 'debility' (2017). Engaging in a theoretically informed examination of soundscapes in Western reportage from those sites will uncover how the media have deployed sound and music to invoke race, ethnicity, and nationality in their coverage of news from abroad. Such (mis-)

DOI: 10.4324/9781003200369-23

250 *James Deaville*

representations through sight and sound connected with favourable audience dispositions at home (e.g. Gabore 2020), which is critical for delivering ratings (Deaville 2006). The chapter concludes with observations regarding how, as the virus ravaged the United States in April 2020, the focus of Western news coverage shifted to the pandemic's effects at home, a move that likewise inflected the soundscapes of COVID-19 news.

Sound in Disaster Reporting

Producers of audiovisual news must display considerable flexibility (and sensitivity) when reporting on breaking and developing crisis situations, which range from human-caused catastrophes like war and insurrection to natural disasters like hurricanes and earthquakes. Their coverage customarily entails both highlighted sound events and contextual ambient sounds for the purpose of conveying realism and credibility to news consumers. As historian Bruce Johnson argues, sound 'produces emotional responses' or 'involuntary affective interpretations' (2017, 16–17), cognitive or pre-cognitive experiences that help to immerse the audience within news items (de Bruin et al. 2022).

The sounds of disaster newscasting obviously vary according to the nature of the event itself, ranging from the uncanny silence of drought to the chaotic tumult of a hurricane or an open battle.[4] Given adequate forewarning and lead time, news networks and stations could commission composers or music-production companies to create themes to fit the event, as they did for the War in Iraq (Deaville 2006).[5] The resulting theme might occur as a stinger at the outset of the newscast,[6] underscoring a graphic title created to signify the disaster while signalling the event's importance and gravity for the broadcast and its network.

In contrast, the sounds from the site(s) of conflict or catastrophe function at the level of the specific news item, diegetically authenticating the report and attempting – in tandem with the moving images and narration – to immerse the audience in the sensory experience of the disaster. As Simon Cottle has argued, reporters in the field engage in communicating the 'experiential ontology of "witnessing," of bodily being there, through the deliberate invocation of bodily senses' (2013, 233). A BBC correspondent wrote the following about his experience covering the Haiti earthquake of 2010: 'you just want to paint a picture for people so that they can smell and hear and see and feel what you are [experiencing] on the ground' (reported in Cottle 2013, 239).

Central to all such visceral narrative storytelling is the soundtrack that brings the latest disaster to the news consumer, primarily comprising the narration of the reporter and any interviewees (witnesses or experts) and the foreground or background sounds from the event itself. The roar of the erupting volcano, the rush of the tsunami wave, the explosion of the mortar attack: these sounds give meaning, life, and presence to the news report (Levy and Pinchevski 2017, 3361). As Engstrom and Larson argue, 'such [news] video also can give viewers a sense of a "you are there" sense of presence and empathy for the people involved in the event' (2019, 107).

Making these types of natural and human-caused disasters come alive for the news audience at home does not require much manipulation of soundbites from the noisy sites of tragedy, but how might one visually and sonically represent the silent lethality of pathogens like the plague or the swine flu? The first global pandemic of the twentieth century, the 1918 influenza, was not covered by newsreels in theatres for several reasons: (1) as the outbreak of the flu coincided with the end of the First World War, the conflict still preoccupied the media (Flecknoe, Wakefield, and Simmons 2018); (2) the authorities did not want to start or spread a panic, and the media censorship of the war carried over into the early days of the pandemic (Enns 2011, 281–310); (3) and lastly, theatres and film production companies

The Sounds and Silence of COVID-19 Quarantine 251

temporarily shut down in accordance with local health requirements (Koszarski 2005, 379).[7] Moreover, the theatrical newsreel at the time was silent, so that the sounds of disaster could only be communicated by intertitles, which meant that if the influenza would have figured in newsreel reports, any associated sounds (sneezing, coughing, etc.) would have had to be acted out (Dixon 2020).[8] Citizens' primary sources of information about the pandemic were the newspapers, which – as Meg Spratt argues – 'supported existing power structures and conferred status on scientific and governmental authorities' (Spratt 2001, 75).

Coverage of the 2009 H1N1 Influenza Pandemic

By the time of the public health catastrophe of the H1N1 virus in 2009, news reporting from the field – whether at home or abroad – had developed into a two- (or more) person operation, involving a reporter and a videographer.[9] During the course of the 2000s, however, ever more media organisations adopted the principles of 'solo journalism' in their news reporting, whereby individual professionals 'are regularly expected to gather information, conduct interviews, write stories, record audio and video elements, and edit it all together into a narrative news story, all by themselves' (Blankenship and Riffe 2021, 41). Such a practice is certainly more efficient than the team approach, although on the one hand it necessitates considerably more labour and skill on the part of the reporter, and on the other it opens the process up to a greater danger of bias. Solo journalism renders the resulting news product more susceptible to the individual reporter's perspective on a given (catastrophic) event, which would be based on their background and interests as well as on the anticipated biases and expectations of audience reactions (Messineo 2015).

The public health disaster of the 2009 H1N1 influenza pandemic served as an immediate predecessor for the COVID-19 pandemic in terms of media reporting,[10] even as news sources and the public tended to erroneously refer to it as the swine flu or Mexican flu, leading to the stigmatisation of Mexican Americans (McCauley, Minsky, and Viswanath 2013, 1116). Within six weeks the virus spread throughout the globe, and the infection became a primary topic for television news coverage, though the mildness of symptoms and the introduction of a vaccine mitigated the effects of the 2009 H1N1 influenza. Some observers accused the media of fostering a 'pandemic of panic' throughout the outbreak (Nerlich and Koteyko 2012, 713), which prompted subsequent studies within the field of academic journalism. In particular, an investigation of one Brazilian television station's overall reporting on the 2009 pandemic concluded with the prescient, open-ended question: 'how to alert the population about an arising virus – . . . whose genetic cause was not well-known and which seemed to be highly lethal – without generating panic?' (da Silva Medeiros, Natércia, and Massarani 2010, 6).

One decisive difference between the coverage of the 2009 H1N1 and the COVID-19 pandemics concerns the perceived transmission pathway (Hoppe 2018): the outbreak of 2009 H1N1 in Mexico, the United States, and Canada was first widely reported at the same time, on 24 April 2009 (McNeil 2009, A13), whereas the spread of the coronavirus was initially reported from Wuhan, China in December 2019, with other global regions to follow in the course of the next two months. Though the swine flu was also called the Mexican flu by some medical authorities and media outlets, when its origins were traced back to central Mexico, Western media attention focused on sites of infection in the United States.[11] In contrast, the North American and European news audience was well aware that the novel coronavirus originated and initially proliferated outside the West, which motivated the Western press during the first months of the pandemic to mobilise constructions of race, ethnicity, and nationality in connecting with the home audience.

252   *James Deaville*

If we consider the decisive initial reporting from the end of April and the month of May, the sights and sounds of the coverage focus on the epidemiology and severity of the 2009 H1N1 virus, as documented and analysed by journalism scholars in a robust body of literature. An important study that aggregated their results concluded that 'media may have – inadvertently – contributed to heightened risk perceptions through a high volume of coverage and an unbalanced emphasis on the threat of H1N1' (Klemm, Das, and Hartmann 2016, 17). Thus a study of the newscasts from five Sydney television stations found that of the coverage, '63.4% ... related to the seriousness of the situation, 12.9% providing advice for viewers and 23.6% involving assurances from government' (Fogarty et al. 2011, 181). Footage from NH News and Sky News Australia, 25 April through 31 May, for example,[12] reveals factual accounts about H1N1 from newscasters based in Australia: they consisted of an alternation between reporter voiceover and commentary by healthcare professionals and government officials, with a flow of relevant images – hospital exteriors, labs, masked pedestrians – under the narration. Under such conditions of informing the local public, there was neither opportunity nor need to present the soundscapes of swine flu outbreaks in distant lands.

Coverage of the COVID-19 Pandemic

In the ten years between the outbreaks of 2009 H1N1 and COVID-19, the news mediascape underwent significant developments that impacted the coverage of the new pandemic. At the time of the 2009 H1N1 outbreak, television broadcasts – terrestrial or cable – remained the primary audiovisual source for news, although social media platforms like Facebook, Twitter, and YouTube were beginning to feature (shared) information about current events; Instagram had yet to launch and Reddit served only a limited audience. The intervening years brought all of these sites to the fore as vehicles for the proliferating distribution of the sights and sounds of news stories, which shifted the production and curation of breaking news about disasters from professional reportage toward the posting of cell phone videos. At the same time, television networks and stations – NBC, CNN, BBC, Sky News, and the like – began posting their own footage online as did major newspapers – for example, the *New York Times* and the *Guardian*, lending their cultural capital to digital news coverage. Even venerable agencies like Reuters and the Associated Press contributed audiovisual reportage to the burgeoning internet market for news about the most current events.

Moreover, the global pathways of transmission of the coronavirus and – given its deadliness – the lockdown and quarantine measures taken at the outset to control its spread warranted a different type of reporting. With the significantly greater and quicker availability of first-hand news video by professional reporters and amateur social media mavens alike, it was possible for news providers to trace infections as they progressed from the initial epicentre in Wuhan, China. Foreign correspondents for the Western press could post absorbing (and terrifying) audiovisual coverage from the exotic sites of the advancing disease, even relying on drones to reveal unprecedented apocalyptic images of major metropolises under lockdown.

Once news of the COVID-19 virus became a lead item in audiovisual reporting, in January 2020, news producers needed to arrive at a fitting sound for the unfolding tragedy.[13] The extradiegetic frame of broadcast news provided an appropriately musical context for pandemic news, through gloomy opening beds and stingers for special coverage.[14] For the news items themselves, however, the coronavirus did not offer any characteristic sounds for its assault on people other than those of the frantic ICU personnel and their life-monitoring and -sustaining apparatuses. The microphones and cameras of reporters were not allowed

The Sounds and Silence of COVID-19 Quarantine 253

to encroach on the space of COVID-19 patients due to quarantine regulations.[15] That said, the earliest human response to the virus' effects did provide soundworthy moments for audiovisual dissemination to the Western public via reports from the distant sites of infection in China and Iran, and then – as it moved westward – Italy. If we comparatively study the visual and sonic environments communicated to the world from China and Italy by news providers, we will discover an underlying ableist and racist neoliberal biopolitics of debility.

China

The outbreak of the novel coronavirus in China caught the attention of global news agencies on 31 December 2019, at which point it was called 'a new form of pneumonia' (Cohen and Normile 2020). As the unnamed, unknown virus spread during January within Wuhan, and eventually led to a full lockdown of public activities there on 23 January 2020, the international news media were eager to transmit the sights and sounds of the new disease to their audiences at home. After all, it was not regarded at the time as representing a threat to the rest of the world but rather an internal Chinese health problem.[16] Other representatives of the Western media, including CNN and the *New York Times*, criticised China's lockdown as 'harsh' (Feng and Chen 2020), 'extreme' (Qin, Myers, and Lu 2020), 'severe' (Griffiths, Gan, and Culver 2020), and 'controversial' (O'Grady 2020). As if to reinforce this rhetoric, Western video news sources showed footage of drones used by authorities to warn citizens who were lockdown-breaking, under headlines like 'Chinese authorities humiliating people for not wearing face masks' (Telegraph 2020), and 'China's massive security state is being used to crack down on the Wuhan virus' (Griffiths and Gan 2020).

As already argued, while disaster reporting has traditionally sensationalised seeing and hearing the hurricane or tornado, or at least seeing the earthquake's aftermath, the question must arise, how do you represent, give visual and sonic life, to an invisible and silent assassin like the coronavirus? Because the direct recipients of its debilitating effects, human bodies, were hardly in evidence for reporters in quarantined Wuhan, the solution of North American and European news media was to fetishise the unprecedented stillness and silence of the incapacitated city.[17] Thus the cityscape revealed by reporters shocked news consumers in the West, but not through the expected pandemonium in hospitals or chaotic street scenes, but rather through the absence of motion and sound within the normally bustling city, which plays into longstanding stereotypes of East Asians (Li and Nicholson 2021). As Eric Hung observes, the labels continually switch 'between model minority and perpetual foreigner/yellow peril,' and the silence enacts the stereotype of 'something scary and incomprehensible' (pers. comm., 4 May 2021).

The media's silent images of empty streets from Wuhan seemed as if ripped from the footage of a post-apocalyptic horror film like *Dawn of the Dead* (Snyder 2004).[18] Quarantine never looked so devastating, never sounded so quietly as constructed by the eyes and ears of the Western press in Wuhan, though such footage would later come to characterise reports from other urban sites of lockdown.[19] Western news media seemed intent on excluding signs of life from inside containment, even though at the time foreign reporters in Wuhan had access throughout the city.[20] Were they so surprised and challenged by what they found – and did not find – that reporters felt compelled to shock viewers by exploiting sensational images and non-sounds of an ostensibly abandoned city? After all, in 2018 the *Guardian* rated the Chinese cities of Guangzhou and Beijing as the two noisiest in the world (Keegan 2018). And an audiovisual production from the state-run YouTube channel New China TV re-appropriated the stereotyping soundscape when the network titled its English-language

254 *James Deaville*

rebirth video from September 2020, *Wuhan: A City Getting Loud Again* (New China TV 2020).

But if we consider the ideological underpinnings for the look and tone of the international coverage in January and February 2020, we discover a deeper layer of possible motivation for the pandemic soundscape. Through the fetishised emptiness and silence, news media in North America and Europe could be said to symbolise, and thus colonise and debilitate, Asian bodies, depriving them of life-affirming physical appearance and voice.[21] For their part, some Western consumers of this news could have regarded the Wuhanese absence from coverage as confirming expectations of the tightly regulated Communist state,[22] in this case, transforming a teeming cityscape to one of a 'ghost city' or 'ghost town.'[23] But also the imaginary construct of neither seeing nor hearing Asians in confinement (or internment) – in accord with the carceral state theorised by Michel Foucault (1975) – aligns with James Kyung-Jin Lee's argument about disappearing or invisible Asians, who assume 'the cultural logic of racial invisibility' (2004, 80). Again, the stereotype is at work that causes the frightening, incomprehensible Asian Other to be banished from consideration.

That logic operates according to the agency of what Naomi Klein has termed 'the shock doctrine' of disaster capitalism (2007), which can be regarded as motivating the Global North's visual/sonic erasure and debilitation of Chinese bodies under quarantine. Lisa Parks and Janet Walker have mobilised the designation 'disaster media' to describe the 'ways that media are … complicit in the amplification of disastrous occurrence' (Parks and Walker 2020, 4).[24] As they observe, such reporting is 'necessarily racialised capitalism, since those who are already disadvantaged often suffer more when disaster strikes,' in particular racialised and disabled people (Parks and Walker 2020, 3). And in the words of Jasbir K. Puar in her study, *The Right to Maim: Debility, Capacity, Disability*, 'the racialized body is in a constant state of becoming disabled' (Puar 2017, 103). Thus under the imperatives of disaster capitalism, the elimination of the quarantined/incarcerated Wuhanese populace in Western news coverage rendered them unprofitable for the neoliberal enterprise, which – as theorised by David Harvey – above all values the maximisation of entrepreneurial freedoms within an institutional framework characterised by … unencumbered markets and free trade' (2007, 22).

So, infectious disease and invalidisation through the quarantine lockdown do not fit easily within this framework of neoliberal capitalism. Robert McRuer's book *Crip Times* exposes how 'vulnerable bodies … become signs of excess that must be trimmed' (2018, 113). However, as Puar has observed, the biopolitics of conditions of incapacitation sustain 'the neoliberal split between the disabled subject as valuable difference … as object of care, and the debilitated body as degraded object' (2017, 92). While people with disabilities can still productively serve the economy, Vandana Chaudhry has argued that 'debilitated bodies … cannot be capacitated within the neoliberal logic of entrepreneurship' (2018). Like the residents of Wuhan, bodies debilitated through quarantine seem incomprehensible and unproductive burdens to the neoliberal capitalist imaginary, and thus they are invisibilised and silenced. If we apply the words of Puar to the early days of the pandemic, the Wuhanese lost their claim to material presence in Western news media through sight and sound, as incarcerated and racialised 'objects of un-care – social pariahs' (2017, 77).

Italy

This work of sonic colonisation and its effect of silencing become all the more apparent when one compares the Western news footage from China with that from Italy in mid-March 2020. The first Italian deaths from the coronavirus were recorded on 22 February,[25] and by the beginning of March the virus had spread throughout the country.[26] Quarantine

The Sounds and Silence of COVID-19 Quarantine 255

measures were introduced by region, but infections had spread so quickly and widely that on 9 March the Prime Minister 'ordered people to stay home and seek permission for essential travel' (BBC 2020). The Italian public was under lockdown and as such was challenged to reach out and relieve the 'quarantedium' (Hartley 2020),[27] which the so-called balcony music (*musica dai balconi*) accomplished, whereby the quarantined populace remained physically separated while viscerally present and socially connected to each other. Necessitated by the summer heat, the residential balcony is a fixture of urban culture in Italy and other warm climes,[28] where they serve as 'liminal spaces that bridge public and private life' (Sheila Crane, quoted in Traverso 2020).[29] Hence the balcony was a logical site for a group of young people from Benevento to perform a 'flash mob' for themselves and their neighbours (Chiu 2020, 7).[30] The video became a viral sensation and initiated similar balcony performances across the country's urban centres, where the close spaces between apartment buildings facilitated such sonic encounters. The resulting balcony musics were not limited to one genre or to individual musicians: typically one to three apartment dwellers sang or performed on various instruments from their balcony, at times alone and at times in chorus with their neighbours.[31]

The resulting extraordinary examples of musicking caught the eyes and ears of the Western press,[32] which disseminated the footage to the audience of news consumers in North America and elsewhere in Europe. The apparently impromptu balcony performances by (white) Italians fulfilled the Global North's sonic bias of the response to the health crisis in the so-called 'land of song' (Bennett 1896).[33] Visually present and sonically resonant, these Italian musicians – amateur and professional – model socially-cohering corona-musicking behaviour in the face of the incapacitating health disaster (Hansen, forthcoming). Here musical sounds fill the media frames, invoking race, ethnicity, and nationality to connect with favourable audience dispositions.

The vocal performances themselves ranged from the Italian national anthem by a flash mob of hundreds in Bari on 13 March (Caters News 2020) to daily operatic excerpts sung by a tenor in Florence (NowThis News 2020). The sounds of instrumental music also contributed to the balcony music, including a flautist and bassoonist playing a dance movement by Bach in Florence on 14 March (Guardian News 2020), a trumpeter playing 'Imagine' in Trapani on 16 March (Maiorana 2020), and a tenor saxophone player interpreting 'Bella Ciao' in Naples on 19 March (Vitale 2020).[34] Originating in the first weeks of Italian lockdown, the plethora of such balcony-music videos disseminated by respected news sources such as US network ABC, CNN, the BBC, the *New York Times*, and the *Guardian* not only added welcome relief to otherwise bleak newscasts (Hansen, forthcoming), but also situated the home audience in a foreign setting populated by individuals who looked like us and performed familiar music.

The phenomenon spread throughout southern Europe in particular, where balcony culture prevails, including Spain and Portugal, but also Havana and New Orleans (Traverso 2020).[35] However, as the progenitor of collective musicking on and from balconies, Italy continued to occupy the attention of the international press, even as the daily rituals lost their novelty for their local practitioners. Closer study of the images might well reveal motivations for the focus on Italian practitioners: white, middle class (and thus at least to a certain extent privileged), able-bodied, and heteronormative-looking, the bodies of these participants could invite identification among audiences of Western news consumers.

Conclusion

With Donald Trump's declaration of a national emergency on 13 March and the government's stay-at-home advisory guidelines from 16 March, the North American news media had good reason to turn its ears and eyes inward, towards the ravages of COVID-19 among its

256   *James Deaville*

own citizenry. However, decisions regarding lockdowns were made by state governments, virtually all of which had their inhabitants isolating themselves by the end of March (Secon 2020). The silence of the ghost city Wuhan became a domestic soundscape of quarantining as presented by sensationalising news coverage across the land, the initial epicentres abroad now forgotten except for the assignment of blame. The silence was not the only soundworthy phenomenon in audiovisual news of COVID-19, however: the frantic soundscapes of hospital ICUs also played out on home screens, a reality check for the coronavirus' severity in our midst and a warning for residents to heed public health measures.[36] As a result, the home audience may have encountered both extremes of pandemic sound and non-sound within the same newscast or even news item on television or online, each serving its purpose, as demonstrated for example by the videos in the MSNBC report from 20 March entitled 'Hundreds Flood NYC Hospital ICU and ERs as Global Coronavirus Cases Top 466,000' (Higgins-Dunn, Mangan, and Stevens 2020).

So that which served as implicit bias in sonic profiling by Western pandemic coverage from foreign parts came home and established itself as the domestic soundscape of lockdown in North America. The uncanny silence of the normally bustling cityscapes and the agitated, even frantic sounds from COVID ICUs (Cecchetto and MacDonald, forthcoming) may have affectively polarised the sonic experience of pandemic news in the first days of isolation, yet either context was dire. As a result, audiovisual newscasters had to turn to the musically and humanly resonant response of the Italians to introduce relief, which was heard and seen at the end of the broadcast.[37] And as time passed and life-saving vaccines were administered, the intended sounds that were foregrounded within the diegesis of screen news reports became the chants of North American protesters, who were envoicing an ideology of disastrous debilitation that was born of ignorant wilfulness, from which no vaccine provides protection.

Notes

1 My thanks extend to Chantal Lemire (Western University) and Adrian Matte (Carleton University) for their assistance in the preparation of this chapter.
2 It should be noted that within traditional North American practices, music is not used during news except for special coverage of a major event.
3 The concept of unintended sounds is problematic, since even environmental sounds can be manipulated or removed in editing. Rather than invoking intentionality, it is more productive to conceive of sounds as foreground and background, as the audience perceives them (Behrendt 2012).
4 See McAlister 2012 for a close study of the audiosphere before, during, and after an earthquake.
5 News providers knew that armed conflict was coming, so they commissioned in advance war themes by (house) composers.
6 Stingers are brief, strongly articulated musical ideas that are intended to capture the attention of the home audience while helping to set the mood for the story to follow.
7 The pages of *Moving Picture World* from October 2018 through January 2019 document the effects of the flu upon the film industry. Richard Koszarski has thankfully complied the reports: the documents make it clear that there never was an order from the federal government to close the theatres; public health measures were regulated locally. The losses to theatre musicians must have been considerable: just for St. Louis, the reports record the layoffs of 2,500 piano players and 150 other musicians (Koszarski 2005, 471).
8 The only existing film source from the period is a British 'film … made by Joseph Best for the Local Government Board (the precursor to the Ministry of Health) for distribution in cinemas during the deadly second wave of the flu outbreak in late 1918' (Dixon 2020).
9 See Tuggle, Carr, and Huffman (2014).
10 There were other epidemics in the twentieth century, like the 1968–1969 flu epidemic, and the SARS outbreak of 2002–2004, but in terms of media coverage, 2009 H1N1 was the most immediate precursor for COVID-19.

The Sounds and Silence of COVID-19 Quarantine   257

11 A few medical reporters did travel to La Gloria, Mexico, to speak with the boy Edgar Hernandez, whom they identified as 'Patient Zero.' However, their accounts read more like human-interest stories than investigative medical reports (CNN 2009).

12 See the YouTube videos 'Swine Flu Australia 2009' (Nhsydneynews 2009) and 'Australia Monitoring 28 for Swine Flu' (Sky News 2009).

13 See Deaville 2006 for a discussion of the factors that informed decisions on music commissions for the Iraq War.

14 Opening beds designate the music at the beginning of the newscast, used as the background (bed) for a brief announcement of the day's top stories, while stingers are short musical ideas or gestures at the beginning of breaking news stories both to capture the audience's attention and to communicate the story's affect. For a more detailed discussion of the terminology for production-music elements used in broadcasting, see Föllmer 2018.

15 Even family members were typically excluded from the coronavirus patient's room.

16 See, for example, ECDPC 2020, which assessed the likelihood of importation of the novel coronavirus to the EU to be low.

17 Disability studies has yet to come to terms with the general use of adjectives like disabled, debilitated, or crippled when they are applied to non-human entities like electronic devices, public and private conveyances, and governmental bodies (including cities).

18 Silence is a favourite device for sound designers in the cinematic subgenre of post-apocalyptic horror, as made explicit in *A Quiet Place* (2018) and *The Silence* (2019).

19 See for example the video 'Coronavirus Outbreak: A Look at Empty Cities around the World' by Canada's Global News, posted on 25 March 2020 (www.youtube.com/watch?v=jwClyd2lHWo). Instead of allowing the metropolis' uncanny silences to accompany the images, the editors felt compelled to add a sober yet rhythmically active musical underscore. As Nuria Lorenzo-Dus and Annie Bryan have argued, the 'strategic use of concrete silence … constitute[s] the best way to capture the intensity of the live moment being shown' (2011, 33).

20 It was not until 10 February that local authorities limited access to Wuhan under the Infectious Diseases Law. See Hooper (2021, 346).

21 Regarding the vivifying or animating role assigned to the human voice in audiovisual media, see Deaville 2019.

22 See Telegraph (2020) and Griffiths and Gan (2020) for these types of critiques.

23 For example, Euronews, 'Coronavirus: Drone Footage Unveils "Ghost Town" Wuhan,' YouTube, 29 January 2020 (www.youtube.com/watch?v=E0fOqXIsMrw), and Reuters, '"Ghost City": Commute through China's Deserted Capital amid Coronavirus,' February 6, 2020 (www.reuters.com/article/us-china-health-beijing-commute-idUSKBN201027).

24 In the case of the Japanese tsunami of 2011, some news agencies provided sensationalizing video titles while enhancing images of the tsunami's devastation by adding an ominous music track. See for example the video entitled 'Japan Tsunami 2011 Massive Wave – Shocking Footage,' posted by stormcentertv on 16 March 2011 (www.youtube.com/watch?v=mcnDdxYNfIk).

25 See Ravizza 2020. Although this article from the major Italian newspaper focusses on Milan, the deaths were widely reported throughout the country.

26 The virus spread so quickly in Italy that within weeks doctors were already required to determine which patients received life-saving care, as reported by Mounk 2020.

27 See the pandemic-related list of neologisms assembled by Hartley 2020.

28 The English chronicler Adam Walker observed the following about Italian balconies in 1792 in his *Remarks Made in a Tour from London to the Lakes of Westmoreland and Cumberland in the Summer of 1791*: 'What business have ITALIAN balconies on the top of an ENGLISH palace? It Italy the climate demands them; but when do we visit the tops of our houses in quest of cool air?' (1792, 14).

29 Traverso's BBC Travel article cites Henri Lefebvre's influential 1992 book *Rhythmanalysis*, which praises the function of the balcony.

30 An Italian source from one year later identified the music as a *tammurriata* (form of tarantella) and the event as a 'jam session' (Bmagazine 2021).

31 Balconies served also as spaces devoted to aural manifestations of gratitude to health care workers. In an article about lockdown rituals, Nanta Novello Paglianti observed how 'Italian applause

258   *James Deaville*

[was] promoted on Facebook by the group "Applaudiamo l'Italia" (Clap for Italy) on 14 March 2020, which prompted people to clap from their balconies at 12 noon for health care workers. Over 300,000 people joined the group proposing flashmobs, performances, flags *tricolore* and singing from their windows or roofs' (Paglianti 2020, 320). See also Catungal 2021, which deals with cultural politics of public appreciation in times of COVID-19.

32   Musicking is a neologism of Christopher Small, used to describe participating 'in any capacity in a musical performance.' For Small, the essence of music lies not in musical works but in taking part in performance, in social action. Music is thus not so much a noun as a verb, 'to music' (1998, 9).

33   Some balcony musical performances did take place in Wuhan, but they did not capture the attention of the Western news media. Thus the *South China Morning Post* posted a video about how 'residents of virus-stricken Wuhan are boosting morale in the city by shouting from their windows and singing patriotic songs,' yet the accompanying images are of gigantic apartment blocks and city office spaces with fixed windows. The sounds are of people shouting responsorially, but the graphic titles note how the officials discouraged the practice for fear 'that it might spread the infection' (South China Morning Post 2020).

34   The last video, of Daniele Vitale, was clearly mixed and edited with added accompaniment in a studio.

35   Traverso's article presents a cultural history of the balcony, positioning it and the coronavirus-inspired phenomenon of balcony music within existing urban practices that invoke space and place.

36   Both government and industry undertook stay-at-home campaigns in the second half of March, releasing videos that urged citizens to 'chill out' and avoid unnecessary risks through public contact. These messages typically incorporated pastel colours, clear graphics, and calming music, in order to encourage public compliance with the stay-at-home orders.

37   Traditional half-hour newscasts in North America typically end with a feel-good human-interest story, which became the placement for the Italian balcony music. See Deaville 2007 for a discussion of music's role in the last items of newscasts.

References

BBC. 2020. 'Coronavirus: Italy Extends Emergency Measures Nationwide.' 10 March 2020. www.bbc.com/news/world-europe-51810673.

Behrendt, Frauke. 2012. 'The Sound of Locative Media.' *Convergence* 18 (3): 283–295. https://doi.org/10.1177/1354856512441150.

Bennett, Joseph. 1896. 'In the "Land of Song."' *The Musical Times* 37: 297–300.

Blankenship, Justin C. and Daniel Riffe. 2021. 'Follow the Leader?: Optimism and Efficacy on Solo Journalism of Local Television Journalists and News Directors.' *Journalism Practice* 15 (1): 41–62. https://doi.org/10.1080/17512786.2019.1695535.

Bmagazine. 2021. 'Iniziava un anno fa la musica dai balconi.' 12 March 2021. www.bmagazine.it/iniziava-un-anno-fa-la-musica-dai-balconi/.

Caters News. 2020. 'Italian Flashmob Singing National Anthem on Balconies.' YouTube, 16 March 2020. www.youtube.com/watch?v=JOy5dmbFNCA.

Catungal, John Paul. 2021. 'Essential Workers and the Cultural Politics of Appreciation: Sonic, Visual and Mediated Geographies of Public Gratitude in the Time of COVID-19.' *Cultural Geographies* 28 (2): 403–408. https://doi.org/10.1177/1474474020978483.

Cecchetto, David and Cameron MacDonald. Forthcoming. 'Listening through a Pandemic: Silence, Noisemaking, and Music.' In *Creative Resilience and COVID-19: Figuring the Everyday in a Pandemic*, edited by Irene Gammel and Jason Wang. London: Routledge.

Chaudhry, Vandana. 2018. 'Capacity, Debility and Differential Inclusion: The Politics of Microfinance in South India.' *Disability Studies Quarterly* 38 (1). https://dx.doi.org/10.18061/dsq.v38i1.5995.

Chiu, Remi. 2020. 'Functions of Music Making under Lockdown: A Trans-Historical Perspective across Two Pandemics.' *Frontiers in Psychology* 11. https://doi.org/10.3389/fpsyg.2020.616499.

CNN. 2009. 'Meet the Boy Believed to Be "Patient Zero."' 29 April 2009. https://edition.cnn.com/2009/HEALTH/04/29/swine.flu.patient.zero/index.html.

Cohen, Jon and Dennis Normile. 2020. 'World on Alert for Potential Spread of New SARS-like Virus Found in China.' *Science (American Association for the Advancement of Science)*, 14 January 2020. https://doi.org/10.1126/science.aba9012.

Cottle, Simon. 2013. 'Journalists Witnessing Disaster: From the Calculus of Death to the Injunction to Care.' *Journalism Studies* 14 (2): 232–248. https://doi.org/10.1080/1461670X.2012.718556.

da Silva Medeiros, Flavia Natércia, and Luisa Massarani. 2010. 'Pandemic on the Air: A Case Study on the Coverage of New Influenza A/H1N1 by Brazilian Prime Time TV News.' *Journal of Science Communication* 9 (3): 1–9. www.arca.fiocruz.br/handle/icict/24970.

de Bruin, Kiki, Yael de Haan, Sanne Kruikemeier, Sophie Lecheler, and Nele Goutier. 2022. 'A First-Person Promise? A Content-Analysis of Immersive Journalistic Productions.' *Journalism* 23 (2) (February): 479–498. https://doi.org/10.1177/1464884920922006.

Deaville, James. 2006. 'Selling War: Television News Music and the Shaping of American Public Opinion.' *Echo* 8 (1): 735–741. www.echo.ucla.edu/Volume8-Issue1/roundtable/deaville.html.

Deaville, James. 2007. 'The Sounds of American and Canadian Television News After 9/11: Entoning Horror and Grief, Fear and Anger.' In *Music in the Post-9/11 World*, edited by Martin Daughtry and Jonathan Ritter, 43–70. New York: Routledge.

Deaville, James. 2012. 'The Changing Sounds of War: Television News Music and Armed Conflicts from Vietnam to Iraq.' In *Music, Politics, and Violence*, edited by Susan Fast and Kip Pegley, 104–126. Middletown: Wesleyan University Press.

Deaville, James. 2019. 'The Moaning of (Un-)Life: Animacy, Muteness and Eugenics in Cinematic and Televisual Representation.' *Journal of Interdisciplinary Voice Studies* 4 (2): 225–245. https://doi.org/10.1386/jivs_00007_1.

Deaville, James and Chantal Lemire. 2021. 'Latent Cultural Bias in Soundtracks of Western News Coverage from Early COVID-19 Epicenters.' *Frontiers in Psychology* 12. https://doi.org/10.3389/fpsyg.2021.686738.

Dixon, Bryony. 2020. 'Silent Film and the Great Pandemic of 1918.' *BFI-Staging*, 3 June 2020. www.bfi-staging.org.uk/news-opinion/news-bfi/features/silent-film-great-pandemic-1918?fbclid=IwAR1qYarS8RmQSrm30_IJBswXSrSWduQhLgqIeKfEupEsd79OA-4433PHOdE.

ECDPC (European Centre for Disease Prevention and Control). 2020. 'Cluster of Pneumonia Cases Caused by a Novel Coronavirus, Wuhan, China, 2020.' 17 January 2020. www.ecdc.europa.eu/en/publications-data/rapid-risk-assessment-cluster-pneumonia-cases-caused-novel-coronavirus-wuhan.

Engstrom, Erika and Gary Larson. 2019. 'Audio and Video Journalism.' In *Convergent Journalism: An Introduction*, edited by Vincent F. Filak, 99–114. New York: Routledge.

Enns, Richard. 2011. *It's a Non-Linear World*. New York: Springer.

Feng, Emily and Amy Chen. 2020. 'Restrictions and Rewards: How China Is Locking Down Half a Billion Citizens.' *NPR*, 21 February 2020. www.npr.org/sections/goatsandsoda/2020/02/21/806958341/restrictions-and-rewards-how-china-is-locking-down-half-a-billion-citizens.

Flecknoe, Daniel, Benjamin Charles Wakefield, and Aidan Simmons. 2018. 'Plagues and Wars: The "Spanish Flu" Pandemic as a Lesson from History.' *Medicine, Conflict, and Survival* 34 (2): 61–68. https://doi.org/10.1080/13623699.2018.1472892.

Fogarty, Andrea S., Kate Holland, Michelle Imison, R. Warwick Blood, Simon Chapman, and Simon Holding. 2011. 'Communicating Uncertainty – How Australian Television Reported H1N1 Risk in 2009: A Content Analysis.' *BMC Public Health* 11 (1): 181. https://doi.org/10.1186/1471-2458-11-181.

Föllmer, Golo. 2018. 'Production and Use of Packaging Elements in Radio: Concepts, Functions and Styles in Transnational Comparison.' In *Transnationalizing Radio Research: New Approaches to an Old Medium*, edited by Golo Föllmer and Alexander Badenoch, 101–116. Bielefeld: transcript Verlag.

Foucault, Michel. 1975. *Surveiller et punir: naissance de la prison*. Paris: Gallimard.

Gabore, Samuel Mochona. 2020. 'Western and Chinese Media Representation of Africa in COVID-19 News Coverage.' *Asian Journal of Communication* 30 (5): 299–316. https://doi.org/10.1080/01292986.2020.1801781.

260 *James Deaville*

Graakjær, Nicolai. 2015. *Analyzing Music in TV Commercials: Television Commercials and Consumer Choice*. New York: Routledge.

Griffiths, James and Nectar Gan. 2020. 'China's Massive Security State Is Being Used to Crack Down on the Wuhan Virus.' *CNN*, 11 February 2020. www.cnn.com/2020/02/10/asia/china-security-police-wuhan-virus-intl-hnk/index.html.

Griffiths, James, Nectar Gan, and David Culver. 2020. 'In Photos: Virus Outbreak Locks Down Chinese Cities.' *CNN*, 30 January 2020. www.cnn.com/2020/01/30/asia/gallery/chinese-cities-lockdown-coronavirus-intl-hnk/index.html.

Guardian News. 2020. 'Coronavirus: Neighbours Play Instruments from Balconies as Italy Stays under Lockdown.' YouTube, 14 March 2020. www.youtube.com/watch?v=k7bobc-vRbQ&t=1s.

Hansen, Niels Christian. Forthcoming. 'Music for Hedonia and Eudemonia during Pandemic Social Isolation.' In *Arts and Mindfulness Education for Human Flourishing*, edited by Tatiana Chemi, Elvira Brattico, Lone Fjorback Overby, and László Harmat. London: Routledge.

Hartley, Anna Pellegrin. 2020. 'Quarantedium, and 16 Other Terms That Should Now Exist in Light of the Coronavirus.' *The Beijinger*, 15 March 2020. www.thebeijinger.com/blog/2020/03/15/quarantedium-and-other-words-should-exist.

Harvey, David. 2007. 'Neoliberalism as Creative Destruction.' *Annals of the American Academy of Political and Social Science* 610 (March): 22–44. https://doi.org/10.1177/0002716206296780.

Higgins-Dunn, Noah, Dan Mangan, and Pippa Stevens. 2020. 'Hundreds Flood NYC Hospital ICU and ERs as Global Coronavirus Cases Top 466,000.' *CNBC*, 25 March 2020. www.cnbc.com/2020/03/25/coronavirus-latest-updates.html.

Hooper, Michael. 2021. 'Fighting a Pandemic According to Law: Examining the Legality of Key Elements of China's Early Covid-19 Response in Wuhan.' *University of Western Australia Law Review* 48 (2): 330–351.

Hoppe, Trevor. 2018. '"Spanish Flu": When Infectious Disease Names Blur Origins and Stigmatize Those Infected.' *American Journal of Public Health* 108 (11): 1462–1464. https://doi.org/10.2105/AJPH.2018.304645.

Johnson, Bruce. 2017. 'Sound Studies Today: Where Are We Going.' In *A Cultural History of Sound, Memory, and the Senses*, edited by Joy Damousi and Paula Hamilton, 1–17. New York: Routledge.

Keegan, Matthew. 2018. 'Where Is the World's Noisiest City?' *Guardian*, 8 March 2018. www.theguardian.com/cities/2018/mar/08/where-world-noisiest-city.

Klemm, Celine, Enny Das, and Tilo Hartmann. 2016. 'Swine Flu and Hype: A Systematic Review of Media Dramatization of the H1N1 Influenza Pandemic.' *Journal of Risk Research* 19 (1): 1–20. https://doi.org/10.1080/13669877.2014.923029.

Klein, Naomi. 2007. *The Shock Doctrine: The Rise of Disaster Capitalism*. Toronto: Alfred A. Knopf Canada.

Koszarski, Richard. 2005. 'Flu Season: Moving Picture World Reports on Pandemic Influenza, 1918–19.' *Film History* 17 (4): 466–485.

Lee, James Kyung-Jin. 2004. *Urban Triage: Race and the Fictions of Multiculturalism*. Minneapolis: University of Minnesota Press.

Levy, Hadar and Amit Pinchevski. 2017. 'Resounding News: The Acoustic Conventions of Israeli Newscasts.' *International Journal of Communication* 11 (January): 3355–3373. https://ijoc.org/index.php/ijoc/article/view/6334.

Li, Yao and Harvey L Nicholson. 2021. 'When "Model Minorities" Become "Yellow Peril" – Othering and the Racialization of Asian Americans in the COVID-19 Pandemic.' *Sociology Compass* 15 (2): e12849, 1–13. https://doi.org/10.1111/soc4.12849.

Lorenzo-Dus, Nuria and Annie Bryan. 2011. 'Recontextualizing Participatory Journalists' Mobile Media in British Television News: A Case Study of the Live Coverage and Commemorations of the 2005 London Bombings.' *Discourse & Communication* 5 (1): 23–40. https://doi.org/10.1177/1750481310390164.

Maiorana, Massimo, via Storyful. 2020. 'Trumpeter Performs "Imagine" on Italian Balcony during Coronavirus Lockdown.' YouTube, 16 March 2020. www.youtube.com/watch?v=Uo6wfP62j1U.

McAlister, Elizabeth. 2012. 'Soundscapes of Disaster and Humanitarianism: Survival Singing, Relief Telethons, and the Haiti Earthquake.' *Small Axe: A Journal of Criticism* 39 (November): 22–38. https://doi.org/10.1215/07990537-1894078.

McCauley, Michael, Sara Minsky, and Kasisomayajula Viswanath. 2013. 'The H1N1 Pandemic: Media Frames, Stigmatization and Coping.' *BMC Public Health* 13 (1): 1116.

McNeil, Donald G. Jr. 2009. 'Unusual Strain of Swine Flu Is Found in People in 2 States: National Desk.' *New York Times*, 24 April 2009, A13.

McRuer, Robert. 2018. *Crip Times: Disability, Globalization, and Resistance*. New York: New York University Press.

Messineo, Dan. 2015. 'Doing It Alone: Do Video Journalists Affect the Quality and Credibility of Television News?' Master's thesis, Colorado State University.

Mounk, Yasha. 2020. 'The Extraordinary Decisions Facing Italian Doctors.' *Atlantic*, 11 March 2020. www.theatlantic.com/ideas/archive/2020/03/who-gets-hospital-bed/607807/.

Nerlich, Brigitte and Nelya Koteyko. 2012. 'Crying Wolf? Biosecurity and Metacommunication in the Context of the 2009 Swine Flu Pandemic.' *Health & Place* 18 (4): 710–717. https://doi.org/10.1016/j.healthplace.2011.02.008.

Nhsydneynews. 2009. 'Swine Flu Australia 2009.' YouTube, 27 April 2009. www.youtube.com/watch?v=c3oTEtXLEdg.

New China TV. 2020. 'Wuhan: A City Getting Loud Again.' YouTube, 24 September 2020. www.youtube.com/watch?v=rHPJwOIHQcM.

NowThis News. 2020. 'Italian Opera Singer Serenades Quarantined Florence Amidst Coronavirus Outbreak | Now This.' YouTube, 16 March 2020. www.youtube.com/watch?v=dhTjGS3QkYE.

O'Grady, Siobhán. 2020. 'China's Coronavirus Lockdown – Brought to You by Authoritarianism.' *Washington Post*, 27 January 2020. www.washingtonpost.com/world/2020/01/27/chinas-coronavirus-lockdown-brought-you-by-authoritarianism/.

Paglianti, Nanta Novello. 2020. 'Rituals during Lockdown: The "Clap for Our Carers" Phenomenon in France.' *Culture e Studi del Sociale-CuSSoc* 5 (1): 315–322.

Parks, Lisa and Janet Walker. 2020. 'Disaster Media: Bending the Curve of Ecological Disruption and Moving toward Social Justice.' *Media + Environment* 2 (1): 1–30. https://doi.org/ 10.1525/001c.13474.

Puar, Jasbir K. 2017. *The Right to Maim: Debility, Capacity, Disability*. Durham: Duke University Press.

Qin, Amy, Steven Lee Myers, and Elaine Yu. 2020. 'China Tightens Wuhan Lockdown in "Wartime" Battle with Coronavirus.' *New York Times*, 6 February 2020. www.nytimes.com/2020/02/06/world/asia/coronavirus-china-wuhan-quarantine.html.

Ravizza, Simona. 2020. 'Coronavirus: primi casi a Milano. Cosa sappiamo dei nuovi contagi in Lombardia, Veneto e Piemonte.' *Corriere della Sera*, 22 February 2020. www.corriere.it/cronache/20_febbraio_22/coronavirus-italia-nuovi-contagi-lombardia-veneto-245e72d4-5540-11ea-8418-2150c9ca483e.shtml.

Reuters. 2020. '"Ghost City": Commute through China's Deserted Capital amid Coronavirus.' 6 February 2020. www.reuters.com/article/us-china-health-beijing-commute-idUSKBN201027.

Secon, Holly. 2020. 'An Interactive Map of the US Cities and States Still under Lockdown – And Those That Are Reopening.' *Business Insider*, 3 June 2020. www.businessinsider.com/us-map-stay-at-home-orders-lockdowns-2020-3.

Sky News. 2009. 'Australia Monitoring 28 for Swine Flu 16th May 2009.' YouTube, 16 May 2009. www.youtube.com/watch?v=sz2MI8TFeV4.

Small, Christopher. 1998. *Musicking: The Meanings of Performing and Listening*. Hanover: University Press of New England.

Snyder, Zach, dir. *Dawn of the Dead*. 2004. Universal Pictures.

South China Morning Post. 2020. 'Wuhan People Chant from Windows to Boost Morale.' Facebook, 28 January 2020, www.facebook.com/scmp/videos/186292952563134/.

Spratt, Meg. 2001. 'Science, Journalism, and the Construction of News: How Print Media Framed the 1918 Influenza Pandemic.' *American Journalism* 18 (3): 61–79. https://doi.org/10.1080/08821127.2001.10739324.

Telegraph. 2020. 'Chinese Authorities Humiliating People for Not Wearing Face Masks.' *The Telegraph*, 19 February 2020. www.telegraph.co.uk/news/2020/02/19/chinese-authorities-humiliating-people-not-wearing-face-masks/.

Traverso, Vittoria. 2020. 'Balconies Have Always Been Designed to Captivate and Inspire the Masses. But amid the Coronavirus Pandemic, They've Taken on a Newfound Importance.' *BBC*, 10 April 2020. www.bbc.com/travel/article/20200409-the-history-of-balconies.

Tuggle, C. A., Forrest Carr, and Suzanne Huffman. 2014. *Broadcast News Handbook: Writing, Reporting & Producing in a Converging Media World*. New York: McGraw-Hill.

Vitale, Daniele. 2020. '"Bella Ciao" – Balcony Sax Performance in Italy.' YouTube, 19 March 2020. www.youtube.com/watch?v=BSWYSww-RlY.

Walker, Adam. 1792. *Remarks Made in a Tour from London to the Lakes of Westmoreland and Cumberland in the Summer of 1791*. London: G. Nicol.

20 Four Sounds against Capitalocene

Lockdown, Music, and the Artist as Producer

Makis Solomos

Lockdown as a Politically Repressive Measure

Listen: https://soundcloud.com/makis-solomos/manifestations-montage-decembre-2019. *Montage of sound ambiences during demonstrations, Paris, December 2019.*

5 December 2019, France: large demonstrations are being held in protest against the pension reform of the Macron government, cities are filled with clamour, slogans, and songs, and the sound ambience is agitated. Then there are clashes with the security forces, and we hear sounds of tear gas and sound grenades being thrown. People try to flee, or to pass by the demonstration, but in the end the song resumes: 'On est là.'[1]

5 March 2020: a successful parade of academics against a proposed neoliberal law concerning higher education,[2] accompanied by the sound of *batucadas* (traditional Brazilian percussion ensembles) and conch orchestras.

8 March 2020: a huge feminist demonstration, organised with a route like a work of art. A group of lawyers performs a haka in protest against pension reform, a dance called Because of Macron is performed in solidarity with migrant women, and there is a mass 'Cry of anger against the "César of shame"'[3] among other manifestations (see Figure 20.1).

In France, an important social movement took place from 5 December 2019 to 8 March 2020, with nine demonstrations at the national level (the last one, undoubtedly, being the most beautiful: a feminist demonstration that converged with the social movement). In these mobilisations, sound, music, and the artists involved played an important role on three levels: first, the movement was accompanied by songs, brass bands, or choreographies that also deserved acclaim for their artistic content, for example as works of mobile art;[4] second, the demonstrations' soundscapes (with the shouts and songs of the crowds, the sound grenades, etc.) show how sounds and sound milieus play an important psycho-political role; finally, musicians and artists also went on strike – opera dancers, musicians from conservatories, and the like – and made their own demands during the social protest.[5]

Then, suddenly, at noon on 17 March 2020, this incredible outpouring of political and artistic energy gave way to a strange silence. The streets were deserted, the public gardens were closed, there were no more cars, trains, or planes. This was the start of the first lockdown in France, the only measure available – due to the government's lack of preparation despite numerous warnings from scientists and environmentalists – to halt the progression of the COVID-19 pandemic. Should lockdown also be taken as a repressive measure against the protest movements? Given that the protests resumed as soon as the first lockdown ceased, and that, despite the subsequent lockdowns that were less rigorous, the demonstrations against other laws proposed by Macron[6] were very large, as were mobilisations in other parts of the world, I would answer that question in the affirmative. The first lockdown not only brought the economy (a certain economy, as we shall see) and people's lives to a standstill, it also cut short the massive mobilisations that were taking place not only in the streets of French cities, but also in the streets of Santiago and other cities in Chile, Hong

DOI: 10.4324/9781003200369-24

Figure 20.1 8 March 2020, Paris. Feminist demonstration as part of the social movement against pension reform (photo taken by the author, Makis Solomos).

Kong, Lebanon, Israel, and many other countries. Indeed, the first lockdown stopped very strong protest movements against neoliberal and/or authoritarian policies, movements as important as the mobilisations of the Arab Spring or the so-called 'place' movements of the early 2010s (see Pleyers and Glasius 2013).

Lockdown is a form of political repression. Another form of political repression is the re-imposition of curfews, which so-called democratic regimes have hardly experienced outside of times of war. Without going so far as to cite Giorgio Agamben's rather ambiguous reflections on the pandemic,[7] we can surely state, without being accused of being conspiracy theorists, that 'the Sars-CoV-2 virus, or COVID-19 … is a non-human intrusion into our social existences. It has exposed a disaster capitalism in which sovereign states coordinate to limit a pandemic by sacrificing political liberty and by differentiated exposure to poverty, illness and death' (Guillibert 2021, 19).

Anthropocene and Capitalocene

Listen: https://soundcloud.com/makis-solomos/journal-de-confinement-mars-2020. *Lockdown diary, Paris, March 2020.*

During the first lockdown, an atmosphere of astonishment, fear, and often despair settled in. For some, the hope was that this would serve as a lesson for the world to come. Sometimes, however, we took pleasure in observing the silence, and the return of birds to the city. From the start of the lockdown, several projects were launched to observe changes in urban sound environments. One example of this was the French project *Silent Cities. Paysages sonores d'un monde confiné*: 'Since the implementation, throughout the world, of measures including the lockdown of populations and social distancing, many press articles have encouraged city dwellers to contemplate what some already present as a renaturation, or even a rewilding, of urbanized spaces' (Challéat et al. 2020). Thus wrote the authors of the project, who ended up bringing together many people from several countries around the world, using a precise protocol for recording sound ambiences from the lockdown. For instance, they compared a recording from the Allée de Brienne in Toulouse on 16 March 2020 at 7:00 p.m. with a recording from the same Allée at the same time on 20 March 2020.

The pandemic clearly marks a new ecological crisis, since it is a zoonosis (see Malm 2020). In this sense, it could be interpreted as the most obvious warning signal yet of the Anthropocene. As such, the disappearance of anthropogenic noise in cities during the first lockdown could be taken as nature taking revenge on human beings, if we continue to distinguish the two rather than speak of the living. We could dream of echoes of the past: as Bernie Krause writes in *The Great Animal Orchestra*,

> It is sixteen thousand years ago, and the plains teem with life. ... It is jam-packed with nonhuman life, whose individual voices coalesce into an intense and collective symphony. ... In this one verdant spot thousands of creatures sing in choruses at all times of the day and night. The visual spectacle is impressive, but the sound is absolutely glorious.
>
> (2012, 3–4)

But why would animals produce what human beings did in the era of tonal music – symphonies, choirs, well-behaved concerts? Why would they not make a noise like the magnificent lion's roar found in Edgard Varèse's *Ionisation*? Today, it is more and more difficult to subscribe to this wilderness ecology, which is nostalgic and that takes us back to the days of the so-called musical sounds, when we grew to love certain forms of noise. Above all, it is an ecology that is insensitive to social issues, to issues of racialisation or to gender issues.

Certainly, we all liked the silence and the birds during the first lockdown, but we should not forget that it was a terrible time – which is far from over – for the most economically precarious populations, for those suffering domestic violence, for students, and for the poorest countries. It is important not to think in terms of environmental ecology alone; Félix Guattari, in his 1989 manifesto, said that there is not only one ecology (the environmental one): there are also 'social' and 'mental' ecologies, and the simultaneous concern for these three ecologies forms what he called 'ecosophy' (2000, 27–28).[8] Increasingly, green political parties are concerned with social issues as well as environmental ones. This is why, with the economist Benjamin Coriat, we could say that analysis of the Anthropocene (with precious and indispensable indicators such as the state of the ozone layer, the melting of glaciers, the warming of the planet, the state of biodiversity, etc.) is useful, but that it has its limits:

> Many of those who refer to it have an apolitical vision of the disruption we are witnessing: for them, it is 'human activity' which is the cause of disturbances. This is where the notion of Capitalocene allows us to go further in the analysis. Based on observations of climate change, it highlights the fact that the state of degradation that we have reached is not linked to 'humanity' – hypostasized and undefined – but to a very particular humanity, organized by a predatory economic system.
>
> (Coriat 2020)

266 *Makis Solomos*

This is also what young militants tell us:

> Today, as capitalism is making the disappearance of humanity as a whole a very real possibility, I know that more and more of us are thinking that there needs to be profound change to this capitalist, patriarchal and racist system that is destroying the planet.
>
> (Mathilde 2021)

Music Is Not Just the World of Consolation

See and listen: www.shutterstock.com/video/clip-10855673-crowded-music-concert-stage-zoom-camera-+10. *Advertisement for a 'crowded-music-concert-stage-zoom-camera.'*[9]

Artists and all those working in the cultural sector suffered tremendously during the successive lockdowns. In 2020, after the concert halls, theatres, cinemas and dance halls closed, there was a succession of cancellations of major summer and autumn festivals. The economic situation of those in France who enjoy the status of freelance workers in show business was dire, even though the government extended their rights for a year. In addition to economic suffering, there were several forms of psychological depression due to cancelled shows, the inability to present work to the public, and the general uncertainty: 'This uncertainty, along with a total lack of any indication as to when it will end, [opened] a chasm, even for professionals experienced in intermittent work and its fallow periods, which necessitate constant self-reinvention' (Merckx 2020).

Nonetheless, during the first lockdown, artists tried not to be defeated. Many of them attempted to overcome this lockdown from its very inception, by doing what they loved: music, dance, theatre, as and where they could, 'looking for forms suited to the moment in order to hold on, to resist, to continue to exist, as well, not to lose a part of their identity which crowns the phase of creation by a presentation, an exchange, a sharing' (Merckx 2020). And the audience seemed to be there, that is to say mainly in front of their computers. Thus, we had ersatz concerts produced by concert halls, by professional orchestras, or by high-level artists who had the means to produce videos to professional standards. In France, we soon had a recording of Ravel's *Boléro* on YouTube performed by the Orchestre National de France,[10] a video that went viral (although whether spectators watched it to the end is another question). In another popular video, the dancers of the Paris Opera – who had been on strike just before the lockdown (see above) – created a *Ballet en visioconférence* as a tribute to caregivers, produced by the filmmaker Cédric Klapisch.[11] The dancers, who were filmed dancing in their homes,

> appeared both natural and professional performing movements whose very nature and exquisite execution stood out against the domestic setting. They transformed a hob, a stairwell, a corridor, the frame of a window, a carpet or a bathtub into a new dance bar, a new workspace.
>
> (Merckx 2020)

As well as these so-called Zoom concerts, we had the famous concerts from windows and balconies at 8:00 pm, which came from Italy. In France, they took a very interesting turn when they became a means of protest: a reminder that we were applauding the very caregivers who had been suppressed by the government when they demonstrated against the break-up of public hospitals.[12] There were many other popular initiatives, for example podcasts combining music or sound creation. Finally, there were all of the initiatives discussed previously, by grassroots musicians and artists (and countless amateurs as well) who tried to resist the desperation of lockdown by playing alone or with others in amateur videos.

Four Sounds against Capitalocene 267

What about Zoom concerts, either by large orchestras or grassroots musicians? Certainly, they brought consolation, but no one was fooled (it is significant that there were far fewer in the following lockdowns). And it is also difficult, with hindsight, to share the enthusiasm of certain journalists, in the early days of the first lockdown, for videos of artists in beautiful houses or splendid gardens playing classical music hits (see for instance the videos of the French violinist Renaud Capuçon).[13] 'Lockdown inspires,' we read in a video posted online by the newspaper *Le Monde* (2020).

The Artist as Producer

Listen: https://soundcloud.com/user-965086014/01-violazioni-della-presenza?in=user-9650 86014/sets/agostino-di-scipio-works-for-strings-and-live-electronics. *Agostino Di Scipio: Violazioni della presenza (excerpt).*[14]

Artists are rightly concerned that Zoom concerts have contributed to the idea that art has to be free and that the artist's work does not have to be remunerated (whereas if there is one thing that should have been free during this health, political, economic, ecological, and social crisis, it would have been internet connection!). On the other hand, as rock musician Dave Grohl wrote, 'the coronavirus pandemic has reduced today's living music to unflattering little windows that look like video surveillance screens, and sound like Neil Armstrong's distorted radio transmissions from the moon, all chopped and compressed' (2020). Zoom concerts do introduce us into a world where everything is recorded, stored, or sold, including our personal data – 'the age of surveillance capitalism' (Zuboff 2019).[15] Most of us don't care, or learn to live with it, until one day we find that Zoom has banned a conference (political censorship), as happened with the well-known Palestinian activist Leila Khaled (see Speri and Biddle 2020). There is no need to invoke Agamben again (or Michel Foucault) to say that capitalism has reached the stage of biopolitics. If you prefer a more moderate, but equally disturbing version, I will cite pianist and researcher Pavlos Antoniadis, who draws a parallel between the live animal markets where COVID-19 was born – where the ground is always wet with runoff blood from slaughtered wild animals and seafood – and the wet markets of music:

> It is the market for streaming digital data between brains connected in networks, the 'wetware' and 'netware' of new computer and communication technologies, complementing the distinction between 'software' and 'hardware' in the context of *cognitive capitalism*.
> (Antoniadis 2022)

The progress of GAFAM (Google-Apple-Facebook-Amazon-Microsoft), of Zoom, and other empires of digital capitalism has been ruthless during the crisis. They are taking over sectors that are still not very industrialised, such as education. Art has been delivered to them for a long time in its commercial version, and it is a prize that will cost the art world dearly: an article in the international edition of the *Guardian* announced in its headline that 'Streaming Platforms Aren't Helping Musicians – and Things Are Only Getting Worse' (Jean 2019). And this is just an example. In fact, to quote the French economist Cédric Durand (2020), we shouldn't be talking about digital capitalism but about 'techno-feudalism,' an economy based on returns, on predation, and the political domination of multinationals. In his book *Techno-Feudalism: A Critique of the Digital Economy*, Durand explains that technological progress conceals

> techno-feudal regression: ... the mutation that comes with the growth of information technology impacts the very foundations of the mode of production. It destabilises

268　*Makis Solomos*

its basic principles … The growth of the digital destroys competitive relationships in favour of relationships of dependence, which derails the whole system and privileges predation over production.

(2020, 227)

The reference to feudalism is thus not a metaphor: analysis of social relationships generated by these changes indicates that the relationship between digital users and multinationals bears a strange resemblance to the links between serfs and lords in feudal societies.

If artists want to avoid being eliminated or instrumentalised – serving as a Trojan horse for techno-feudalism – by this development, it is important to think about the means of production they use and the social relationships they induce. Art and especially music are sometimes steeped in a form of idealism whereby, with the pretext of focusing on the music itself, musicologists leave this reflection to sociologists or politically engaged people. The mutation that we are witnessing, for which in many ways the crisis serves as a gigantic territory for experimentation, will surely change the face of music. Let us become a little materialistic, in order to think of the artist as a producer, and quote the end of Walter Benjamin's lecture from the 1930s, 'The Author as Producer': 'You may have noticed that the chain of thought whose conclusion we are approaching only presents the writer with a single demand, the demand of reflecting, of thinking about his position in the process of production' (Benjamin 1970, 95).

This evolution is not inevitable. It was possible that a new notion of the commons may develop, as many claim. The writer Alain Damasio (2020) said:

The GAFA[M] will perish on their own when we understand that the Internet must become a public service. It must be part of the commons, just like air and water. A global commons. We have to work on that as much as on a solidarity economy.

We can also imagine cultural short circuits that would enable us, 'as in agriculture, to do without intermediaries, while taking into account the reduction of the ecological footprint and the constraints of social distancing' (Laurent Jeanpierre, quoted in Correai 2020). We can still think of degrowth (even though the word may be frightening) not only in terms of living with less technology, but also in terms of developing new technologies, which would allow us to move towards autonomy, which is 'emancipation from dependence on alienating and heteronomous systems' (Deriu 2015, 125). Thinking of the artist as a producer would perhaps open the doors to many other concrete, 'real utopias,' to quote sociologist Erik Olin Wright (2010), and would certainly not distance us from the art world – quite the contrary.

Notes

1 The song 'We Are Here' is emblematic of the so-called yellow vests protests. Originally, it was a football song, sung by supporters of Lens' Racing Club: 'We are here, we are here, in misfortune or glory, we are here, for the love of the jersey you wear on your back, in misfortune or glory, we are here,' which was then taken up during the railway strike in 2018: 'We are here, we are here, even if you don't want us, we are here, for the honour of the railway workers and the future of our kids, even if you don't want us, we are here.' The yellow vests' version, taken up during the demonstrations that were part of the social movement from December 2019 to March 2020 is: 'We are here, we are here, even if Macron does not want us, we are here, for the honour of the workers and for a better world, even if Macron doesn't want us, we are here.'

Four Sounds against Capitalocene 269

2 The LPPR (Multiannual Research Programming Law). The government ended up passing the law after the first lockdown, in September 2020, despite very strong opposition from the entire university community.

3 The 'César of shame' refers to the fact that, despite rape accusations, Roman Polanski had just been awarded the 2020 César.

4 In recent years there has been increased research into mobile musical or listening practices, often in connection with mobile listening devices (see Gopinath and Stanyek 2014), and we could also refer to soundwalks. On musical practices during demonstrations, see the interview with the Fanfare Invisible (Collective 2023b).

5 Opera dancers led an important strike against pension reform, from December 2019 to February 2020. Some of their actions were highly publicised, such as the performance of an extract from *Swan Lake* on the steps of the Opéra Garnier on 24 December 2019 (see Collective 2020). As for musicians from conservatories, see for instance the journal *La Crécelle* (subtitled 'The Crackling Newspaper of the Struggling Conservatory'), https://documentations.art/La-Crecelle-Journal-du-Conservatoire-en-lutte. For documentation of the role that music, sound, and artists played during this social protest, see Collective 2023a.

6 In particular the global security law (autumn 2020). And of course we must also mention the protests against the health passport (summer 2021).

7 Agamben's first article on the COVID-19 pandemic was published in *Il Manifesto* on 26 February 2020. In it, he makes this point, among others: 'We have to consider the state of precarity and fear that has been in recent years systematically cultivated in people's minds – a state which has resulted in a natural propensity for mass panic, for which an epidemic offers the ideal pretext. ... Limitations on freedom are thus being willingly accepted, in a perverse and vicious cycle, in the name of a desire for security – a desire that has been generated by the same governments that are now intervening to satisfy it' (quoted in Agamben 2021). Its title is 'The Invention of the Pandemic.' It was criticised because it underestimated the scale of the pandemic; but it is true that it was published just before the major spread of the pandemic in Europe. In a subsequent article ('New Reflections,' published on 27 April 2020 in *Neue Zürcher Zeitung*), he wrote:

> The hypothesis that we are experiencing the end of a world – the world of bourgeois democracy that is built on rights, parliaments, and the division of powers – is now spreading widely. That world is being replaced by a new despotism that, with the pervasiveness of its controls and with its suspension of all political activity, will be worse than the totalitarianisms we have known thus far. Political commentators call it the 'Security State' – in other words, a state where 'for security reasons' (in this instance for the sake of 'public health,' a term that recalls the Reign of Terror's infamous 'Committee of Public Safety') there's no limit to the repression of individual freedoms.

> (quoted in Agamben 2021)

8 'The Earth is undergoing a period of intense techno-scientific transformations. If no remedy is found, the ecological disequilibrium this has generated will ultimately threaten the continuation of life on the planet's surface. Alongside these upheavals, human modes of life, both individual and collective, are progressively deteriorating. Kinship networks tend to be reduced to a bare minimum; domestic life is being poisoned by the gangrene of mass-media consumption ... It is the relationship between subjectivity and its exteriority – be it social, animal, vegetable or Cosmic – that is compromised in this way, in a sort of general movement of implosion and regressive infantilisation. [We need then] an ethico-political articulation – which I call *ecosophy* – between the three ecological registers (the environment, social relations and human subjectivity)' (Guattari 2000, 27–28).

9 This is what can be found by typing 'zoom concert background' into your browser. Of course, you can try putting the sound on but it doesn't make any difference, since it's up to you to create the sound!

10 www.youtube.com/watch?v=Sj4pE_bgRQI.

11 www.youtube.com/watch?v=NiM-x4fPFRI.

12 See the chapter by Alessandro Greppi and Diane Schuh in this volume.

13 See for instance the video in the following link: www.youtube.com/watch?v=MnCNkClwxYA.

270 *Makis Solomos*

14 Agostino Di Scipio is a composer who is very attentive to the material conditions of music production. See his recent important book (2021).

15

> Surveillance capitalism unilaterally claims human experience as free raw material for translation into behavioral data. Although some of these data are applied to product or service improvement, the rest are declared as proprietary *behavioural surplus*, fed into advanced manufacturing processes known as 'machine intelligence,' and fabricated into *prediction products* that anticipate what you will do now, soon, and later.
>
> (Zuboff 2019, 8)

References

Agamben, Giorgio. 2021. *Where Are We Now? The Epidemic as Politics*. Translated by Valeria Dani. London: Eris.

Antoniadis, Pavlos. 2022. 'Performance et crise sanitaire.' In *Arts, Ecologies, Transitions*, edited by Roberto Barbanti, Isabelle Ginot, Makis Solomos, and Cécile Sorin. Unpublished manuscript.

Benjamin, Walter. 1970. 'The Author as Producer.' Translated by John Heckman. *New Left Review* I/62 (July–August): 83–96.

Challéat, Samuel, Nicolas Farrugia, Amandine Gasc, and Jérémy Froidevaux. 2020. 'Silent·Cities. Paysages sonores d'un monde confiné.' https://laboratoireparallele.com/2020/04/17/silent%C2%B7cities-paysages-sonores-dun-monde-confine.

Collective. 2020. '"Si des bourgeois viennent nous voir, nous ne sommes pas responsables": rencontre avec des danseuses de l'opéra en grève.' *Acta.Zone*, 25 January 2020. https://acta.zone/si-des-bourgeois-viennent-nous-voir-nous-ne-sommes-pas-responsables-entretien-avec-des-danseuses-en-greve-de-lopera-de-paris.

Collective. 2023a. 'Paris, décembre 2019 – mars 2020. Documenter le movement social avec la musique et le son.' *Filigrane. Musique, esthétique, sciences, société* 28, forthcoming.

Collective. 2023b. 'Interview avec la Fanfare Invisible.' *Filigrane. Musique, esthétique, sciences, société* 28, forthcoming.

Coriat, Benjamin. 2020. 'L'âge de l'anthropocène, c'est celui du retour aux biens communs.' *Mediapart*, 16 May 2020. www.mediapart.fr/journal/economie/160520/benjamin-coriat-l-age-de-l-anthropocene-c-est-celui-du-retour-aux-biens-communs.

Correai, Mickael. 2020. 'La culture comme bien commun.' *Mediapart*, 5 May 2020. www.mediapart.fr./journal/culture-idees/050520/la-culture-comme-bien-commun.

Damasio, Alain. 2020. 'Pour le déconfinement, je rêve d'un carnaval des fous, qui renverse nos rois de pacotille.' Interview by Hervé Kempf. *Reporterre. Le quotidien de l'écologie*, 28 April 2020. https://reporterre.net/Alain-Damasio-Pour-le-deconfinement-je-reve-d-un-carnaval-des-fous-qui-renverse-nos-rois-de-pacotille.

Deriu, Marco. 2015. 'Autonomie.' In *Décroissance. Vocabulaire pour une nouvelle ère*, edited by Giacomo D'Alisa, Federico Demaria, and Giorgos Kallis, 121–128. Paris: Le passager clandestin.

Di Scipio, Agostino. 2021. *Circuiti del Tempo. Un percorso storico-critico nella creatività musicale elettroacustica e informatica*. Lucca: Libreria Musicale Italiana.

Durand, Cédric. 2020. *Techno-féodalisme. Critique de l'économie numérique*. Paris: Éditions de la Découverte.

Gopinath, Sumanth and Jason Stanyek, eds. 2014. *The Oxford Handbook of Mobile Music Studies*. 2 vols. Oxford: Oxford University Press.

Grohl, Dave. 2020. 'The Day the Live Concerts Return.' *Atlantic*, 11 May 2020. www.theatlantic.com/culture/archive/2020/05/dave-grohl-irreplaceable-thrill-rock-show/611113.

Guattari, Felix. 2000. *The Three Ecologies*. Translated by Ian Pindar and Paul Sutton. London: Athlone Press.

Guillibert, Paul. 2021. *Terre et capital. Pour un communisme du vivant*. Paris: Éditions Amsterdam.

Jean, Evet. 2019. 'Streaming Platforms Aren't Helping Musicians – And Things Are Only Getting Worse.' *Guardian*, 20 November 2019. www.theguardian.com/culture/2020/nov/14/streaming-platforms-arent-helping-musicians-and-things-are-only-getting-worse.

Krause, Bernie. 2012. *The Great Animal Orchestra: Finding the Origins of Music in the World's Wild Places*. London: Profile Books.

Le Monde. 2020. 'Coronavirus: le confinement inspire les musiciens, stars ou inconnus.' 18 March 2020. www.youtube.com/watch?v=m4QZS6hSiHE.

Malm, Andreas. 2020. *La Chauve-souris et le Capital. Stratégie pour l'urgence chronique*. Translated by Étienne Dobenesque. Paris: La Fabrique.

Mathilde, Irène. 2021. 'On veut une université ouverte aux enfants d'ouvriers, aux travailleurs, aux étrangers!' *Révolution permanente*, 11 December 2021. www.revolutionpermanente.fr/Irene-On-veut-une-universite-ouverte-aux-enfants-d-ouvriers-aux-travailleurs-aux-etrangers.

Merckx, Ingrid. 2020. 'Le spectacle vivant au temps du Corona.' *Mediapart*, 5 November 2020. www.mediapart.fr/journal/culture-idees/051120/le-spectacle-vivant-au-temps-du-corona/.

Pleyers, Geoffrey and Marlies Glasius. 2013. 'La résonance des "mouvements des places": connexions, émotions, valeurs.' *Socio. La nouvelle revue des sciences sociales* 2: 59–80. https://doi.org/10.4000/socio.393.

Speri, Alice and Sam Biddle. 2020. 'Zoom Censorship of Palestine Seminars Sparks Fight Over Academic Freedom.' *Intercept*, 14 November 2020. https://theintercept.com/2020/11/14/zoom-censorship-leila-khaled-palestine.

Wright, Erik Olin. 2010. *Envisioning Real Utopias*. London: Verso.

Zuboff, Shoshana. 2019. *The Age of Surveillance Capitalism: The Fight for the Future at the New Frontier of Power*. London: Public Affairs.

Afterword

Coping with Crisis through Coronamusic

Niels Chr. Hansen

Setting the Scene

For many Danes, an iconic soundbite of the pandemic year 2020 was Prime Minister Mette Frederiksen's firm assertion on 11 March that 'now we must stand together by keeping a distance from each other' (Frederiksen 2020; see Lantz 2021 for a rhetorical analysis). On the historic night of the WHO's pandemic declaration, her government had called an urgent press conference to announce the almost complete shutdown of Danish society, starting from the following morning. Along with New Zealand's Jacinda Ardern, Norway's Erna Solberg, Germany's Angela Merkel, Finland's Sanna Marin, Iceland's Katrín Jakobsdóttir, and Taiwan's Tsai Ing-wen, the Danish prime minister was a member of an exquisite ensemble of female heads of state who acted swiftly, stringently, and empathically during the early days of the novel coronavirus pandemic – erring on the side of caution – to successfully shield their most vulnerable populations.[1]

Paradoxically, Frederiksen simultaneously encouraged social cohesion while prohibiting the most commonly used means for achieving it. Closure of schools, childcare facilities, restaurants, nightclubs, concert venues, indoor leisure activities, non-critical public workplaces, private associations, and religious gatherings effectively removed crucial nodes in the social infrastructure of this northern European welfare state (Birk 2017). Like many other communities worldwide, the Danes were requested to sacrifice soft, social needs for hard, human health.

Yet, unbeknownst to the political level at the time, music offers a valuable means to connect people safely and seamlessly with exceptional speed across long distances despite societal shutdowns – and this capacity increases manifold when mediated by modern information technology. In fact, recent epidemiological modelling of download count time series data established that the electronica music genre has a basic reproduction number ($R0$) of 3,430 (Rosati et al. 2021, 10), making it about 190 times as 'contagious' as measles. This revitalises the strong imagery of 'viral tunes' promoted in public discourse and serious science (e.g., Fink et al. 2021).

Soundbites from political speeches like the one offered at the outset of this chapter were soon accompanied by even more memorable musical ones. Researchers have defined 'coronamusic' as the 'cultural products resulting from engaging with music during COVID-19 lockdown in ways that make either direct or indirect reference to pandemic life circumstances or the coronavirus itself' (Hansen 2022; see also Hansen et al. 2021). In this afterword, an overview of the Danish context will first exemplify how people coped with the crisis through corona-themed music. Next, a review follows of the emerging research literature on coronamusic creation and consumption as a global strategy for tackling the psychological struggles that arise in the wake of government-sanctioned home confinement.

DOI: 10.4324/9781003200369-25

An Infectious Case of Musical Revivals

In pandemic Denmark, a nerdy – albeit highly competent – choir enthusiast and chief conductor of the Danish National Girls' Choir, Phillip Faber, became a national hero, arguably on equal footing with the chief virologists and epidemiologists advising the government on pandemic management. A widely watched Facebook livestream developed into daily morning singing sessions on national television, led by Faber from a dated grand piano in his line manager's living room. Under normal circumstances, such a niche programme would only have reached a marginal audience – if allowed to air at all. Yet, the Danish population enthusiastically embraced the initiative (Sørensen, Baunvig, and Andersen 2021), and within the span of two weeks, morning singing was upgraded into a weekly, one-hour, prime-time show on Friday nights on DR1, the main channel of the Danish Broadcasting Corporation.

This participatory music making programme achieved the highest viewership of any television programme for six weeks in a row, spellbinding more than one fifth of the population.[2] Tellingly, the ratings for this show surpassed those of the more passive, presentational – and hitherto vastly successful – X Factor show, occupying the same timeslot on a competing nationwide channel during the preceding weeks (Ulfstjerne 2020). Throughout the pandemic year 2020, Danes continued celebrating major events like midsummer, Queen Margrethe's eightieth birthday, the postponed UEFA Championship, and the centenary of the reunification of Northern Schleswig and Denmark, with sing-along programmes on national television.

When other programmes finally took over the leading position, nationally nostalgic themes – such as Danish wildlife and Northern Schleswig – remained in focus. The release of the nineteenth edition of the Folk High School Songbook (*Højskolesangbogen*) in November further cultivated the national song treasure. For the first time in its 126-year history, this songbook reached the publishers' bestseller lists (Baunvig 2020). National nostalgia was sometimes wrapped in self-deprecating humour and ironic distance – the hallmarks of Danish informality. With their quirky paraphrases of patriotic repertoire – replacing public pride with the private trivialities of lockdown life – the singing satire group MAGT, for example, received nearly a quarter million YouTube views.[3]

The aforementioned Danish prime minister rode this wave with opportunistic cunningness, posting videos of herself and her family singing sentimental Danish 1980s hits, while peeling potatoes accompanied by the national sing-along broadcast. While many political leaders' poll numbers increased as populations worldwide rallied around their respective flags (Baekgaard et al. 2020; however see also Kritzinger et al. 2021), Frederiksen's Instagram followership grew dramatically (Mehlsen 2000). This ultimately offered her a direct channel of communication to more than 20 per cent of young Danes.

Celebrities from the musical domain also partook in building community resilience. An example from Denmark was the coronakoncerter.dk initiative, wherein 20 concerts with famous musicians were televised daily from the young bassist Mathæus Bech's Copenhagen living room, from 12 March onwards, culminating in a four-day, virtual mini-festival in April. While celebrity musicians were featured prominently in televised shows from their private homes, the participatory element nearly always took centre stage. Specifically, songs were drawn almost exclusively from the well-known repertoire, and karaoke-style lyrics were displayed on-screen. Clips with neighbours singing joyfully from outdoor settings, compliant with restrictions, were frequently interspersed with family videos from social media, to encourage amateur-focused audience participation.

Summing up, during the lockdown in Denmark, musical activities achieved a more recognised position in public discourse, associated with widespread participation in nationally televised sing-along programmes. Participation was active and extended across traditional age and gender gaps (Sørensen, Baunvig, and Andersen 2021) – and also to amateurs, whose involvement in music making has traditionally been stigmatised (Stebbins 2020). The resulting group identity centred around nostalgic themes, mediated in part by a revival of the national treasure of song. Communal singing may indeed have provided a non-religious form of collective ritualised behaviour, which was suitable for replacing ceased rituals – church ceremonies, concerts, and sports games alike – in an increasingly secular world (Sørensen, Baunvig, and Andersen 2021). Notably, the fact that this resurgence towards flow TV and live-streamed concerts occurred during a period when streaming platforms were otherwise winning over market shares suggests that the contemporaneity that live transmission offers in comparison to taped formats may have increased its overall potential as a means of coping (Sørensen, Baunvig, and Andersen 2021), possibly through enhancing social connection (Swarbrick et al. 2021).

Public figures such as the Danish prime minister capitalised upon music's pandemic popularity, and may even have benefitted from it when soliciting support for difficult political decisions. It seems relevant to speculate whether music-induced social cohesion could have played a causal role in the high compliance rates and testing willingness in Denmark (Jørgensen et al. 2021). By funding a research project on a related topic, The Independent Research Fund Denmark indeed acknowledged the potential of music-based, virtual rituals in everyday coping behaviour (Baunvig 2020). The topical interest in the effects of musical behaviours on wellbeing offers a unique opportunity for researchers to establish and share an empirical evidence base for future policy making. Consequently, what role pandemic-instigated musical activities and corona-themed musical repertoires played in worldwide, collective and individual psychological coping during the pandemic lockdowns of 2020 will now be discussed.

Mental Wellbeing during Lockdown

Musical coping behaviours were adopted in response to the mental health epidemic that arose in the wake of extensive societal lockdowns (Howlin and Hansen, forthcoming). Throughout the world, once the virus arrived, mass gatherings were banned, cultural and sporting events were cancelled, travel and commuting were restricted, and restaurants, bars, shops, and schools were temporarily shut down (Hale et al. 2021; Porcher 2020). These limitations in personal liberty, mobility, and social life were associated with increases in stress, anxiety, fear, depression, and sleep disorders, alongside general boredom, uncertainty, financial insecurity, and emotional disturbance (Hossain et al. 2020). While depressive symptoms may have decreased slightly as restrictions eased over the northern hemisphere summer, they remained consistently high, with averages in the mild depression range (Fancourt, Steptoe, and Bu 2021).

Certain sociodemographic, professional, and contextual factors predisposed individuals to the severity of the deterioration in their wellbeing. People who were young (Kiernan et al. 2021; H. Wang et al. 2020); female (Torales et al. 2020; C. Wang et al. 2020); single, separated, or divorced (Torales et al. 2020); or possessed lower education levels (Mazza et al. 2020) were especially affected. Other risk factors included excessive exposure to COVID-19-related news and social media content; negative coping styles (H. Wang et al. 2020); low levels of physical exercise (Kiernan et al. 2021); limited confidence in health providers; restricted access to personal protective equipment (C. Wang et al. 2020); and certain chronic and mental health comorbidities (Torales et al. 2020). Occasional positive effects of reduced social pressure were also observed in some populations (Pan et al. 2021). Finally, people employed in the healthcare sector (Benfante et al. 2020), and those who regularly came into

Afterword: Coping with Crisis through Coronamusic   275

contact with COVID-19 patients (Benfante et al. 2020; H. Wang et al. 2020) suffered more deeply.

Yet despite these individual differences, mental health ailments affected a large proportion of the population – not just those directly infected with the virus. Due to urgent crisis management, governments understandably did not initially prioritise the wellbeing needs of their citizens. This created an extensive demand for safe, widely accessible, and highly individualised coping strategies (Hansen 2022). Along with crafts, digital arts, writing, and reading for pleasure, music offered a highly prized means of fulfilling these needs (Mak et al. 2021).

Musical Coping Behaviours

More than half of the general public used musical engagement for psychological coping (Cabedo-Mas, Arriaga-Sanz, and Moliner-Miravet 2021; Fink et al. 2021; Kiernan et al. 2021), making it the potentially most frequent solitary, leisure activity during lockdown (Finnerty et al. 2021). Music was widely considered the most effective creative activity for enhancing mental health (Granot et al. 2021; Kiernan et al. 2021), rated at least as effective as exercise and sleep (Vidas et al. 2021). Coping efficacy was, however, mediated by prior importance assigned to music (Fink et al. 2021; Granot et al. 2021). Unlike music, pandemic social media usage and television watching were overall negatively associated with wellbeing (Krause et al. 2021).

Importantly, consumer-based user statistics and self-reports provide starkly divergent pictures of pandemic music listening (Hansen 2022). When asked directly, most people reported increased listening during lockdown (Cabedo-Mas, Arriaga-Sanz, and Moliner-Miravet 2021; Carlson et al. 2021; Ferreri et al. 2021; Fink et al. 2021; Ziv and Hollander-Shabtai 2021; but see also Krause et al. 2021). Yet, as everyday listening situations like commuting and public exercising decreased, so did aggregated weekly streaming counts of the top 200 songs on Spotify (Sim et al. 2022). As playlist followership exhibited similar declines, these changes could not merely be explained by migration towards more specialised niche repertoires. Nevertheless, subjective and objective measures converge on simultaneously increasing video-based consumption on platforms like YouTube (Carlson et al. 2021; Sim et al. 2022). These findings suggest that adapting musical activities for coping purposes involved more attentive and committed listening modes.

These active modes of engagement extended to music making. While many bands and artists cancelled concerts and postponed releases of new music during lockdown (Messick 2020), amateurs took the stage as collective co-creators. Instead of fully ceasing musical activities deemed effective for psychosocial coping (Kiernan et al. 2021), 83 per cent of choirs, 86 per cent of dancing groups, and 60 per cent of instrumental ensembles adopted virtual formats (Draper and Dingle 2021). Despite technological challenges, and lower ratings of group identification and psychological needs satisfaction in virtual compared to physical formats (Draper and Dingle 2021), in many cases, collective music making online succeeded in maintaining identity, sense of purpose, empowerment, and social anchoring during difficult times, in amateurs (Morgan-Ellis 2021) and professionals (MacDonald et al. 2021) alike.

The switch of performance location from traditional rehearsal spaces and music venues to home and outdoor settings (see Hansen et al. 2021), introduced considerable degrees of informality into popular music culture. By performing from their living rooms in casual attire for television shows, and large-scale, virtual charity events like *One World: Together at Home* (18 April 2020) and *One Love Asia Concert* (27 May 2020), celebrity figures allowed

276   *Niels Chr. Hansen*

the public into their more intimate spheres (McIntosh 2021). Musical idols may have thus attained an instrumental role in strengthening global unity, and promoting healthy behaviour in the public at large (Hansen et al. 2021).

Typical technological solutions adopted by musicians involved different degrees of simultaneity between performers and audience (Onderdijk et al. 2021; Swarbrick et al. 2021). For example, virtual concerts that were live-streamed could typically be played back afterwards. Audience members were occasionally visible on-screen, could sometimes make song requests, and/or provide live-chat commentary. Virtual choirs and ensembles used a post-production combination of individual offline recordings, muted singing via teleconferencing, live-broadcasting of singing along to live-streams, and joint singing via low-latency software (Morgan-Ellis 2021). While availability and skills largely determined the solutions adopted, research reveals intriguing relationships between technology and coping potential. Specifically, social connection was felt more strongly by those affected by pandemic-induced stress, and could be enhanced by wearing virtual-reality headsets, or by streaming live (Onderdijk et al. 2021; Swarbrick et al. 2021).

People listened to both nominally happy and sad music during lockdown. While some self-reporting studies found increased happy music and decreased sad music listening (Ferreri et al. 2021), the latter remained prominent in younger age groups (Vidas et al. 2021), who – as we recall – experienced greater impediments to their wellbeing overall (Kiernan et al. 2021; H. Wang et al. 2020). This makes sense in light of findings that people experiencing increased negative emotions used music for solitary emotion regulation, whereas those experiencing increased positive emotions used it as a proxy for social interaction (Fink et al. 2021). The latter positive emotion group may overlap with those producing coronamusic that emphasises positive emotions like happiness, humour, togetherness, and being moved (Hansen et al. 2021), and those 25–50 per cent who discovered new styles and artists during lockdown (Cabedo-Mas, Arriaga-Sanz, and Moliner-Miravet 2021; Ferreri et al. 2021; Fink et al. 2021). The former negative group may, on the other hand, overlap with those 36–68 per cent turning towards nostalgic repertoires (Fink et al. 2021; Gibbs and Egermann 2021). Operationalizing this construct as release dates older than three years, Yeung (2020) showed lockdown-related increases in nostalgic music listening, in 17 trillion Spotify plays across six European countries.

A Special Role for Corona-Themed Music

The preceding review clearly shows how music became a widely used and effective pandemic coping tool. Playing turned more informal, collective, and amateur-driven, and listening turned more attentive, individualised, and multimodal. As intimacy and proximity were progressively mediated through digital means, musical arts were reconfigured to reflect and accommodate pandemic life circumstances (Ulfstjerne 2020). This resulted in the emergence of the coronamusic phenomenon (Hansen 2022; Hansen et al. 2021).

Multiple survey studies substantiate the prominence of coronamusic during lockdown. In France, Germany, Italy, the United Kingdom, the United States, and India, for example, 57 per cent developed a moderate-to-extreme interest in coronamusic (Fink et al. 2021). Nearly 90 per cent of Israelis were exposed to coronamusic in the shape of humorous clips, splitscreen recordings, or performances recorded in empty concert halls or musician's homes (Ziv and Hollander-Shabtai 2021). More than half of Spaniards resided in neighbourhoods with balcony singing, 31 per cent followed coronamusic initiatives, and another 24 per cent had actively participated in some capacity (Cabedo-Mas, Arriaga-Sanz, and Moliner-Miravet 2021). Some people may also have engaged more with live-streaming media (Carlson

Afterword: Coping with Crisis through Coronamusic   277

et al. 2021), especially to overcome social isolation (Ferreri et al. 2021). Google search trends reveal that coronamusic interest took off in mid-February 2020, peaked in mid-March, and may have declined to nearly pre-pandemic levels by June of the same year (Hansen 2022). Therefore, survey studies with early data collection times (i.e., April–May) – such as those by Cabedo-Mas, Arriaga-Sanz, and Moliner-Miravet (2021), and Fink et al. (2021) – may warrant particular attention with regards to the coping efficacy of coronamusic (see Hansen, Wald-Fuhrmann, and Davidson 2022).

Corona-themed music may in fact have played an especially important role in pandemic coping. Using light gradient-boosted regressor models, Fink et al. (2021) found their respondents' interest in other people's coronamusic behaviours to be the strongest predictor of how much music helped them regulate emotions, and feel connected to others. In other words, those who actively engaged with corona-themed repertoires benefitted most noticeably from music as a socio-emotional coping tool. Contextually tailored musical behaviours may thus be adaptive for managing societal crises, which, in turn, makes a deeper understanding of coronamusic scientifically pertinent.

Key Characteristics of Coronamusic

In their Coronamusic Database, Hansen et al. (2021) used citizen science and retrospective Google searches of YouTube to source 465 music videos, and 254 news reports from all around the world that were shared online between 8 February and 27 July 2020. This crowdsourced collection features ample examples of diverting balcony singing, livestreamed concerts, virtual raves, and listening parties, alongside more reflective gratitude and resilience songs composed in honour of frontline workers to promote community spirit. The Danish case as well as the global research literature reviewed above, and the varied videos from the Coronamusic Database (Hansen et al. 2021) converge in a preliminary model of common features of corona-themed repertoires and behaviours (see Figure A.1). Overall, coronamusic seems to diverge from pre-pandemic music behaviours in terms of modes of engagement, temporality, agency, identity, interaction, location, topicality, and sentiment.

First, musical engagement during lockdown emphasised creation over consumption. People who were active concertgoers prior to the pandemic were especially more likely to make music for coping (Fink et al. 2021). Music selection overall became a more active process; rather than passive following of expert-curated hit lists, many turned towards individualised playlists, and acquainted themselves with new technology, resulting in multi-modal engagement patterns with integration of visual and sensorimotor elements, such as video-based streaming (Carlson et al. 2021) and dance (Hansen et al. 2021). Second, widespread use of live transmission, and a resurgence of flow-TV formats indicated an increased focus on temporal simultaneity, although reminiscence also occurred via nostalgic content. Third, the collective took over from the individual as the primary musical agent. This was exemplified by the prominence of interaction and unity themes in coronamusic videos (Fraser, Crooke, and Davidson 2021; Hansen et al. 2021). Fourth, these co-creators often identified as amateurs rather than as professionals. In Denmark, this was evident from the emphasis on communal singing over X Factor shows. Fifth, interactions of coronamusic performers with their audiences were remarkably relaxed. Mistakes and technical glitches were widely tolerated, and informal dress codes were adopted by musical guests on television shows and by orchestral musicians featured in splitscreen videos. Sixth, musical performances inhabited physical spaces that are traditionally conceived of as private, intimate, inaccessible, and possibly even anti-social – such as balconies and homely living rooms, including those belonging to celebrities (Hansen et al. 2021; Ulfstjerne 2020). Seventh, instead of traditional

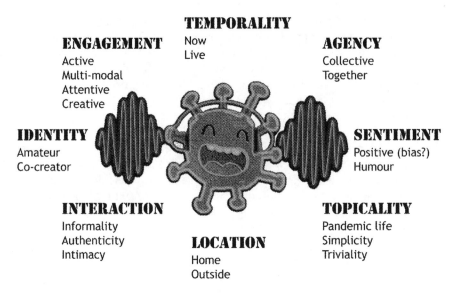

Figure A.1 The putative characteristics of coronamusic – open for future investigation. Overall, coronamusic seems to differ from pre-pandemic music in terms of its active, multimodal, creative, and attentive modes of engagement; its temporal focus on the here and now; the agency it assigns to the collective with amateur and co-creator identities; its positive and humorous sentiment; its informal, authentic, and intimate interaction patterns; its location in home and outdoors settings; and its topical focus on the simplicities and trivialities of life in lockdown.

themes of love, lust, party, and politics (Ruth 2018), pertinent topics in coronacover lyrics touched upon pandemic life circumstances and the simplicities of life at home. In fairness, however, topical escapism may also have been present in pandemic listening to nostalgic songs, and exploration of new artists and repertoires (Cabedo-Mas, Arriaga-Sanz, and Moliner-Miravet 2021; Ferreri et al. 2021; Fink et al. 2021; Gibbs and Egermann 2021). Eighth and finally, optimistic emotions such as happiness, humour, hope, togetherness, and being moved dominated coronamusic content (Hansen et al. 2021). Suggestive of a positive affective bias, this sentiment diverged remarkably from the general public sphere during the pandemic (Chandrasekaran et al. 2020), but also from traditional commemorative disaster songs responding to rapid-onset events such as earthquakes, plane crashes, industrial accidents, or terrorist attacks (Sparling 2017).

Pandemic balcony singing provides an apt example of a musical phenomenon captured by the model in Figure A.1. Such events were typically amateur-driven, involving active participation in the collective co-creation of informal and multimodal live music and dance, with a positive sentiment, in home and outdoor settings. Interviews by Calvo and Bejarano (2020) established that motivations for participating were multifaceted, spanning from emotion regulation – that is, relieving stress, battling loneliness, and entertaining kids – to social cohesion and identity formation – that is, connecting with neighbours, celebrating birthdays, solidifying cultural or national identities, commemorating the dead, meeting challenges, and recruiting followers for social media, as well as fulfilling one's duty as an artist. Participation in balcony singing varied widely between countries (from 36 per cent in Spain to 4 per cent in Mexico: see Granot et al. 2021), and even in Italy – whose balconies dominated the early pandemic media landscape (Deaville and Lemire 2021; Hansen et al.

Afterword: Coping with Crisis through Coronamusic   279

2021) – overall self-reported increases in singing, playing, and composing music at home were more pronounced than those for balcony singing in particular (Fink et al. 2021). Yet, associations with stronger feelings of togetherness (Granot et al. 2021) hint at the considerable socio-emotional coping potential of pandemic balcony singing.

Needless to say, significant caveats remain in the model proposed here. First, coronamusic repertoires were exceptionally diverse (Hansen et al. 2021), and therefore likely served a multitude of both related and unrelated functions. Second, substantial individual differences moderated pandemic coping behaviour, such as employment situation (Cabedo-Mas, Arriaga-Sanz, and Moliner-Miravet 2021), reward sensitivity (Ferreri et al. 2021), and emotional experience (Fink et al. 2021), to name a few. Third, despite certain cross-cultural efforts (e.g., Ferreri et al. 2021; Fink et al. 2021; Granot et al. 2021), cultural biases and assumptions remain (see Deaville and Lemire 2021), not least in crowdsourced initiatives such as the Coronamusic Database (Hansen et al. 2021). Fourth and finally, as of yet, not many elements of this model have been tested empirically. These factors obviously limit its universal applicability; therefore, it should be regarded first and foremost as a hypothesis-generating tool to guide future research endeavours.

Future Musicovid Research

When coronamusic videos started proliferating across the social media landscape during the early days of the COVID-19 pandemic, music scientists from around the world conceived of new research endeavours investigating people's engagement with music under the imposed lockdowns (Hansen, Wald-Fuhrmann, and Davidson 2022). My own interactions with colleagues on Twitter, for example, led to large-scale, international collaborations (i.e., Fink et al. 2021; Hansen et al. 2021). While the emergence of these new prospects was overall exciting, a concern grew amongst some of us that the music science community – confined to isolation – would produce an abundance of under-powered survey studies which were all designed to answer the same basic questions: did people listen to more music during lockdown? And if so, what exactly motivated them to do so? Therefore, to prevent redundancy, and unnecessary expenditure of limited resources, as well as to allow the field to progress towards more sophisticated and nuanced research questions, Melanie Wald-Fuhrmann and I founded the MUSICOVID network, which quickly came to comprise more than 400 researchers from over 250 universities, in about 45 countries from all six inhabited continents. At the two virtual inauguration events on 19 May 2020, manifold research ideas were nurtured, and international collaborations were drawn up to improve scientific quality and increase our potential to tackle the societal challenges arising in the wake of the pandemic. Many of the findings reviewed above were published in the special issue resulting from this coordinated effort (Hansen, Wald-Fuhrmann, and Davidson 2022).

As the novel coronavirus moves towards achieving its predicted endemic status (Veldhoen and Simas 2021), researchers with an interest in the coping capacity of music may proceed to formally test the validity, generalisability, and underlying psychological mechanisms giving rise to the coronamusic model presented in the previous section. Particularly, science and society at large have an interest in better understanding the role of thematically tailored musical innovations in human coping behaviour, on historical and evolutionary timescales. For example, why does cheerful coronamusic differ so fundamentally from sad disaster songs (see Sparling 2017)? To solve these challenges in the digital era, the scientific community will need to develop novel analytical tools and suitably sized corpora for understanding music-related behaviour in the online and offline spheres.

280   *Niels Chr. Hansen*

If musical initiatives can indeed become more readily available components of the societal crisis toolbox, grassroots initiatives such as those that arose during the COVID-19 pandemic could be supplemented by more strategic undertakings driven from the top down. This raises pertinent questions about what role music could play in tackling the climate crisis, natural disasters, and migration (arising naturally and in their wake), the biodiversity crisis, ongoing struggles for social justice and equality, the crisis of fake news and democracy, political polarisation, and future war and conflict. Moreover, musically mediated social cohesion obviously entails exclusionary potential. For example, when music brought the Danish population closer together around a national identity, were others then simultaneously excluded? What about immigrants, asylum seekers, and international students who did not necessarily identify with the national-conservative emphasis on a nostalgic heritage of which they did not feel a part? Regardless, a statement made on 26 March 2020 by a close colleague of the Danish prime minister, Joy Mogensen, Minister for Culture at the time, puts this into perspective. Specifically, Mogensen expressed that she would regard it as 'inappropriate' if she talked about culture during an imminent health crisis (Borup 2020). The emerging research on coronamusic suggests that this view may have been indeed misguided.

In conclusion, the musical sounds of the pandemic were not merely noise in a troubled system; rather, they constituted a much needed lubricant for enhanced mental wellbeing. The sonic accompaniment of life in isolation played an intricate role in psychological coping, and its musical characteristics, lyrics, and forms of expression were directly tailored to best serve this purpose.

Notes

1 See www.forbes.com/sites/avivahwittenbergcox/2020/04/13/what-do-countries-with-the-best-coro navirus-reponses-have-in-common-women-leaders.
2 Weekly viewership data are available from Kantar at http://tvm.gallup.dk/tvm/pm/. As described by Baunvig (2020), the 'Fællessang – Hver for sig' television programme and its viewership's enthusiastic engagement with it were also positively reviewed, and heavily discussed in national Danish media during the time.
3 For a good example, see www.youtube.com/watch?v=Zu6OJE8xl-8.

References

Baekgaard, Martin, Julian Christensen, Jonas Krogh Madsen, and Kim Sass Mikkelsen. 2020. 'Rallying around the Flag in Times of COVID-19: Societal Lockdown and Trust in Democratic Institutions.' *Journal of Behavioral Public Administration* 3 (2): 1–12. https://doi.org/10.30636/jbpa.32.172.

Baunvig, Katrine Frøkjær. 2020. 'Forestillede fællesskabers virtuelle sangritualer.' *Tidsskriftet Sang* 1: 40–45.

Benfante, Agata, Marialaura Di Tella, Annunziata Romeo, and Lorys Castelli. 2020. 'Traumatic Stress in Healthcare Workers during COVID-19 Pandemic: A Review of the Immediate Impact.' *Frontiers in Psychology* 11. https://doi.org/10.3389/fpsyg.2020.569935.

Birk, Rasmus H. 2017. 'Infrastructuring the Social: Local Community Work, Urban Policy and Marginalized Residential Areas in Denmark.' *Environment and Planning A* 49 (4): 767–783. https://doi.org/10.1177/0308518X16683187.

Borup, Birgitte. 2020. 'Jeg ville opfatte det som upassende, hvis jeg stod og talte om kultur lige nu.' *Berlingske Tidende*, 26 March 2020. www.berlingske.dk/aok/joy-mogensen-jeg-ville-opfatte-det-som-upassende-hvis-jeg-stod-og-talte-om.

Afterword: Coping with Crisis through Coronamusic   281

Cabedo-Mas, Albeto, Cristina Arriaga-Sanz, and Lidon Moliner-Miravet. 2021. 'Uses and Perceptions of Music in Times of COVID-19: A Spanish Population Survey.' *Frontiers in Psychology* 11. https:// doi.org/10.3389/fpsyg.2020.606180.

Calvo, Kerman and Ester Bejarano. 2020. 'Music, Solidarities and Balconies in Spain.' *Interface: A Journal for and about Social Movements* 12 (1): 326–332.

Carlson, Emily, Johanna Wilson, Margarida Baltazar, Deniz Duman, Henna-Riikka Peltola, Petri Toiviainen, and Suvi Saarikallio. 2021. 'The Role of Music in Everyday Life during the First Wave of the COVID-19 Pandemic: A Mixed-Methods Exploratory Study.' *Frontiers in Psychology* 12. https://doi.org/10.3389/fpsyg.2021.647756.

Chandrasekaran, Ranganathan, Vikalp Mehta, Tejali Valkunde, and Evangelos Moustakas. 2020. 'Topics, Trends, and Sentiments of Tweets About the COVID-19 Pandemic: Temporal Infoveillance Study.' *Journal of Medical Internet Research* 22 (10). https://doi.org/10.2196/22624.

Deaville, James and Chantal Lemire. 2021. 'Latent Cultural Bias in Soundtracks of Western News Coverage from Early COVID-19 Epicenters.' *Frontiers in Psychology* 12. https://doi.org/10.3389/ fpsyg.2021.686738.

Draper, Grace and Genevieve Dingle. 2021. ' "It's Not the Same": A Comparison of the Psychological Needs Satisfied by Musical Group Activities in Face to Face and Virtual Modes.' *Frontiers in Psychology* 12. https://doi.org/10.3389/fpsyg.2021.646292.

Fancourt, Daisy, Andrew Steptoe, and Feifei Bu. 2021. 'Trajectories of Anxiety and Depressive Symptoms during Enforced Isolation Due to COVID-19 in England: A Longitudinal Observational Study.' *The Lancet Psychiatry* 8 (2): 141–149. https://doi.org/10.1016/ S2215-0366(20)30482-X.

Ferreri, Laura, Neomi Singer, Michael McPhee, Pablo Ripollés, Robert J. Zatorre, and Ernest Mas-Herrero. 2021. 'Engagement in Music-Related Activities during the COVID-19 Pandemic as a Mirror of Individual Differences in Musical Reward and Coping Strategies.' *Frontiers in Psychology* 12. https://doi.org/10.3389/fpsyg.2021.673772.

Fink, Lauren, Lindsay Warrenburg, Claire Howlin, William Randall, Niels Chr. Hansen, and Melanie Wald-Fuhrmann. 2021. 'Viral Tunes: Changes in Musical Behaviours and Interest in Coronamusic Predict Socio-Emotional Coping during COVID-19 Lockdown.' *Humanities and Social Sciences Communications* 8 (180). https://doi.org/10.1057/s41599-021-00858-y.

Finnerty, Rachael, Sara A. Marshall, Constance Imbault, and Laurel Trainor. 2021. 'Extra-Curricular Activities and Well-Being: Results from a Survey of Undergraduate University Students during COVID-19 Lockdown Restrictions.' *Frontiers in Psychology* 12. https://doi.org/10.3389/ fpsyg.2021.647402.

Fraser, Trisnasari, Alexander Hew Dale Crooke, and Jane W. Davidson. 2021. ' "Music Has No Borders": An Exploratory Study of Audience Engagement with YouTube Music Broadcasts during COVID-19 Lockdown, 2020.' *Frontiers in Psychology* 12. https://doi.org/10.3389/ fpsyg.2021.643893.

Frederiksen, Mette. 2020. 'Pressemøde om COVID-19 den 11. marts 2020.' 11 March 2020. www.stm. dk/presse/pressemoedearkiv/pressemoede-om-covid-19-den-11-marts-2020/.

Gibbs, Hannah and Hauke Egermann. 2021. 'Music-Evoked Nostalgia and Wellbeing during the United Kingdom COVID-19 Pandemic: Content, Subjective Effects, and Function.' *Frontiers in Psychology* 12. https://doi.org/10.3389/fpsyg.2021.647891.

Granot, Roni, Daniel Spitz, Boaz Cherki, Psyche Loui, Renee Timmers, Rebecca Schaefer, Jonna Vuoskoski, Ruth-Nayibe Cárdenas-Soler, João F. Soares-Quadros Jr., Shen Li, Carlotta Lega, Stefania La Rocca, Isabel Cecilia-Martìnez, Matìas Tanco, Marìa Marchiano, Pastora Martìnez-Castilla, Gabriela Pérez-Acosta, José Darìo Martìnez-Ezquerro, Isabel M. Gutiérrez-Blasco, Lily Jiménez-Dabdoub, Marijn Coers, John Melvin Treider, David M. Greenberg, and Salomon Israel. 2021. ' "Help! I Need Somebody": Music as a Global Resource for Obtaining Wellbeing Goals in Times of Crisis.' *Frontiers in Psychology* 12. https://doi.org/10.3389/fpsyg.2021.648013.

Hansen, Niels Chr. 2022. 'Music for Hedonia and Eudaimonia During Pandemic Social Isolation.' In *Arts and Mindfulness Education for Human Flourishing*, edited by Tatiana Chemi, Elvira Brattico, Lone Overby Fjorback, and László Harmat. London: Routledge. Preprint, https://doi. org/10.31234/osf.io/s9jf6.

Hansen, Niels Chr., Melanie Wald-Fuhrmann, and Jane W. Davidson. 2022. 'Editorial: Social Convergence in Times of Spatial Distancing: The Role of Music During the COVID-19 Pandemic.' *Frontiers in Psychology* 13. https://doi.org/10.3389/fpsyg.2022.910101.

Hansen, Niels Chr., John Melvin G. Treider, Dana Swarbrick, Joshua S. Bamford, Johanna Wilson, and Jonna K. Vuoskoski. 2021. 'A Crowd-Sourced Database of Coronamusic: Documenting Online Making and Sharing of Music during the COVID-19 Pandemic.' *Frontiers in Psychology* 12. https://doi.org/10.3389/fpsyg.2021.684083.

Hale, Thomas, Noam Angrist, Rafael Goldszmidt, Beatriz Kira, Anna Petherick, Toby Phillips, Samuel Webster, Emily Cameron-Blake, Laura Hallas, Saptarshi Majumdar, and Helen Tatlow. 2021. 'A Global Panel Database of Pandemic Policies (Oxford Covid-19 Government Response Tracker).' *Nature Human Behaviour* 5: 529–538. https://doi.org/10.1038/s41562-021-01079-8.

Hossain, Md Mahbub, Samia Tasnim, Abida Sultana, Farah Faizah, Hoimonty Mazumder, Liye Zou, E. Lisako J. McKyer, Helal Uddin Ahmed, and Ping Ma. 2020. 'Epidemiology of Mental Health Problems in COVID-19: A Review.' *F1000Research* 9: 636. https://doi.org/10.12688/f1000research.24457.1.

Howlin, Claire and Niels Chr. Hansen. Forthcoming. 'Music in Times of Covid-19.' In *Musik und Medizin/Music and Medicine*, second edition, edited by Gunter Kreutz and Günther Bernatzky. Berlin: Springer. Preprint, https://doi.org/10.31234/osf.io/z94fq.

Jørgensen, Frederik, Alexander Bor, Marie Fly Lindholt, and Michael Bang Petersen. 2021. 'Public Support for Government Responses against COVID-19: Assessing Levels and Predictors in Eight Western Democracies during 2020.' *West European Politics* 44 (5–6): 1129–1158. https://doi.org/10.1080/01402382.2021.1925821.

Kiernan, Frederic, Anthony Chmiel, Sandra Garrido, Martha Hickey, and Jane W. Davidson. 2021. 'The Role of Artistic Creative Activities in Navigating the COVID-19 Pandemic in Australia.' *Frontiers in Psychology* 12. https://doi.org/10.3389/fpsyg.2021.696202.

Krause, Amanda, James Dimmock, Amanda Rebar, and Ben Jackson. 2021. 'Music Listening Predicted Improved Life Satisfaction in University Students during Early Stages of the COVID-19 Pandemic.' *Frontiers in Psychology* 11. https://doi.org/10.3389/fpsyg.2020.631033.

Kritzinger, Sylvia, Martial Foucault, Romain Lachat, Julia Partheymüller, Carolina Plescia, and Sylvain Brouard. 2021. '"Rally Round the Flag": The COVID-19 Crisis and Trust in the National Government.' *West European Politics* 44 (5–6): 1205–1231. https://doi.org/10.1080/01402382.2021.1925017.

Lantz, Prins Marcus Valiant. 2021. 'Affecting Argumentative Action: The Temporality of Decisive Emotion.' *Argumentation* 35: 603–627. https://doi.org/10.1007/s10503-021-09546-2.

MacDonald, Raymond, Robert Burke, Tia DeNora, Maria Sappho Donohue, and Ross Birrell. 2021. 'Our Virtual Tribe: Sustaining and Enhancing Community via Online Music Improvisation.' *Frontiers in Psychology* 11. https://doi.org/10.3389/fpsyg.2020.623640.

Mak, Hei Wan, Meg Fluharty, and Daisy Fancourt. 2021. 'Predictors and Impact of Arts Engagement during the COVID-19 Pandemic: Analyses of Data from 19,384 Adults in the COVID-19 Social Study.' *Frontiers in Psychology* 12. https://doi.org/10.3389/fpsyg.2021.626263.

Mazza, Cristina, Eleonora Ricci, Silvia Biondi, Marco Colasanti, Stefano Ferracuti, Christian Napoli, and Paolo Roma. 2020. 'A Nationwide Survey of Psychological Distress among Italian People during the COVID-19 Pandemic: Immediate Psychological Responses and Associated Factors.' *International Journal of Environmental Research and Public Health* 17 (9): 3165. https://doi.org/10.3390/ijerph17093165.

McIntosh, Heather. 2021. 'Charity Benefit Concerts and the *One World: Together at Home* Event.' *Rock Music Studies* 8 (1): 76–82. https://doi.org/10.1080/19401159.2020.1852773.

Mehlsen, Camille. 2020. '10 ting, du (måske) ikke vidste om influencere.' *Kommunikations Forum*, 10 December 2020. www.kommunikationsforum.dk/artikler/Unges-SoMe-brug.

Messick, Kyle J. 2020. 'Music Industry in Crisis: The Impact of a Novel Coronavirus on Touring Metal Bands, Promoters, and Venues.' In *The Societal Impact of Covid-19: A Transnational*

Perspective, edited by Veysel Bozkurt, Glenn Dawes, Hakan Gülerce, and Patricia Westernbroek, 83–98. Turkey: Istanbul University Press.

Morgan-Ellis, Esther. 2021. '"Like Pieces in a Puzzle": Online Sacred Harp Singing During the COVID-19 Pandemic.' *Frontiers in Psychology* 12. https://doi.org/10.3389/fpsyg.2021.627038.

Onderdijk, Kelsey, Dana Swarbrick, Bavo Van Kerrebroeck, Maximillian Mantei, Jonna K. Vuoskoski, Pieter-Jan Mae, and Marc Leman. 2021. 'Livestream Experiments: The Role of COVID-19, Agency, Presence, and Social Context in Facilitating Social Connectedness.' *Frontiers in Psychology* 12. https://doi.org/10.3389/fpsyg.2021.647929.

Pan, Kuan-Yu, Almar AL Kok, Merijn Eikelenboom, Melany Horsfall, Frederike Jörg, Rob A. Luteijn, Didi Rhebergen, Patricia van Oppen, Erik J. Giltay, and Brenda W.J.H. Penninx. 2021. 'The Mental Health Impact of the COVID-19 Pandemic on People with and without Depressive, Anxiety, or Obsessive-Compulsive Disorders: A Longitudinal Study of Three Dutch Case-Control Cohorts.' *The Lancet Psychiatry* 8 (2): 121–129. https://doi.org/10.1016/S2215-0366(20)30491-0.

Porcher, Simon. 2020. 'Response2covid19, a Dataset of Governments' Responses to COVID-19 All Around the World.' *Scientific Data* 7. doi:10.1038/s41597-020-00757-y

Rosati, Dora P., Matthew H. Woolhouse, Benjamin M. Bolker, and David J. D. Earn. 2021. 'Modelling Song Popularity as a Contagious Process.' *Proceedings of the Royal Society A* 477 (2253). https://doi.org/10.1098/rspa.2021.0457.

Ruth, Nicolas. 2018. '"Where Is the Love?" Topics and Prosocial Behavior in German Popular Music Lyrics from 1954 to 2014.' *Musicae Scientiae* 23 (4): 508–524. https://doi.org/10.1177/1029864918763480.

Sim, Jaeung, Daegon Cho, Youngdeok Hwang, and Rahul Telang. 2022. 'Virus Shook the Streaming Star: Estimating the COVID-19 Impact on Music Consumption.' *Marketing Science* 41 (1): 19–32. https://doi.org/10.1287/mksc.2021.1321.

Sparling, Heather. 2017. '"Sad and Solemn Requiems": Disaster Songs and Complicated Grief in the Aftermath of Nova Scotia Mining Disasters.' In *Singing Death: Reflections on Music and Mortality*, edited by Helen Dell and Helen M. Hickey, 90–104. London: Routledge.

Stebbins, Robert A. 2020. *The Serious Leisure Perspective: A Synthesis*. Cham, Switzerland: Springer Nature.

Swarbrick, Dana, Beate Seibt, Noemi Grinspun, and Jonna K. Vuoskoski. 2021. 'Corona Concerts: The Effect of Virtual Concert Characteristics on Social Connection and *Kama Muta*.' *Frontiers in Psychology* 12. https://doi.org/10.3389/fpsyg.2021.648448.

Sørensen, Jesper F., Katrine F. Baunvig, and Peter B. Andersen. 2021. 'Håndvask og fællessang. Ritualer og ritualiserede handlinger i Coronaens tid.' *Religionsvidenskabeligt Tidsskrift* 72: 116–136. https://doi.org/10.7146/rt.vi72.126503.

Torales, Julio, Marcelo O'Higgins, João Mauricio Castaldelli-Maia, and Antonio Ventriglio. 2020. 'The Outbreak of COVID-19 Coronavirus and Its Impact on Global Mental Health.' *International Journal of Social Psychiatry* 66 (4): 317–320. https://doi.org/10.1177/0020764020915212.

Ulfstjerne, Michael Alexander. 2020. 'Songs of the Pandemic.' *Anthropology in Action* 27 (2): 82–86. https://doi.org/10.3167/aia.2020.270213.

Veldhoen, Marc and J. Pedro Simas. 2021. 'Endemic SARS-CoV-2 Will Maintain Post-Pandemic Immunity.' *Nature Reviews Immunology* 21: 131–132. https://doi.org/10.1038/s41577-020-00493-9.

Vidas, Dianna, Joel L. Larwood, Nicole L. Nelson, and Genevieve A. Dingle. 2021. 'Music Listening as a Strategy for Managing COVID-19 Stress in First-Year University Students.' *Frontiers in Psychology* 12. https://doi.org/10.3389/fpsyg.2021.647065.

Wang, Cuiyan, Riyu Pan, Xiaoyang Wan, Yilin Tan, Linkang Xu, Roger S. McIntyre, Faith N. Choo, Bach Tran, Roger Ho, Vijay K. Sharma, and Cyrus Ho. 2020. 'A Longitudinal Study on the Mental Health of General Population during the COVID-19 Epidemic in China.' *Brain, Behavior, and Immunity* 87: 40–48. https://doi.org/10.1016/j.bbi.2020.04.028.

Wang, Huiyao, Qian Xia, Zhenzhen Xiong, Zhixiong Li, Weiyi Xiang, Yiwen Yuan, Yaya Liu, and Zhe Li. 2020. 'The Psychological Distress and Coping Styles in the Early Stages of the 2019

Coronavirus Disease (COVID-19) Epidemic in the General Mainland Chinese Population: A Web-Based Survey.' *PLOS ONE* 15 (5): e0233410. https://doi.org/10.1371/journal.pone.0233410.

Yeung, Timothy Yu-Cheong. 2020. 'Did the COVID-19 Pandemic Trigger Nostalgia? Evidence of Music Consumption on Spotify.' *Covid Economics* 44. https://cepr.org/sites/default/files/news/MusicConsumption.pdf.

Ziv, Naomi and Revital Hollander-Shabtai. 2021. 'Music and COVID-19: Changes in Uses and Emotional Reaction to Music under Stay-at-Home Restrictions.' *Psychology of Music* 50 (2): 475–491. https://doi.org/10.1177/03057356211003326.

Index

Note: Page locators in **bold** and *italics* represents tables and figures, respectively.

ABC network 255
Abejon, Emilio 82
acoustic changes, in domestic space 31, 37
acoustic sensitivity 22
Acoustic Spatiality 33
acoustic sustainability 244
adda 40
Adele 197
aesthetic work 70
Against Modern Football movement 78
Agamben, Giorgio 264
agency, sense of 104, 203
Ahmed, Sarah 34
air traffic reduction 16
Alberti, Leon Battista 88
Aldir Blanc Emergency Bill 6, 163, 165, 167, 168, 171, 172, 173n10
Also sprach Zarathustra 232
Anderson, Benedict 37n5
#*andràtuttobene* 177
Aniceto, Raimundo 164
Annalisa **186**
anthropogenic sounds, in urban environments 69, 238, 243, 265
anti-COVID-19 rules 53, 184
anti-government protests 51
anti-Gypsyism 136
anti-police slogan 53
anti-social 277
Antoniadis, Pavlos 267
Antunes, Carlinhos 170–1
anxiety 4, 6, 7, 22–3, 25, 196–7, 214
Aoki, Steve 182
applauders 54–6
Apple: experimentation 69; Podcast + subscription service 69
Arcade Boyz 186
Ardern, Jacinda 272
Arduini, Luca 186
Arendt, Hannah 51, 59
Arés-Muzio, Patricia 103
Armstrong, Dan 219
Armstrong, Neil 267

Around the World with Guests project 170, 174n18
Arriaga-Sanz, Cristina 173n4, 277
Arshad-Ayaz, Adeela 36
Assad, Badi 170
Assange, Julian 51
Assuniri, Puraké 164
Asterism 152
Athawale, Ramdas 43
atmospheres 66, 68, 70, 77–8, 81; *atmospheric-medial mis-en-scène* 68; atmospheric practices 81; atmospheric transfer 78, 81; atmospherology 70, 73n8
audience members 7, 152, 195, 199, 276
audio devices 69
audiotactile music 4, 71, 74n22
audio-vision 77
audiovisual: news 249–50, 256; producers 249–50, 256; product 65, 81–2, 85n24
auditory involvement 18
Auslander, Philip 5, 178
automatic content identification tools 118–19
Autonomous Sensory Meridian Response (ASMR) 70
auto-poietic systems 153
Av3ry (2019) 152–4
Avraamov, Arseny 59
axiosemiotic object 17
Azules (Los Angeles) 111

Bachhan, Amitabh 43
background listening 242
Baeza, Gilberto Salvador Perez 101–2, 105
Bains, Nagina 203
balcony singing 3, 193–6, 198, 202–3, 232, 276–9
banging vessels 42–3
banner 51–4, *54*, 56
Barbanti, Roberto 140
Barenboim, Daniel 232
Barnett, Jack 48n11
Baroni, Edoardo **186**
Barreto, Luisa Marques 167
Bastos, Cristovão 173n6

286 *Index*

batucadas sound 263
Baunvig, Katrine Frøkjær 280n2
BBC 250, 252, 255, 257n29
Bech, Mathæus 273
Beck, Nora 228
Bee Gees 196
Beethoven's melody 231
Befera, Luca 6
Being involved with something 178–9
Bejarano, Ester 278
Belo Horizonte 170
Benjamin, Walter 67, 71, 268
Berger, Peter Ludwig 15
Berger-Tal, Oded 244n1
Berio, Luciano 180
Bernaldina, Vovó 164
Berque, Augustin 50
Berry, David 99
Besseler, Heinrich 197–8
Bevilacqua, Carmine 140
Bijsterveld, Karin 31–2
Birdsall, Carolyn 32
birds chirping, spectrogram of *241*
birdsongs 22, 243
Biswas, Srijita 7
Blanc, Aldir 163, 165n6
Blast Theory/BeAnotherLab 149
Blesser, Barry 89
Boccaccio, Giovanni 226, 227, 231, 234
Bocelli, Andrea 186
Bodner, Ehud 208n14
Boléro on YouTube 232, 266
Bolsonaro, Jair 164, 167, 173n10
Bon Jovi 196
Born, Georgina 10n14, 35
Borsetti, Luca 188n8
Bosco, João 173n6
Bratus, Alessandro 6, 150, 187n1
Bravi, Paolo 129–30
Brazilian music world 165
Brazilian pandemic: consideration of 172–3;
    emergency bill, negotiating 165–8; LAB
    projects, and impacts 169–72; musicians
    163–73
Brilli, Stefano 10n13
Brueck, Laura 42
BTS 182
Buccino, Rocco 137
Buch, Esteban 8
Bull, Michael 4

Cabedo-Mas, Alberto 173n4, 277
Calabre, Lia 166
Caldeira, Izael 163
Calderón, Doris Elena Pinos 101
Caliandro, Alessandro 6, 187n1
Calvo, Kerman 278
Camus, Albert 226
Canada, H1N1 virus in 251

Canto della Verbena 9
cantu a tenore 125–7, 133n3
Cantzonis 6, 129–30, 132, 134n14
Canzonetta 183–4
Capitalocene 9n8; Anthropocene 264–6; artist
    as producer 267–8; music 266–7; as politically
    repressive measure 263–4
Caporaletti, Vincenzo 71, 74n20
Carousel 206
Carpenter, Edmund 47n4
Carr, Forrest 256n9
Caruso, Fulvia 6, 187n1, 188n16
Casamonica, Vittorio 145n14
casserolade 7, 57
Castoriadis, Cornelius 32
Cavazza, Damiano 139
central policy makers 239
Ceravolo, Flavio Antonio 6, 187n1
César of shame 263, 269n3
Chadabe, Joel 217
Chakrabarty, Dipesh 42, 43, 47n2
Chatterjee, Nakshatra 7, 40
Chaudhry, Vandana 254
China, sounds/silence of COVID-19 quarantine
    253–4
Chinni, Riccardo 72n2
Chun, Wendy 91
Cirese, Alberto 129
civic/religious calendars 8
clapping *see* applauders
Classroom users 66
Clément, Jean Baptiste 52
closed village 30
cloud storage 217
CNN News 252, 253, 255
Cognitive Maps in Rats and Men 241
Cohen, Susanna 208n2
Colombian sound systems 110, 115
Combs, Luke 196
common-sense knowledge 15
communication strategy 4, 66
conditio humana 200
confusion, perception of time 95
Congress of Brazil 164
Consolatio philosophiae 200
consumer-based user statistics 275
Contagion (2011) 233
Conte, Giuseppe 137
Control, A Perfect Circle 156
conventional rite 56
Cooley, Timothy J. 133n2
coping, with crisis: coronamusic, characteristics
    of 277–9; corona-themed music, special role
    276–7; future musicovid research 279–80;
    mental wellbeing, during lockdown 274–5;
    musical coping behaviours 275–6; musical
    revivals 273–4; through coronamusic 272–80
Corbin, Alain 47n5
Coriat, Benjamin 265

corona contrafacta 196, 197, 199, 202–4
coronamusic 272, 274, 276–7; adaption/response, in challenging times 214; characteristics of 277–9; collage performances 221; communication/production/collaboration, new instruments of 216–18; consolation 200–2; coping, with crisis 272–80; creative/ innovative mindset 221–2; database 193, 277; differentiated expertise 223; expression, through remotivity 218; income stream disruption 216; internet, performances 194–5; live music performances 194; live streaming, from home 219–20; living/working, in isolation 216; lockdown policies 193; mediated consolation, practice of 199; musical consolation 202–6; music practitioners, response of 215; new performance platforms 220–1; parody songs 196; playlists 195–6; production/consumption 215; putative characteristics of *278*; remote/home working, history of 214; remotivity, practice of 221–3; repertoire creation 195–7, 277; stable client/ audience base 222–3; studios relocating, to home 218–19; technologically enabled global network, disruption of 215–16; typology 193; Umgangsmusik 197–9; videos 277, 279
coronamusicking 7, 193, 197, 199
coronavirus 163, 233, 251, 272, 279; in China 253; communication strategy, fundamental trait of 4; coping capacity 279; mediatising reality 5–6; musicking, as technology of the self 6–8; North American and European news on 251; rapid spread of 3; time/memory 8–9
#CortègeDeFenêtres 7, 51–4, 56
cortège de tête 51
Cottier, Russel 219
Cottle, Simon 250
Couldry, Nick 10n14
COVID-19 pandemic 15, 32, 36, 41, 45, 66, 87, 105, 130, 132, 180, 182, 195, 199, 233, 249, 263; during Bolsonaro Regime 163–73; crisis of 173; dance in the middle of 110; emergency 186; Indian cities, *aural topography* of 41; infections 30; Janata Curfew 42–4; live music 115–19; lockdown 237, 272; *media economy* 115–19; music/community, digital spaces of 99–107; noise 44–5; outbreak, impact of 40; outset of 102; privileged affective experiences 67; restrictions 81; Richard TV's list, of videos *120*; self-evident nature of 15; silence, rhapsody of 45–6; Slovakia/Czech Republic *see* COVID-19 pandemic, in Slovakia/ Czech Republic; social stigma 112–15; *sonidero digital channels* 119–21; *sonideros* 115–19; SonTube channels 101–7; South Asia, sounding 41–2; Tempo Reale 180; YouTube channels 99
COVID-19 pandemic, in Slovakia/Czech Republic: mutual trajectories 32–5; overview

of 30–1; privatised public, sounds of 35–7; remembering sound 31–2
COVID-19 quarantine 102; China, sounds/ silence 253–4; hybrid auditory realities 87–90; Italy, sounds/silence 254–5; sounds/silence *see* sounds/silence, of COVID-19 quarantine
COVID-19 survivor 204
COVID-19 symptoms 196
COVID-19 trains 45
COVID-19 virus 44
COVID-19 wave 1, 228; Australian survey 228; music, as 'most helpful' activity 228
Covid/Back 204
COVID warriors 42
Crawlers graphics 152–3
Crutzen, Paul 9n8
Cubase VST Connect 217
cultural memory 24, 31
cultural practitioners 110
cultural production, platformisation of 99
cultural sector 266
Cultura Viva 166
culture advisers 168
cumbia sonidera 111, 112; in Mexico 110; videos 6
Cunha, Cristiano 170
curfew, by clapping 42
Curva Fiesole 77
Czocher, Anna 28n18

Dace, Acelino 164
da Luz, Maria Auxiliadora 164
Damasio, Alain 268
Daft Punk 196
dance tempo 90
Darbietungsmusik 197–8
da Silva, Luiz Inácio 165
da Trindade, Vitor 171
Davidson, Jane 208n2, 208n15
Davis, Owen 219
Dawn of the Dead (film) 253
Deadmau5 182
Deaville, James 5, 8
Decameron 226–8
de Castro, Jeanne 169
DeLIVEry 181
de Nieve, Bola 170
DeNora, Tia 201
de Oliveira, Martinho Lutero 164
depression 6, 165, 197, 199, 201, 214, 266, 274
Deriggi, Ana 169
De Rosa, Maria 137
Diabelli Variations 232
Diamond, Neil 205
Dicuonzo, Antonella 6, 136
digital communication 87, 89, 94–5, 148, 157
digital ethnography 5, 125, 131–3, 185
Digital Ethnography 131
Digital Love 196

288 *Index*

digital-mediated communication 151
disaster capitalism 249, 254, 264
disaster newscasting 250
distance learning 21, 70
DJing 113
DJ mix 43
DJs moving 230
Doidge, Mark 83n1
Do-It-Yourself (DIY): approach 220; culture 100; media channels 6
Domeniche d'Essai 181
domestic space 18; acoustic changes 31; transformation of 19
Donizetti, Gaetano 179
Douglas, Mary 2
Draaisma, Douwe 32
Dropbox 217
Ducheneaut, Nicolas 128
Duncan, Zélia 170
Durand, Cédric 267

Earwitnessing: *Sound Memories of the Nazi Period* 32
economic/social hardships 61
Eerola, Tuomas 201
Eichenbaum, Howard 245n6
Eidsheim, Nina Sun 88
Elbphilharmonie 155–6
Electronic Arts 80
embodiment, sense of 104
emojis, in 'race critical code studies' 105, 106
English Premier League (EPL) 77
Engstrom, Erika 250
Erkkilä, Jaakko 201
ethical issues, of virtual audio 81–3
ethical-political correctness 79
European live music scenes 181
European Roma Institute for Arts and Culture (ERIAC) 140, 141
EU's Environmental Noise Directive 2002/ 49/ EC 240
Everybody Hurts 196
Exeter, Mike 219

Faber, Phillip 273
Facchinetti, Roby 177
Facebook 57, 110, 113, 127, 130, 137–9, 183, 184; Facebook Live 219, 222; pages 183, 184
face-to-face social life 87
Faith and Tradition page 184
fans' agency 82
Federación de Accionistas y Socios del Fútbol Español (FASFE) 82
Feghali, Jandira 168, 173n10
female dancer 120
Fernandinho Beatbox 170
Ferreira, Juca 165, 166
Festival Khamoro 36
FIFA series 80

Fike, Dominic 221
financial crisis (2007–2008) 2
Fink, Lauren K. 202, 277
flat line effect 237
Fleck, Gábor 35
Flores, Hector Manuel Delgado 101, 104
football chants, appropriation/commodification of 82
football television sounds 79, 81
Forsyth, Iain 48n11
Fortnite 220, 222
Foucault, Michel 254
Frankfurter Allgemeine Zeitung 202
Franklin, Malvin F. 214
Frank, Sybille 76
Frederiksen, Mette 272, 273
Freire, Paulo 170
French population 58
French social movements, during lockdown 50–61
French sociologist claims 17
Freud, Ahmed 34
Friends of Cefferino Gimenez Malla, known as 'EL PELÉ' 137
functional daytime activities 115

Gabrielli, Robert 137
Gabrielli, Scen 138
Gaga, Lady 195
Galdino, Regina 170
games 79; anthropological points of focus 79; audio rewriting of 79; interface 157
Gan, Nectar 257n22
Garda, Michela 6
Garibaldi, Giuseppe 129
GavBroadcast: chats 101, 105; streamed queues 103
Gaye, Marvin 214
Gay, Leslie 133n2
Gebrauchsmusik 198
Gelder, Rudy Van 214
Gelem, gelem 36
Gemini, Laura 10n13
Genesis 149; game interface *149*; virtual reality *158*
German music 197
gesture, social/political manipulations of 56
Gialluisi, Pipo 169
Gibson, James 89
Gil, Gilberto 166, 169, 174n16
Ginsborg, Jane 208n2
Gintoli, Eliana 139, 141
Girard, René 2
Gitano, Rocco 137, 140, 142, 143
Giuliani, Francesca 10n13
Glennon, Mike 208n5
global market 223–4
global pandemic 15, 87, 95, 250
'Go Corona Go' chants 43

Goehr, Lydia 232
Goldman Sachs 215, 224
Good Friday video 127
Google 66, 120
Google-Apple-Facebook-Amazon-Microsoft
    (GAFAM) 267
Gopinath, Praseeda 43
Graakjær, Nicolai Jørgensgaard 79, 80, 81
gradient-boosted regressor models 277
Great Animal Orchestra, The 265
Greece, first lockdown in 238
Greppi, Alessandro 7
Griffero, Tonino 70, 83n8
Griffiths, James 257n22
Grohl, Dave 267
Grondin, Simon 94, 95
Groten, Raphaela 104
Grusin, Richard 10n14
Guardian 202, 252–3, 255
Guattari, Félix 265
Gubitsch, Tomás 230
Gugolati, Maica 188n16
Guinga 173n6

H1N1 virus 251
Haas, Kira 203
Hagood, Mack 33
half-things 71
Hamburg International Music Festival (2020) 154
Hammer, MC 196
Hansen, Niels Christian 7, 8, 10n12, 208n2,
    208n15, 233
Hanser, Waldie E. 201
Hartley, Anna Pellegrin 257n27
Harvey, Trevor S. 133n2
head-space 21
healthcare sector 60, 164, 274
Heider, Wally 214
Hepp, Andreas 10n14
Hergenhahn, Baldwin 241
Hernandez, Edgar 257n11
Herzfeld, Michael 33, 36, 38n6
Hesmondhalgh, David 99
Hilder, Thomas R. 133n2
Hilmes, Michele 31
Hinduism, religious practices 44
Histórias, Marcelo das 173n10
Hoffmann, Carl-John 159n1, 159n3
Hoffman, Steve 214
Hofman, Ana 34
Hollander-Shabtai, Revital 173n4
Holy Cross Confraternity 126
Holy Trinity 183; pilgrimage 183; shrine in
    Vallepietra 9
Holy Week 126, 127
#HOMEPLAYING 180
home studio 214

home working, history of 214
Horwitz, Heinrich 159n1
HTML framework 92
Huapanguitos en un domingo de cuarentena 102,
    104, 105
Huffman, Suzanne 256n9
human-computer interaction (HCI) 148
human habits 66
Hung, Eric 253
hybrid auditory realities 87–90, 92, 94–5

Ice Cold 233
Icon series 220
Ile-de-France, sensitive neighbourhoods of 60
Indian cities, *aural topography* of 41
Indian Sound Cultures, Indian Sound Citizenship
    (2012) 42
indifferentiating process 2; anthropological
    process of 2; pandemic, powerful impact of 2
influenza of 1918 216
Instagram 138
institutionalisation, lack of 112
intermedia practices, during pandemic:
    aesthetic references 152–4; digital trespassing
    157–9; institutional aspects 154–6; mapping
    Schubert's latest production 151–2; overview
    of 148–50; theoretical background 150–1; into
    unknown 156–7
International Roma Day 36, 138, 140, 143
interviewees 169
interviews 278
Ionisation 265
#iorestoacasa 186
Irabién, Raúl 197
Italian pandemic lockdown 65
Italy sounds/silence, of COVID-19 quarantine
    254–5

Jacobsen, Kristina 132
Jakobsdóttir, Katrín 272
Jamaican sound systems 110
Janata Curfew 7, 41–6
Janatā, Diego 171
JavaScript 92
John, Elton 196, 220
Johnson, Bruce 250
Jones, Ellis 99
Jovanovi, Jarko 145n3

Kaingáng, Artemínio Antônio 164
Kalinak, Kathryn 93
Kaye, Lenny 229
Kealey, Edward 218
KeepOn Live 181
Kethane Movement 139, 140, 143–4
keynote sounds 80, 84n12, 238–9
Khaled, Leila 267
Kilten, Konstantina 104
King, Eoin 244n5

290 *Index*

Klapisch, Cédric 266n11
Klein, Naomi 249, 254
Klien-Thomas, Hannah 188n16
Koszarski, Richard 256n7
Krause, Amanda E. 173n4
Krause, Bernie 265
Kun, Josh 112

LaBelle, Brandon 33
Lacey, Kate 36
Lange, Barbara Rose 133n2
La Repubblica 230
Larson, Gary 250
latent acoustic learning: auditory cognitive maps
    242–3; basic terminology 238–9; blackbirds
    238, 243–5; continuous noise/salami sound
    239–41; COVID-19 lockdown 237; latent
    learning/cognitive maps 241–2; noise pollution
    239; silent urban soundscapes 237–8
Latin American cultures 111
La Tortolita 106
Latour, Bruno 25
leakage effect 88, 89
Lee, James Kyung-Jin 254
Lefebvre, Henri 257n29
Leib 67
Lei Emergencial Aldir Blanc (LAB) *see* Aldir
    Blanc Emergency Bill
Lemire, Chantal 256n1
Le Parisien 54
Le temps des cerises *52*
Lévi-Strauss, Claude 47n4
Levy, Gabriel 171
Levy, Hadar 249
Lima, Deborah Rebello 166
Lipa, Dua 182, 220, 221, 222
Lippman, Alexandra 112
*Listening Publics: The Politics and Experience of
    Listening in the Media Age* 36
live music events 115, 181
Living Cultures (Cultura Viva) 166
lockdown 16, 33, 216; in 2020 7; ambivalent
    silence 21–3; anti-state positions 53;
    assumptions/categories/methodology 16–18;
    beginning of 60; clapping 54–5; consensual
    ritual, hijacking 56–8; digital liveness 178–80;
    French social movements 50–61; in Greece
    238; happy/sad music 276; honking horns
    58–60; hybrid auditory realities 88; listening
    modality 89; mental wellbeing 274–5;
    multiplied schizophonia 18–21; negative
    impacts 163; parisian suburbs 60–1; as
    politically repressive measure 263–4; political
    repression 264; public forums 51; radio cars
    23, 24; re-inventing virtual communities,
    on YouTube 185–7; restrictions 18, 177;
    silence and immobility 243; soundmark
    23–5; soundscape of 17; streaming music
    performances 232

Lord Alge, Chris 223
Losiak, Robert 4
Lost Nutcracker, The (2020) 180
Lotis, Theodoros 5
Loutte, Bernard 229
Luckmann, Thomas 15
L'ultimo a morire 143
L'ultimo concerto? 1, 2, 4–6, 8, 182
Lutzu, Marco 6, 133n1
Lysloff, René 133n2

Macchiarella, Ignazio 6, 133n1
Machado, Regina 6, 173n10
Maestro Ziikos 208n11
Maintenance of Cultural Spaces 171
Mak, Kathy 203, 205
Mallica, Marcello 126
Malm, Andreas 9n8
Manga, Mario 169
Mann, Chris 197, 204, 209n19
Mantellini, Massimo 69–70
Maracinescu, Mara 73n11
Marcel Zaes 90
Marchi, Valerio 83n1
marginalised group identity 36
Margolies, Daniel 5
Marin, Sanna 272
Márquez, Gabriel García 28n21
Marsola, Mônica 169
martial law 24
Masaoka, Miya 226, 233
Matte, Adrian 256n1
Mattos, Livia 170
MC Kunumi 171
McAlister, Elizabeth 256n4
McCue, Kevin 85n24
McDowell, John H. 100
McLuhan, Marshall 47n4
McRuer, Robert 254
mediation 9, 10n14, 78, 84n19, 88, 89, 90, 94, 95,
    112, 119, 148, 150, 152, 153, 159, 160n5, 179;
    remediation 82
mediatisation 5, 66–8, 76–7, 79, 83, 115
Meek, Joe 214, 223
Meizel, Katherine 133n2
Melis, Davide 126
memory and repetition relation 91
Mendoza-Duran, Esteban 94, 95
mental wellbeing, during lockdown 274–5
Merkel, Angela 272
Mexican music: artists 110; and cultural
    practitioners 110; media coverage 110;
    musicians 5; musicians to pandemic 5
Mexican stringed instruments 106
Mexico, H1N1 virus 251
Miceli, Francesco 179
Microsoft: Microsoft Teams 66; Soundscape
    project 69
Middle Ages 51

migrant workers 44; in India 46; jobless 44
Milleddu, Roberto 134n14
Miller, Marcus 221
Mingus, Charles 214
Ministry of Culture (MinC) 166
Mintert, Svenja 83n1
Modi, Narendra 7, 42, 43
Mogensen, Joy 280
Moliner-Miravet, Lidon 173n4, 277
money, for public hospital 52
Monson, Ingrid 91
mood regulation strategies 201
Moore, Allan 160n5
Moravčíková, Dominika 4
most valuable player (MVP) 232
Motta, Claudio Augusto 164
Mounk, Yasha 257n26
Movimento Artigo Quinto 170
Movimento Kethane – Rom e Sinti per l'Italia see Kethane Movement
Mowitt, John 31
MP3, digital distribution 218
Munakata, Naomi 164
Murphy, Enda 244n5
Música de Tzitzio 30 de agosto año COVID 19 106
musical community 4
musical instrument digital interface (MIDI) 217
musical performance, during/after COVID-19 pandemic 176–87; digital liveness 178–80; re-inventing virtual communities, on YouTube 185–7; streaming liveness, in pandemic time 180–2; suspended rites/media compensation 183–5
musical revivals 273–4
musicians' tools 216
Music in the Air 215
MUSICOVID network 4, 7, 279
music practitioners 217, 221
music psychological studies 201
music, role of 177, 178
music streaming, on TV/games consoles 215

Nadel, Lynn 245n6
Nanzer, Nate 221
Naseem, M. Ayaz 36
National Council of Culture 166
National Plan for Culture 166, 173n8
natural/human-caused disasters 250
Nazi-Fascism 139
NBC 252
Negus, Keith 218
Není nutno 35–6
neo-auratic encoding 71–2
Nessun concerto! 2
Netflix 145n12, 233
Neves, Jonathan 164
Newstatesman 226
New Yorker 229

New York Times 44, 252, 253, 255
NH News 251
Ninth Symphony 232
noise emission 239, 244
noise pollution 21, 237–8, 243
Novak, David 45
Nürnberger, Laura 208n6

O Bom e Velho – A Day in Tropicália (2021) 170
Odio song 140
O'Grady, Pat. 217
O'Keefe, John 245n6
Olaniyan, Tejumola 41, 47n4
Oliveros, Pauline 188n4
Olson, Matthew 241
one-to-one communication 159
Ong, Walter 47n4
Ong, Wendy 220
online communication 223
online services 66, 213
Ortega, Lena 70, 72n2
#otherbeats 9, 87, 90–2, 91, 92; idiosyncratic playback system and user interface 95; music compositional 94; playback system/user interface 92–5; screenshot of 91
#otherbeats website, screenshot of *91*

Padovani, Izabel 169
Paglianti, Nanta Novello 257n31
Paiakan, Paulinho 164
Palgi, Yuval 208n14
Palm Sunday Mass 127
pandemic-related listening modalities 90
pandemic soundscaping 5, 68; aesthetics/ atmospheres/new aura 70–2; *Hic et Nunc*, of unique experience 67–8; high-resolution technology 68–70; immersive mediatisation 66–8; investigative perspective 72; surprisingness 65
pandemic time 103; announcements 23; auditory 19; domestic soundscape of 20; opponents of 22; silence 23
Pani, Antonio 130–2, 133n1
Pani, Diego 128
Paris Commune 52
Paris Opera 266
Parks, Lisa 254
Parliamentary Inquiry Commission (CPI) 164
parody songs 193, 195–8, 203, 208n7
participatory location rhythm performances, digital archive 87–95
participatory music making 9, 194, 198, 273
participatory performance, categories of 198
Partition of 1947 9
pathic sphere 67, 71
Paulo Gustavo Bill (2021) 172
Pavlovic, Dijana 140
people-resistance 136
Perfect Circle, A (2019) 152

292   *Index*

performance 102, 104, 110–11, 116, 127, 165, 169, 176–81, 187, 194, 199, 214–16, 219, 220, 222; aural topography 41; balcony singing/splitscreen 195; collage 221; concept 150; digital liveness 178–80; during/after COVID-19 pandemic 176–87; evolution of 150; festival 103; folk music 9; Gelem, gelem/Není nutno 36; *gosos* scheme 126; group 36; internet 194–5; John, Elton 220; live music 194; live performances, in Brazil 165; livestreamed 102, 104, 222; during lockdown 232; Mawaca 171; media 153; musical 5, 31, 35, 43, 137, 176–87, 197, 277; musicians' live broadcasts 144; music recordings/presentational 110, 199; National Public Radio 220; raw material 34; re-inventing virtual communities, on YouTube 185–7; rhythm 87–95; Scott, Travis 222; solo 7, 170; *son huasteco* 101; *sonidero*-oriented media coverage 113; Sos Battor Colonnas 128; streaming liveness, in pandemic time 180–2; suspended rites/media compensation 183–5; theatrical/computer-based interaction 159; video, from home 110; video share, on Facebook 132; viral sensation 255; virtual 127, 220, 231; vocal 255; on YouTube 149
Performance of Secrecy: Domesticity and Privacy in Public Spaces, The 33
Peroni, Ludovico 4, 5
Pevarello, Eva 144, 145n13
Philippe, Louis, King 57
Phone Sex 234
phonographic recording 71
physical/virtual realities 157
Pianos on the Street initiative 35
Pianto delle Zitelle 184
Piasere, Leonardo 144n1
Piekut, Benjamin 88, 95
Pietrobruno, Sheenagh 99, 100
Pilosu, Sebastiano 126
Piña, Celso 111
Pinchevski, Amit 249
Pink, Sarah 131, 132
plague-spreaders 139
playback system 92–3, 95
Polanski, Roman 269n3
political balconism 58
Pollák, Peter 30, 34, 37n1
Pollard, Jane 48n11
Possi, Zizi 170
post-pandemic cumbia 110
power of *cantu a tenore* 126
pre-COVID gatherings 103
presentational performances 194, 198
Pretto, Marcelo 170
Procurade 'e moderare 125
Programme for Cultural Action (PROAC) 168
ProTools/Logic 217
'provincialize' Europe 43

provoking anxiety 24
Puar, Jasbir K. 249, 254
public health crisis 36
public social life, *internalisation* of 19
Pucci, Magda 171, 172
Puruborá, Elizer Tolentino 164
Purushothaman, Priya 48n11

quarantedium 255
quarantine *see* COVID-19 quarantine
QuerrequeFilms 101, 103–5; stream queue 104
Quiet Comfort: Noise, Otherness, and the Mobile Production of Personal Space 33
quilombo community 168

Radano, Ronald 41, 47n4
Rainbow, Randy 197
Rajendran, Lakhsmi 47n1
Ramos-Kittrell, Jesús A. 111
Rauh, Andreas 99
Ravel, Maurice 266
Ravizza, Simona 257n25
REABRE program 113
re-enchantment 153
Regulation 206/2012 240
Regulation 626/2011 240
Reily, Suzel Ana 6, 133n2
remote working, history of 214
Renou, Aymeric 54
rhythm, tree leaf graphic representation of *94*
Richard TV 113, *114*, 115, 120; channel 119; content 120; oldest videos 120; signature 121; videos 121; YouTube video *117*, *118*
Right to Maim: Debility, Capacity, Disability, The 254
Rioux, Pier-Alexandre 94, 95
road traffic reduction 16, 22
Robin, William 214
Roca, Irene T. 244n1
Rocco, Giulia Di 137, 142, 145n7
Rogers, Maggie 229
Roma community, in Campobasso 140
Romani communities, on Italian soil 142
Roma/Sinti musicians: becoming visible 136–43; first wave, stay home 137–42; Romani flag 139; second wave/beyond 142–3
Rosa, Maria De 137
Rousseff, Dilma 164–6
Rüfenacht, Annie 93
Rughini, Cosima 35

Saarikallio, Suvi 201, 208n1, 208n13
Sagesser, Marcel 9
salami sound 239; constant noise 237; spectrogram of *241*
Salter, Linda-Ruth 89
Sanborn, David 221
Sanborn Sessions 221
Sandunes 48n11

Sanremo Festival 142
Santini, Alexandre 173n10
São Paulo 168, 170
Sarbadhikari, Sukanya 44
Sardinian anthem, of 1794 revolutionary
    movements 125
Sardinian traditional music: cantzonis
    129–30; during COVID-19 pandemic
    125–32; imaginary polyphonies 125–6; media
    strategies, for multipart singing 126–7; Singing
    a Cuncordu in Time of Pandemic 127;
    social media platforms 125; 'su baballoti,' by
    Antonio Pani 130–2; virtual multipart singing
    127–9
Sarmah, Uddipan 46
Sarno, Giulia 9
SARS-CoV-2 coronavirus 130, 163, 195, 264
Savoretti, Jack 186, **186**
Scanu Montiferru's group 128
Schafer, Raymond Murray 10n10, 17, 19, 20,
    27n3, 27n8, 27n12, 32, 37n3, 42, 50, 69,
    84n12, 238, 239, 244
schizophonia 18–21, 25, 27n7
Schubert, Alexander 6, 149, 150, 151, 152, 153,
    154, 156, 159, 159n1, 160n7, 160n12; aesthetic
    references 156; homepages of *154–6*; strategies
    150; virtual environments 153; works,
    prominent features of *151*
Schuh, Diane 7
Schulke, Hans-Jürgen 76
Schulze, Holger 34
Schütz, Alfred 179
Scicolone, Ugo 143
Scott, Travis 220, 221, 222
Scudamore, Richard 78
Self-Isolation Home Studio Setup 219
self-location, sense of 104
self-quarantined communities 227
Semaine sanglante 52
SEM Student News 128
Sen, Jai 47
Sennett, Richard 146n17
Şerban, Shirley 205
Sheckler, Harrison 206
sheltering-in-place locations 90
Shifriss, Roni 208n14
Shramik Specials 45
shruti 44
shutting down, sensory technology 33
silence: ambivalent 21–3; birds return to city,
    during lockdown 264; calm-bringing 26; in
    city 22; of COVID-19 quarantine 243–56;
    listening 4–5; during lockdown 45, 243; minds
    associations with death and disaster 22;
    pandemic 4–5; pandemic silence 22; positive
    valuation 22, 23; rhapsody of 45–6; rhythm of
    1; *space of possibility* 5; on television 79
Silvério, Artur Camilo 164
Singh, Jivraj 48n11

Sinti Nel Mondo 138
Sirigu, Efisio Pintor 129
Sky Deutschland 80
Sky News 252
Slater, Mel 104
Slovak ethnic majority 35
Slovakia/Czech Republic, COVID-19 pandemic
    see COVID-19 pandemic, in Slovakia/Czech
    Republic
Slovak media 30
Small, Christopher 257n32
smartphone 69, 70, 85, 115, 117, 126, 128, 130
Smith, Jacob 42
Smith, Robert B. 214
'social' and 'mental' ecologies 265
social distance 45
social life 5, 19, 50, 87, 125, 155, 274
social media 5, 6, 50, 56, 65–6, 69–70, 83, 119,
    125, 130–2, 137–8, 171, 217, 230; anti-social
    environments 185; comments, from audience
    2; Facebook 252; Instagram 113; musical
    performance 176; national archive 172; online
    streaming 156; royalties 165; Sardinian music
    makers 128; smartphones 115; Twitter 40, 252;
    YouTube 100, 252
social movement 50; feminist demonstration
    264; in France 263
social networks *see* social media
social organisation 34, 136
social regulation 35
social solidarity 36
social surrogate 201–2
sociodemographic 274
socio-emotional coping potential 199
socio-musical culture 111
Soderbergh, Steven 233
Solberg, Erna 272
Solomos, Makis 8
sonic citizenship 41, 45
sonic cocoon 20
*Sonic Color Line: Race and the Cultural Politics
    of Listening, The* 31
sonic environment 17, 20, 42, 50, 237–40, 242–4,
    253
Sonic Paradigm of Urban Ambiances, A 17
sonic sensibility 59
sonic turn 41
sonic world 93
sonidero Latino TV's YouTube video *116*
sonideros 110; audiovisual producers 119;
    broadcaster's revenue 119; community 113;
    culture 6, 111, 113, 115, 117, 119; oriented
    media coverage 113; scene 114; videos 119;
    YouTube channels 118, 121
SonTube channels 100–2, 106; GavBroadcast
    101; streamed queues 103; streamed queues,
    during COVID-19 101–7
Sos Battor Colonnas 128
sound-as-memory 31

294   *Index*

sound environment 21; improvement 22; lockdown restrictions 18
Soundhouse: Intimacy and Distance 180
soundmark 23–5, 27n12, 58, 69; pandemic change 25
sound of fandom 78
Sound of Music, The 203, *204*
soundscape: acoustic stimuli 242; children's games and rhymes 33; COVID-19 news 250; *cumbia rebajada* 111; domestic 9, 256; *du Capitole* 58; game's 80; hermeneutical tool 78; high-resolution technology 68–70; holistic 242; lockdown 21; during lockdown 16; lo-fi/hi-fi 69, 238; low-information high-redundancy 240; Microsoft's Soundscape project 69; perceptual construct 32; salami sound 240; stereotyping 253; towns/suburbs in protest 58; urban areas 3, 21, 237–8, 243; Wroclaw's 16, 26
*Soundscape: Our Sonic Environment and the Tuning of the World, The (*Raymond Murray Schafer*)* 238
soundscaping 15, 32, 33, 42, 68, 71
sounds, neighbourhood 19
Sounds of Brazil Project 170
Sounds of the Pandemic: International Online Conference 3
Sound Souvenirs: Audio Technologies, Memory and Cultural Practices 31
sounds/silence, of COVID-19 quarantine: China 253–4; disaster reporting 250–1; H1N1 influenza pandemic (2009), coverage of 251–2; Italy 254–5; overview of 249–50; pandemic coverage 252–5
soundtrack 23, 25, 57, 58, 85n24, 94, 250
Sowodniok, Ulrike 34
space of appearance 51, 59
Spada, Max 137
Spano, Canon Giovanni 129
Spanu, Michaël 6, 110
Spinelli, Gennaro 139, 144
Spinelli, Santino 139, 144, 145n7, 145n11, 145n16
'spiritual' songs 138
Spiro, Neta 208n2
Spoonful of Clorox, A 197
Spotify consumption 229
Stabat Mater 126–7
Stanyek, Jason 88, 95
Stay Home and Make Music 140
Steets, Silke 76
Steingo, Gavin 41
Sterne, Jonathan 31, 41
Sting 221
Stobart, Henry 133n2
Stoermer, Eugene 9n8
Stoever, Jennifer Lynn 31, 35
Stokes, Martin 36
Stoltz, Chuck 208n6

Strub, J. A. 5
'Su Baballoti' lyrics **131**, 132
subaltern counterpublics 59
Sunday Times 77
supporters' cultural traditions 82
Surui, Warini 164
Survivors Rhapsody 142
Suter, Alice H. 244
Swami Jr. 170
Sweet Caroline 205
Syed, Nasir 133n2
Sykes, Jim 41
Symphony of Factory Sirens (1922) 59
szczekaczki 24

taali bajao 42, 44
Tagore, Rabindranath 40, 47n3
Talk Radio and Home Video 187
Tańczuk, Renata 4, 28n21
Tan, Shzr Ee 133n2
Tapscott, Don 217
Techno-Feudalism: A Critique of the Digital Economy 267
Teixeira, Nisio 208n2
telecommunications, to supply social closeness 148
Telegram 66
televised football: in pandemic times 78–81; sounds 79
Temer, Michel 167
temporal distortion 90
Tenores Sardinia Association 125
terrorist attack, on twin towers in 2001 2
testimonial-informative strategies 80
thaali/taali bajao 42, 44
Thibaud, Jean-Paul 17, 26, 27n3
Thompson, Emily Ann 32, 47n5
Thorley, Mark 6, 213
three-dimensional (3D) sound 69
Tikuna, Djuena 171
Timberlake, Justin 204
Times of India 46
Timóteo, Aginaldo 163
Timson, Zach 209n21
Tine mal 142
Tiny Desk Concert 220
Tjay, Lil 233
Togetherness 196
Toritos de Petate Morelia Carnaval 2021 vs Covid nos vemos en el 2022 ánimo Morelia 102
Torres, Camarera 104
Torres, Maria del Carmen Camarena 101
Tropicália Movement 169
Trentino-Alto Adige 137
Truax, Barry 24, 26n2, 47n5, 242
Trump, Donald 255
Tsai Ing-wen 272
Tuggle, C. A. 256n9
Tukker, Sofi 182

Tuning of the World, The (1977) 42
Turino, Célio 166, 167, 173n10
Turino, Thomas 71, 197, 198
TV crews 30
Twitch 110

Uber drivers 58
Uhlíř, Jaroslav 35
#ultimoconcerto 1
ultras 77, 78, 80, 82, 83n3, 84n11, 84n17
Umgang 198
umgangsmäßig 199
Umgangsmusik 197–9
Umgangssprache 198
United States, H1N1 virus in 251
Unity Switch (2019) 152, 156
unpleasant impression 24
urban experience, somatic-sensory dimension of 17
user comments *204*; Kathy Mak's parody *205*; *Sound of Music* parody, positive aesthetic evaluations of *204*

van Dijck, José 31, 32
Van Gelder, Rudy 214
Varèse, Edgard 265
Vastano, Mino 140
Veloso, Caetano 174n16
Venega, Gregório 164
Venegas, Cristina 101
Vera, Gabino 101
Verma, Neil 42
Victoroff, David 55
video calls 66
videoconciertos 115
video recording *93*; documenting 65; of police cars 25
videos 1; broadcast, on YouTube 103; channel title **186**; conferencing systems 89; heritage value of 121; streaming 119
Vieira, Maria Antônia 164
Vietnam War 214
virtual audio 81; ethical issues 81–3; solutions 77; strategies 81
virtual ethnography 50, 183
virtual images 82
virtual journey 68
virtual performances 194, 220, 231
virtual reality 6, 67, 152, 155, 158, 276
Vistula River 24
Vitale, Daniele 257n34

Voegelin, Salomé 59, 91
V-Pop hit 'Ghen' 197
VR glasses 149, 153
VST Transit 217
vulgar anti-clerical songs 25
vulnerable workers, categories of 163

Wald-Fuhrmann, Melanie 7, 10n12, 208n2, 208n15
Walker, Adam 257n28
Walker, Janet 254
Web Audio API 92
webcam, of images sharing 68
Webex 66
WebTV 179
Weheliye, Alexander 91
wellbeing 44, 101, 229, 230, 232, 233, 274–6, 280
Western culture 88
WeTransfer 217
WhatsApp 66, 132, 139, 144
Wieczorek, Sławomir 4
Wikinomics 217
Wiki-Piano.Net 152–3
Williams, Anthony D. 217
Wirz, Mirjam 111
Wood, Abigail 133n2
Wright, Erik Olin 268
Wrocław's soundscape 16

Xerente, João Sõzê 164

Yawalapiti, Aritana 164
Yellow Vests, The 50, 56
Yeung 276
YouTube 99, 110, 113, 119, 120, 127, 132, 149, 185, 186, 202, 219, 222, 232, 277; algorithms 99, 103; channels 99, 127, 143, 253; chats 104; commercial interests 99; consumption 229; ContentID 119; DIY character of 100; Mexico City 6; page *114*; pandemic-era 101; quantitative metrics 185; re-inventing virtual communities, in lockdown 185–7; role 120; *sonidero* channels 121; video 40, 99, *114*
YouTubers 113, 185–6

Zigana Evangelical Mission 138
Ziv, Naomi 173n4
Zizek, Slavoj 244
Zoom 66, 182, 267; concerts 266, 267; videoconferencing platforms 217

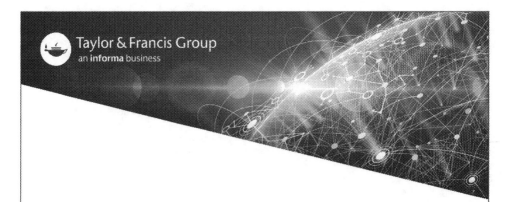

Taylor & Francis eBooks

www.taylorfrancis.com

A single destination for eBooks from Taylor & Francis with increased functionality and an improved user experience to meet the needs of our customers.

90,000+ eBooks of award-winning academic content in Humanities, Social Science, Science, Technology, Engineering, and Medical written by a global network of editors and authors.

TAYLOR & FRANCIS EBOOKS OFFERS:

- A streamlined experience for our library customers
- A single point of discovery for all of our eBook content
- Improved search and discovery of content at both book and chapter level

REQUEST A FREE TRIAL
support@taylorfrancis.com

Printed in the United States
by Baker & Taylor Publisher Services